Aerosol Remote Sensing

Jacqueline Lenoble, Lorraine Remer and Didier Tanré (Editors)

Aerosol Remote Sensing

 Springer

Published in association with
Praxis Publishing
Chichester, UK

Professor Jacqueline Lenoble
LOA
Université Lille 1
Villeneuve d'Ascq
France

Dr Lorraine Remer
JCET UMBC
Baltimore
Maryland
USA

Dr Didier Tanré
CNRS
LOA
Université Lille 1
Villeneuve d'Ascq
France

SPRINGER-PRAXIS BOOKS IN ENVIRONMENTAL SCIENCES

ISBN 978-3-642-17724-8   ISBN 978-3-642-17725-5 (eBook)
DOI 10.1007/978-3-642-17725-5
Springer Heidelberg New York Dordrecht London

Library of Congress Control Number: 2011943628

Cover design: Jim Wilkie
Project copy editor: Rachael Wilkie
Typesetting: David Peduzzi

Printed on acid-free paper

Springer is part of Springer Science+Business Media (www.springer.com)

To the memory of
Yoram Kaufman

Contents

List of authors

Colette Brogniez, Laboratoire d'Optique Atmosphérique, Université Lille1, France

Brian Cairns, NASA Goddard Institute for Space Studies, New York, USA

Alain Chédin, Laboratoire de Météorologie Dynamique, Paris, France

Oleg Dubovik, Laboratoire d'Optique Atmosphérique, Université Lille1, France

Maurice Herman, Laboratoire d'Optique Atmosphérique, Université Lille1, France

Christina Hsu, NASA Goddard Space Flight Center, Greenbelt, Maryland, USA

Brent Holben, NASA Goddard Space Flight Center, Greenbelt, Maryland, USA

Ralph Kahn, NASA Goddard Space Flight Center, Greenbelt, Maryland, USA

Michael King, Laboratory for Atmospheric and Space Physics, University of Colorado, Boulder, USA

Kevin Leavor, Hampton University, Hampton, Virginia, USA

Michel Legrand, Laboratoire d'Optique Atmosphérique, Université Lille1, France

Jacqueline Lenoble, Laboratoire d'Optique Atmosphérique, Université Lille1, France

Patrick McCormick, Hampton University, Hampton, Virginia, USA

Michael Mishchenko, NASA Goddard Institute for Space Studies, New York, USA

Clémence Pierangelo, Centre National d'Etudes Spatiales, Toulouse, France

Lorraine A. Remer, JCET University Maryland Baltimore County, Maryland, USA

Glenn Shaw, University of Alaska (retired)

Piet Stammes, Royal Netherlands Meteorological Institute (KNML), De Bilt, The Netherlands

Didier Tanré, Laboratoire d'Optique Atmosphérique, Université Lille1, France

Omar Torrès, NASA Goddard Space Flight Center, Greenbelt, Maryland, USA

List of figures

List of tables

List of symbols and acronyms

Latin symbols

a_n, b_n	Mie coefficients
b_l, g_l, a_l, z_l, d_l	coefficients of expansion of the reflectance matrix
B	blackbody radiance
c	speed of light in a vacuum
C_e, C_a, C_s	extinction, absorption, scattering cross sections
d	depolarization factor
E_0	extraterrestrial solar irradiance
$\mathbf{E_0}$	extraterrestrial solar irradiance vector
$\mathbf{E_l}, \mathbf{E_r}$	orthogonal components of the complex electric vector
f_i	instrument spectral response function for channel i
$\mathbf{F}(\Theta)$	Scattering matrix
$\mathbf{F_R}(\Theta)$	Rayleigh scattering matrix
F_{ij}	elements of the scattering matrix
F^+, F^-	upward, downward flux
g	asymmetry factor
h	Planck's constant
I, Q, U, V	Stokes parameters
$\mathbf{I} = (I\ Q\ U\ V)^T$	Stokes vector, associated to any radiative quantity I
i	incidence angle
J, J_{em}, J_{sc}	source function, due to emission, due to scattering
$\mathbf{J}$	source matrix
k	Boltzmann's constant
$\mathbf{k}$	wave vector
$k = |\mathbf{k}| = 2\pi/\lambda$	wave number
$\mathbf{k_0}$	wave vector for the sun direction
$l\ (l = 0, L)$	order of the expansions into Legendre polynomials or functions
L	radiance
L^+, L^-	upward, downward, radiance
$\mathbf{L} = (I\ Q\ U\ V)^T$	radiance matrix

L_0	path radiance
$L = \sigma_e/\sigma_b$	lidar ratio
$m = m_r + i\, m_i$	complex refractive index
M	airmass; ratio of the slant optical path to the vertical optical path
$n(r) = dN/dr$	number size distribution
$n(\ln r) = dN/d\ln r$	number size distribution on the logarithmic scale
N	total number of particles in a given volume
$p(\Theta)$	phase function
P_l	Legendre polynomial of order l
P_m^l	standard associated Legendre functions
P_{mn}^l	generalized spherical functions
P	degree of linear polarization
$\mathbf{P}(\Theta)$	phase matrix
p	pressure
P_0	surface pressure
Q_e, Q_a, Q_s	extinction, absorption, scattering efficiency factors
r	particle radius
r_m	mode radius
r_{eff}	effective radius
$r_l, r_r, r_F,$	Fresnel reflectance coefficients
R	reflectance coefficient of a Lambert surface (in Chapter 10, R represents the range of the Lidar beam)
$R(\mu, \varphi, \mu_0, \varphi_0)$	bidirectional reflectance distribution function (BRDF)
$\mathbf{R}(\mu, \varphi, \mu_0, \varphi_0)$	reflectance matrix
R_s^l and T_s^l	functions associated to Pmnl
$s\ (s = 1, S)$	order of the expansion into Fourier series of azimuth
S_1, S_2	Mie series
$S\,(\tau^*, \mu, \varphi, \mu_0, \varphi_0)$	atmosphere reflectance
$S\,(\tau^*, \mu)$	atmosphere flux reflectance
S	spherical albedo
S_L	line strength of the Doppler profile
sig	width of the size distribution
T	Temperature
$T\,(\tau^*, \mu, \varphi, \mu_0, \varphi_0)$	Atmosphere diffuse transmittance
$T\,(\tau^*, \mu)$	Atmosphere flux diffuse transmittance
t	transmittance of an atmospheric layer
t_v	spectral transmittance
$T_{tot}(\tau^*, \mu)$	atmosphere flux total transmittance
$\mathbf{T}(\chi)$	Stokes rotation matrix
$v(\ln r)$	volume size distribution
v_{eff}	effective variance
w	wind speed
$x = kr$	size parameter for a spherical particle of radius r
z	altitude over (mean) sea level
Z_t	tangent altitude

Greek symbols

α	Ångström coefficient
α_D	Doppler width
α_L	Lorentzian Full-Width Half-Max
β_l	coefficients of expansion of the phase function
$\beta_l, \gamma_l, \alpha_l, \zeta_l, \varepsilon_l, \delta_l$	coefficients of expansion of the phase matrix
Δ	Dirac function
δ	phase difference between the two components of the electric field
ε_0	dielectric constant of a vacuum
ε_{surf}	surface emissivity
η	fine mode weighting
θ	zenith angle
θ_0	solar zenith angle
θ_n	normal zenith angle
Θ	scattering angle
λ	wavelength
$\mu = \cos\theta$	cosine of the view zenith angle
$\mu_0 = \cos\theta_0$	cosine of the solar zenith angle
$\mu_n = \cos\theta_n$	cosine of the normal zenith angle
ρ	atmospheric reflectance or normalized radiance
σ_e	extinction coefficient
σ_{ag}	gaseous absorption coefficient
σ_a	absorption coefficient
σ_b	backscatter coefficient
σ_s	scattering coefficient
σ_{sR}	Rayleigh scattering coefficient
σ	width of the size distribution, defined by $\ln \sigma = sig$
τ	optical thickness or optical depth (AOT or AOD)
τ_e	extinction optical thickness or depth
τ_a	absorption optical thickness or depth
τ_s	scattering optical thickness or depth
τ_{aer}	aerosol optical thickness
τ_{Ray}	Rayleigh optical depth
τ^*	total atmospheric optical depth
τ^{SP}	slant path optical thickness (SPOD)
Φ	spectral shape factor from Lorentzian broadening
φ	azimuth angle
φ_0	solar azimuth angle
ϖ	single scattering albedo (SSA)
Ω	ozone amount

Acronyms and abbreviations

AAI	Absorbing Aerosol Index
AAOD	Absorbing Aerosol Optical Depth
AATS	Ames Airborne Tracking Sunphotometer
AATSR	Advanced Along-Track Scanning Radiometer
ACE	Aerosol Characterization Experiment
ADEOS	Advanced Earth Observation Satellite
AEM	Applications Explorer Mission
AERI	Atmospheric Emitted Radiance Interferometer
AEROCAN	Canadian aerosol sunphotomer network, a federated member of the AERONET project
AERONET	Aerosol Robotic Network
AHSRL	Arctic High Spectral Resolution Lidar
AI	Aerosol Index
AIRS	Atmospheric Infrared Sounder
ALISSA	Atmosphere par Lidar Sur SAliout
AMSU	Advanced Microwave Sounding Unit
AOD	Aerosol optical depth
AOT	Aerosol optical thickness
APS	Aerosol Polarimetric Sensor
ARM	Atmospheric Radiation Measurement
ASOS	Advanced Sustainable Observation System
ATBD	Algorithm Theoretical Basis Document
ATSR	Along-Track Scanning Radiometer
AVHRR	Advanced Very High Resolution Radiometer
BAPMON	Background Air Pollution Monitoring Network
BOREAL	Boreal Ecosystem Atmosphere Study
BRDF	Bidirectional reflectance distribution function
BT	Brightness temperature
CAI	Cloud Aerosol Imager
CALIOP	Cloud-Aerosol Lidar with Orthogonal Polarization
CALIPSO	Cloud-Aerosol LIdar and Pathfinding Satellite Observation
CANDAC	Canadian Network for Detection of Atmospheric Change
CARL	Climate Research Facility Raman Lidar
CCD	Charge-Coupled Device
CCN	Cloud Condensation Nuclei
CIMEL	Construction d'Instruments de Mesures Electroniques (French company who provide the sunphtometers used in AERONET)
CLAES	Cryogenic Limb Array Etalon Spectrometer
CLAMS	Cheasapeake Lighthouse & Aircraft Measurements for Satellites
CLIMAT	Conveyable Low-noise Infrared radiometer for Measurements of Atmosphere and ground surface Targets
CloudSat	Cloud Satellite
CTP	Cloud top pressure

CNES	Centre National d'Etudes Spatiales
CRF	Climate Research Facility
CZCS	Coastal Zone Color Sensor
DDA	Discrete dipole approximation
DIAL	Differential Absorption Lidar
DISORT	Discrete Ordinates Radiative Transfer Program for a Multi-Layered Plane-Parallel Medium
DLR	Deutsches Zentrum für Luft- und Raumfahrt
EARLINET	European Aerosol Research Lidar Network
EBCM	Extended Boundary Conditions Method
Envisat	Environmental Satellite (ESA)
EOF	Empirical Orthogonal Functions
EOS	Earth Observing System (NASA)
EP	Earth Probe (NASA)
ERBS	Earth Radiation Budget Satellite (NASA)
ERS	European Remote Sensing satellite (ESA)
ERTS	Earth Resources Technology Satelllite (NASA)
ESA	European Space Agency
EUMETSAT	European Organisation for the Exploitation of Meteorological Satellites
FDTDM	Finite Difference Time Domain Method
FEM	Finite Element Method
FOV	Field Of View
FTIR	Fourier Transform InfraRed spectroscopy
GACP	Global Aerosol Climatology Project
GALION	GAW Aerosol Lidar Observation Network
GASP	GOES Aerosol/Smoke Product
GAW	Global Atmosphere Watch
GEISA	Gestion et Etude des Informations Spectroscopiques Atmosphériques
GIMMS	Global Inventory Modeling and Mapping Studies
GLAS	Geoscience Laser Altimeter System
GLI	Global Imager
GMS	Geostationary Meteorological Satellite
GOES	Geostationary Operational Environmental Satellite
GOM	Geometrical optics method
GOME	Global Ozone Monitoring Experiment
GOMOS	Global Ozone Monitoring by Occultation of Stars
GOSAT	Greenhouse gases Observing Satellite
GSFC	Goddard Space Flight Center
HALOE	Halogen Occultation Experiment
HIRDLS	High Resolution Dynamics Limb Sounder
HIRS	High Resolution Infrared Sounder
HITRAN	High-resolution Transmission molecular absorption database
HSRL	High Spectral Resolution Lidar
IASI	Infrared Atmospheric Sounding Interferometer

ICESat	Ice, Cloud,and land Elevation Satellite (NASA)
IDDI	Infrared Difference Dust Index
IFOV	Instantaneous field of view
IGBP	International Geophysical Biological Program
IIR	Imaging InfraRed radiometer, on CALIPSO
IR	Infrared
ISAMS	Improved Stratospheric and Mesospheric Sounder
ISCCP	International Satellite Cloud Climatology Project
ISLSCP	International Satellite Land Surface Climatology Project
ISRF	Instrument spectral response function
JAXA	Japan Aerospace Exploration Agency
KA	Kirchhoff approximation
Landsat	Earth Resources Technology Satellite (NASA)
LaRC	Langley Research Center
LER	Lambert Equivalent Reflector
LITE	Lidar In space Technology Experiment
LOA	Laboratoire d'Optique Atmosphérique
LOS	Line of sight
LSM	Least Squares Method
LTE	Local Thermodynamic Equilibrium
LUT	Look Up Tables
MAIAC	Multi-Angle Implementation of Atmospheric Correction
MAN	Marine Aerosol Network
MERIS	Medium Resolution Imaging Spectrometer
METEOSAT	European Geostationary Meteorological Satellites (EUMETSAT)
Metop	Meteorological Operational satellite programme (EUMETSAT)
MILAGRO	Megacity Initiative – Local and Global Research Observations
MISR	Multiangle Imaging SpectroRadiometer
MLM	Maximum likelyhood method
MLO	Mauna Loa Observatory
MODIS	MODerate resolution Imaging Spectroradiometer
MODTRAN	MODerate resolution atmospheric TRANsmission
MOM	Method of Moments
MPLnet	Micropulse Lidar Network
MSG	Meteosat Second Generation (EUMETSAT)
MSS	Multi Spectral Scanner
MVIRI	Meteosat Visible Infra-Red Imager
NASA	National Aeronautics and Space Administration
NCEP	National Center for Environmental Prediction
NDVI	Normalized Difference Vegetation Index
Nd-YAG	Neodymium Yttrium AluminumGarnet
NESDIS	National Environmental Satellite, Data, and Information Service
NILU	Norwegian Institute for Air Research
NOAA	National Oceanic and Atmospheric Administration
NPOESS	National Polar-Orbiting Environmental Satellite System

NPP	NPOESS Preparatory Project
NWP	Numerical Weather Prediction model
OMAERO	Aura OMI Aerosol Data Product
OMAERUV	Aura OMI Aerosol Data Product using two wavelengths in the near UV
OMI	Ozone Monitoring Instrument
OPAC	Optical Properties of Aerosol and Clouds database
OrbView-2	Satellite carrying the SeaWiFS sensor for ocean color (Geoeye)
OS	Method of successive orders of scattering
OSIRIS	Optical Spectrograph and Infrared Imaging System
PARASOL	Polarization and Anisotropy of Reflectances for Atmospheric Science Coupled with Observations from a Lidar
PATMOS	Pathfinder-ATMOSphere
PBL	Planetary Boundary Layer
PFR	Precision Filter Radiometers
PHOTONS	Photométrie pour le Traitement Opérationnel de Normalisation Satellitaire (French Component of AERONET)
PMD	PhotoMultiplier Detector
PMOD-WRC	Physikalisch Meteorologisches Observatorium Davos - World Radiation Center
POAM	Polar Ozone and Aerosol Measurement
POES	Polar Operational Environmental Satellite (NOAA)
POLDER	Polarization and Directionality of the Earth's Reflectances
PSC	Polar Stratospheric Cloud
PSD	Particle size distribution
QA	Quality Assurance
RASA	Russian Aviation and Space Agency
RSP	Research Scanning Polarimeter
RT	Radiative Transfer
RTE	Radiative Transfer Equation
SAD	Surface Area Density
SAGE	Stratospheric Aerosol and Gas Experiment
SAM	Stratospheric Aerosol Measurement
SAMUM	Saharan Mineral Dust Experiment
SCAR	Sulfate clouds and radiation
SCIAMACHY	Scanning Imaging Absorption Spectrometer for Atmospheric Chartography
SeaWiFS	Sea-viewing Wide Field of view Sensor
SEM	Scanning Electron Microscope
SEVIRI	Spinning Enhanced Visible and Infrared Imager
SM	Superposition Method
SKYNET	Observing network for aerosol-radiation-cloud interaction (Japan)
SOLRAD-NET	Solar Radiation Network
SPOD	Slant Path Optical Depth
SPOT	Satellite Pour l'Observation de la Terre

SRC	Spectral Regression Coefficient
SSA	Single scattering albedo
SVM	Separation of variables method
TIGR	Thermodynamical Initial Guess Retrieval
TIR	Thermal Infrared
TIROS	Television Infrared Observation Satellite (NOAA)
TM	Thematic Mapper
TMM	T-Matrix Method
TOA	Top Of Atmosphere
TOMS	Total Ozone Mapping Spectrometer
TRMM	Tropical Rainfall Measuring Mission (NASA/JAXA)
UARS	UpperAtmosphericResearch Satellite (NASA)
UMBC	University of Maryland Baltimore County
USVI	United States Virgin Islands
UV	UltraViolet
UVAI	UltraVioletAerosol Index, the same as AAI
VAAC	Volcanic Ash Advisory Center
VHRR	Very High Resolution Radiometer
VIE	Volume integral equation
VIIRS	Visible Infrared Imager Radiometer Suite
VIRS	Visible and Infrared Scanner
VISSR	Visible and Infrared Spin Scan Radiometer
VSWF	Vector Spherical Wave Functions
WMO	World Meteorological Organization
WODRC	World Optical Depth Research Center

Foreword

Throughout history, aerosols have been associated with local pollution of the atmosphere and the outbreak of severe episodes that threaten human health. In reality, aerosols are and always have been an integral part of the atmosphere, and they play an important role in a wide range of weather and climate phenomena on local, regional, and global scales. No consideration of climate change can avoid taking into account the radiative and hydrological effects of aerosols within the climate system. Yet it is only in the past decade or so that general circulation models have begun to incorporate aerosols interactively in climate simulation and forecasting studies. Despite this progress, there is a long path ahead. Due to the complex nature of aerosols and the variety of ways they interact with the climate system, there is still considerable uncertainty on the magnitude, and even the sign, of their contribution to past changes in global climate, hence the need for a text to consolidate what we now know and provide guidance for continuing investigation. The present text is focused on the observational aspects of aerosols and the application of radiative transfer theory for remote sensing from ground-based and satellite-based radiometers.

Aerosols are a non-gaseous part of the atmosphere that play a critical role in cloud formation and precipitation processes. They exist in a variety of forms: solid matter (soluble or insoluble), liquid matter (chemicals dissolved in water), or a mixture of the two. They are subject to continuous transformation – the result of chemical, photochemical and hydrological reactions – and they affect the weather and climate system directly through their effect on the radiation field, and indirectly through their interaction with clouds and precipitation.

Aerosols are generally the result of direct emission into the atmosphere or the product of interactions involving the surface (including the biosphere), the atmosphere, and radiation. Although science is often a pursuit of knowledge for its own sake — one may call it curiosity — atmospheric science has a practical aspect that is of great importance to human society. We need to be able to forecast changes in the atmosphere and oceans on a wide range of time scales and, in the present era, where humans can contribute significantly to change, one needs to distinguish between "natural " and "anthropogenic" aerosols. This could not always be done unambiguously. Perhaps it is easy to make the distinction with respect to sources of direct emission. But from the climate point of view, the main difference between an anthropogenic aerosol and a natural aerosol is not where it came from so much as the extent to which human activity contributed to its production. This is where the varied and complex interactions with the climate system come in. A simple but purely hypothetical example would be the reclamation of an arid area to grow trees. In this case, the natural, wind-blown mineral dust is replaced by organic aerosols emitted by vegetation that might, under other circumstances, be considered natural, but in this case considered anthropogenic. Thus, one can see how the distinction between natural and anthropogenic can be a bit fuzzy.

Because aerosols play such an important role in atmospheric processes and because the sources of aerosols are highly varied, it has become essential to develop and deploy a global monitoring system to map their spatial distribution, to track changes over time, and to study their interaction with the other components of the atmosphere. The monitoring relies on radiation measurements and inversion algorithms. This is where Dr Yoram Kaufman, to whom this textbook is dedicated, made major contributions that greatly advanced our knowledge of aerosols and measurement methodology. Like aerosols, Dr Kaufman had a direct and an indirect impact on the science. In addition to seminal papers on remote sensing and interpretation of observations, his affable disposition and gregarious nature led him to be a much sought-after collaborator on many projects. His interaction with colleagues the world over permitted his ideas and insights to spread and have a multiplying effect on scientific progress.

In tribute to his catalytic role in fostering science, the Atmospheric Sciences Section of the American Geophysical Union established the *Yoram J. Kaufman Unselfish Cooperation in Research Award*, given annually to honor senior scientists, like him, who have a "broad influence in atmospheric science through exceptional creativity, inspiration of younger scientists, mentoring, international collaborations, and unselfish cooperation in research". For his accomplishments over a career that was brought to a sudden, premature end in 2006 by a tragic accident, Dr Kaufman received many awards, including the NASA Medal for Exceptional Scientific Achievement in 2001 and the Verner E. Suomi Award of the American Meteorological Society in 2007.

Perhaps Dr Kaufman's greatest contribution to the science of aerosols is providing the motivation for the development of the aerosol monitoring system now in place worldwide. The system consists of satellite-based and ground-based components. The satellite-based sensors provide global coverage, while the ground-based sensors provide local measurements that can be related to *in situ* collections of aerosol samples. A paper published in 1994, of which Dr Kaufman is senior author, demonstrated the feasibility of retrieving many of the properties of aerosols from an instrument that combines a sun photometer that measures direct sunlight and a sky radiometer that measures sky radiance. Although the instrument was portable and manually-controlled, with measurements recorded by hand, it served as the proof of concept for what is now the largest ground-based system for monitoring aerosols: **AE**rosol **RO**botic **NET**work (AERONET). Today, AERONET includes approximately 500 automated instruments distributed around the world.

The satellite-based component of the aerosol monitoring system consists of a variety of sensors on multiple platforms, including the **MOD**erate Resolution **I**maging **S**pectroradiometer (MODIS) deployed on the Terra and Aqua satellites of NASA's Earth Observing System (EOS) and a lidar system and imaging radiometer on the Cloud-Aerosol Lidar and

Infrared Pathfinder Satellite Observations (CALIPSO), a joint mission of NASA and the French CNES. Dr Kaufman was a member of both the MODIS and CALIPSO Science Teams and served as Project Scientist of Terra during its development. He worked closely with atmospheric scientists at NASA Goddard Space Flight Center and collaborators from academic institutions around the world to develop and implement the algorithms for processing the satellite data to produce a rich aerosol data archive. The authors of this text are amongst the principal contributors.

The data archive consists of a continuous space-time record of aerosol properties based on MODIS, beginning in year 2000, supplemented by more detailed but less extensive observations derived from other satellite sensors and the ground-based AERONET. MODIS, for example, provides a near-global, vertically integrated picture, while CALIPSO provides samples of vertically resolved information. All of these data are freely available for research by the international community. As Terra Project Scientist, Dr Kaufman was particularly involved in ensuring that the data from MODIS and other sensors on Terra are easily accessible for display and downloadable directly from the NASA web site. This global picture of aerosols allows us to test theoretical models and parameterizations that are now being incorporated into weather and climate models.

Aerosols are no longer the mysterious constituent that contributes "noise" to atmospheric processes. They are an important and integral constituent of the atmosphere.

This text provides the basic principles of how one extracts information on aerosols from radiation measurements, serving as an essential source for educating graduate students in this subject, and at the same time it is a convenient accounting of the present state-of-art that can serve as a platform for future research.

Albert Arking, Johns Hopkins University

Preface

Today, as we write this Preface, there are more than 20 separate satellite sensors in orbit around Earth that are able to retrieve information about atmospheric aerosols. In addition, ground networks and other suborbital instruments are systematically using remote sensing techniques to measure and derive aerosol products and making those products freely available to the public. There has been a surge of interest in these products, and they are widely used. A search of Science Citations yields nearly 5000 hits for the simple search of satellite AND aerosol*.

The retrieval algorithms for each sensor are based on rigorous remote sensing theory, and each sensor brings different and often complementary capabilities to the table. To understand the products, their sensitivities and limitations, one must return to the basics of the physics and begin from first principles. Any beginning scientist at the onset of their career who is interested in using remote sensing tools to understand aerosol effects on climate, on the hydrological cycle or on human health needs to see the progression from the theoretical to the practical.

Today, a graduate student has a wide choice of texts to choose from if interested in the theory of radiative transfer. Some of these books give examples that touch on remote sensing, and some may even use aerosol retrievals as an example. However, a radiative transfer text is not focused on aerosol remote sensing, and will not provide the specifics and intercomparison of all current and historical retrieval methods. A student might piece together the knowledge by accessing individually published papers from scientific journals or on-line accessible technical reports written by different instrument teams. These papers and reports are immensely valuable, but such a literature search is not the same as having the information in one easily accessible volume. There is also a disconnect between a theoretical radiative transfer text and the literature that describes individual practical solutions to retrieve aerosol information from remote sensing. We felt that the time had come for a book that provides an easy path from theory to practical algorithms.

Aerosol Remote Sensing begins from first principles. It does not substitute for an introductory text on radiative transfer, but reviews the basics and sets off from there. Both passive and active sensors and methods are addressed, different types of inversions are described, ground and satellite applications are included, and a historical context to today's products is provided. The remote sensing products produced today will be archived and used repeatedly long past the time the orbiting satellites that produced those data are no longer functioning. As succeeding "generations" of graduate students discover the need for aerosol remote sensing products and access these climate data records for their research, they will want and need a text to learn the science behind the archived data. We intend for *Aerosol Remote Sensing* to be that text.

We could not have produced this book without the efforts of many people who we would like to thank here. Foremost among them are the individual authors of the chapters, who worked diligently to meet our expectations and deadlines, and put much effort into developing the depth and breadth of information that comprehensively covers the scope of each chapter. These individuals are Colette Brogniez, Brian Cairns, Alain Chedin, Oleg Dubovik, Maurice Herman, Brent Holben, N. Christina Hsu, Ralph Kahn, Michael King, Michel Legrand, Kevin R. Leaver, M. Patrick McCormick, Michael Mischenko, Clemence Pierangelo, Glenn Shaw, Piet Stammes and Omar Torres. Some of these individual authors took on much more work than they originally signed up for, contributing to multiple chapters and sometimes making uncredited contributions by providing editorial comment on other authors' work. Thank you, all. *Aerosol Remote Sensing* is very much a cooperative effort.

We would also like to give special thanks to Albert Arking for writing the Foreword, and to Jean-Luc Deuze, Richard Ferrare and Tom Eck who read early versions of the text and provided very valuable suggestions for improvement. Marines Martins created the compiled Bibliography from the reference lists of each chapter, and we are grateful for her effort. We would also like to thank the editorial and production staff at Praxis who patiently waited for the finished manuscript through several delays and then provided excellent cover art, proofing and typesetting. We would especially like to thank Romy Blott, Clive Horwood, David Peduzzi, Rachael Wilkie and Jim Wilkie.

Finally, although his name does not appear as an author on any chapter, Yoram Kaufman has made a major contribution to producing this text. Yoram's influence in the field of aerosol remote sensing permeates every topic covered in this book. Yoram was involved in early discussions between the editors in defining the need for a comprehensive book that became *Aerosol Remote Sensing*, and if Yoram had not died tragically in 2006, he would have contributed to this work as an editor and an author. We missed his contibution as we compiled this book. More importantly, we miss his scientific contributions to the field, and we miss him personally. This text is dedicated to his memory.

1 Introduction

Jacqueline Lenoble, Lorraine A. Remer, Didier Tanré

1.1 What are aerosols?

Aerosols are the solid and liquid particles suspended in the atmosphere. They can be seen as the amorphous haze that decreases visibility on polluted days, or the well-defined plumes of particles that rise out of a burning fire. These particles can be either natural (lofted desert and soil dust, sea salt particles, volcanic emissions, wildfire smoke, biogenic emissions) or anthropogenic (industrial emissions, biomass burning for agriculture, or land use changes that accelerate erosion and evaporation of lakes). As most aerosols are produced at the Earth's surface, they are generally concentrated in the lower layers of the troposphere, and near the production sources. However particles can reach higher levels (4–6km) and can be transported over long distances. Aerosols are removed from the atmosphere by dry deposition, scavenging by precipitation, and evaporation. At the stratosphere level, they are abundant only after major volcanic eruptions, and they are mainly formed by gas-to-particle conversion. Although much less numerous than tropospheric aerosols, they may have an important impact, due to their long residence time and to their spread all around the globe.

In this book, discussion of aerosols excludes water droplets and ice crystals that are the constituents of clouds. There exists some controversy as to the separation of aerosols from cloud particles (Koren et al., 2007; Charlson et al., 2007), but here we focus on those particles that are generally smaller than cloud droplets or crystals, and whose origins differ from the cloud generation processes of condensation and sublimation of water vapor. However, much of the radiative theory developed in this book to describe the light scattering by small particles can also be applied to cloud droplets and crystals.

While atmospheric gases such as oxygen (O_2), nitrogen gas (N_2) or minor constituents such as water vapor (H_2O), ozone (O_3) are clearly defined by their chemistry, the term "aerosol" is generic for a wide variety of substances, not all well defined. Even subclassifications of aerosols such as dust or smoke are themselves generic terms for many combinations of minerals or organics, respectively. At best we can discuss aerosol types. Table 1.1, reproduced from Chin et al. (2009), presents the main types of aerosol, as represented by current aerosol transport models. These models generate aerosol from estimates of source emissions, then allow the particles to be transported by modeled meteorology, transformed from modeled chemical processes and removed from the atmosphere by modeled wet and dry deposition. The fourth column gives the median value and range from an ensemble of different models of the total mass burden of aerosols in the atmosphere. Note that by mass, the mostly natural sources of aerosol that include dust and sea salt represent the overwhelming mass burden of the atmosphere. Also note the wide range of model results. Estimating the distribution of the aerosol loading of the atmosphere across specific aerosol types is highly uncertain.

Aerosol type	Total source[1] (Tg/yr[1])	Life time (day)	Mass loading[1] (Tg)
	Median (Range)	Median (Range)	Median (Range)
Sulfate[2]	190 (100–230)	4.1 (2.6–5.4)	2.0 (0.9–2.7)
BC	11 (8–20)	6.5 (5.3–15)	0.2 (0.05–0.5)
POM[2]	100 (50–140)	6.2 (4.3–11)	1.8 (0.5–2.6)
Dust	1600 (700–4000)	4.0 (0.03–7)	20 (5–30)
Sea Salt	6000 (2000–120000)	0.4 (0.03–1.1)	6 (3–13)

[1] Tg (teragram) = 10^{12} g or 1 million metric tons

[2] The sulfate aerosol source is mainly SO_2 oxidation plus a small fraction of direct emission. The organic matter source includes direct emission and hydrocarbon oxidation

Table 1.1 Estimated source strengths, lifetimes, mass loadings of major aerosol types. Statistics are based on results from 16 models examined by the Aerosol Comparisons between Observations and Models (AeroCom) Projects (Textor and Kinne et al., 2006). BC=Black Carbon; POM = particulate organic matter.

One of the reasons for applying remote sensing techniques to retrieve aerosol characterization and loading is to provide strong observational constraints on model depictions of the global aerosol distribution. While all types of observations provide important information and help to constrain models, only satellite remote sensing has the ability to observe and quantify the aerosol system on a global scale.

We will see in Chapter 2 that the aerosol optical or radiative characteristics, which determine the possibilities of remote sensing, depend on the particles' chemical composition, their size, and their shape. Composition is represented by the particles' complex refractive indices, with the real part contributing to the determination of the particles' light scattering properties and the imaginary part of the refractive index contributing to the determination of the particles' light absorption properties. In turn, these physical characteristics depend on the aerosol origin and transport. Moreover, an aerosol layer is generally an ensemble of particles of different composition, different sizes, and different shapes. Even a single particle may not be homogeneous, but contain inclusions of different materials; in this last case, we speak of "internal mixture" as opposed to "external mixture" for an ensemble of different homogeneous particles.

To further complicate the situation, measurement of particles' size and complex refractive indices in the laboratory or with in situ instrumentation in the field will characterize the dry properties of the particles, but not necessarily represent the ambient properties observed by remote sensing in the atmosphere. The reason is that several of the substances that form atmospheric aerosols are hygroscopic, meaning they attract water vapor that condenses onto the dry substance, forming a suspended liquid particle that could be a highly diluted solution of the original chemical with greatly altered physical properties. Hygroscopic coatings of mineral dust may even cause these typically nonspherical particles to uptake water smoothing out their rough edges.

Table 1.2 lists the physical properties of the aerosol particles necessary to determine the particles' optical properties.

Roughly, in the visible, the real part of the refractive index is around 1.53 for most dry particles, with a small imaginary part of $0.5–1.0 \ 10^{-2}$. Hygroscopic particles that absorb water vapor dilute these values, causing the actual ambient refractive indices to draw towards the value of pure water: 1.33 (real) and 0.00 (imaginary). For example, oceanic particles, mostly hygroscopic sea salt, have almost no absorption and a real refractive index slightly larger than pure water, around 1.38. Ambient sulfate particles are also strongly hygroscopic with a representative complex refractive index of $1.45 + 0.0035i$. Soot or black carbon have a very high absorption, with a refractive index of approximately $1.75+0.45i$ (Radiation Commission, 1986), although black carbon is most often found internally mixed with other particle types, especially other organic material, which dilutes these values. Depending on composition, particle refractive indices may have a strong spectral dependence. For example, mineral dust containing some hematite, an iron compound, absorbs strongly in the shortwave, but less so towards the midvisible and higher wavelengths, causing dust particles to appear "reddish" to the eye. This is represented by spectral shifts in the dust-imaginary part of the refractive index.

Liquid particles are approximately spherical, whereas solid aerosol may present very different shapes. Modeling generally assumes the simplest case of spherical particles, but theory and measurements are now available for different nonspherical shapes (see Chapter

2, Section 2.7). It has been shown in some cases that introducing nonspherical particles in modeling significally improves agreement with experimental data (Herman et al., 2005).

The aerosol sizes vary from a few molecules (10^{-3} μm) to a few tens of micrometers in radius, but most of the aerosol mass is contained in the size range 0.05 to 10 μm; that corresponds also to the aerosol major impact on radiation. Different particle types favor different size ranges. Particles derived from combustion sources gas-to-particle conversion such as sulfates and most organics, tend to be an order magnitude smaller (0.05–0.40 μm) than those particles originating from wind-driven processes such as sea salt (~1 μm) and dust (0.40–10 μm). Long-lived volcanic particles tend to fall between these two size ranges.

For the purpose of modeling, it is convenient to introduce mathematical expressions of the size distribution; different expressions are used, but one of the most popular is

$$n(r) = \frac{dN}{dr} = \frac{N}{rsig\sqrt{2\pi}} \exp\left[-\frac{\ln^2 r/r_m}{2sig^2}\right], \tag{1.1}$$

Physical property	Sulfate	BC	POM	Dust	Sea salt
Real part of refractive index – dry	~ 1.53	~1.75	~1.53	~1.53	~1.50
Imaginary part of refractive index – dry	~ 0.005	~ 0.440	~0.006	~ 0.008	~0.000
Real part of refractive index – wet	1.35–1.45		1.35–1.45	~1.48	1.35–1.45
Imaginary part of refractive index – wet	~ 0.002		~ 0.003	~0.005	0.000
Size distribution (effective radius in μm)	r_{eff}= 0.1–0.2		r_{eff}= 0.1–0.2	r_{eff}= 1.5–3	r_{eff}= ~ 1
Shape	sphere	*not sphere	sphere	not sphere	sphere

* Black carbon (BC) originates as long nonspherical chains, but those chains collapse and are often incorporated in other organic material (POM) that adds coatings over the collapsed BC chains. Rarely is BC seen in its pure state in the real atmosphere.

Table 1.2 Approximate typical physical properties for different aerosol types. Refractive index values are a function of wavelength. Here they are given for 0.55 μm.

where $n(r)$ is the number of particles with a radius between r and $r+dr$; N is the total number of particles.

On a logarithmic scale, Eq. (1.1) writes as

$$n(\ln r) = \frac{dN}{d\ln r} = rn(r) = \frac{N}{sig\sqrt{2\pi}}\exp\left[-\frac{\ln^2 r/r_m}{2sig^2}\right];$$ (1.2).

Eq. (1.2) is a normal distribution on the logarithmic scale, callled lognormal distribution (LND); r_m is the mode radius, i.e. the radius corresponding to the maximum of $n(\ln r)$; sig measures the width of the distribution; many authors introduce the notation

$$\ln\sigma = sig.$$ (1.3)

More often, an aerosol is defined by a bimodal size distribution, which is the sum of two equations similar to Eq. (1.1), with N_1, N_2, r_{m1}, r_{n2}, sig_1, sig_2, for component 1 and 2. A coarse mode with large particles (r_m around 0.5 to 1.0 μm) is associated to a fine or accumulation mode (r_m around 0.02 to 0.17 μm) (Dubovik et al., 2002). In general each individual mode corresponds to a specific aerosol type with a common origin.

Equations (1.1) or (1.2) define a number size distribution; they are used in Chapter 3. Sometimes, a volume size distribution is preferred

$$v(\ln r) = dV/d\ln r = \frac{4\pi r^3}{3}n(\ln r);$$ (1.4)

$v(\ln r)$ can be derived from $n(\ln r)$ defined in Eq. (1.2); this distribution is no more lognormal, and its maximum is reached for

$$r_v = r_m\exp(3sig^2)$$ (1.5)

If the choice is to have a lognormal volume size distribution, as done in Chapters 5 and 9, it is directly written as

$$v(\ln r) = \frac{C}{sig\sqrt{2\pi}}\exp\left[-\frac{\ln^2 r/r_m}{2sig^2}\right];$$ (1.6)

C is a constant depending on the total amount of particulate matter, and r_m is a mode radius for the volume size distribution. The number size distribution then derives from Eq. (1.4), and it is not LND.

An important factor for radiative effects is the effective radius (Hansen and Hovenier, 1974)

$$r_{eff} = \frac{\int_0^\infty r^3 n(r)\,dr}{\int_0^\infty r^2 n(r)\,dr} \; ; \tag{1.7}$$

the associated effective variance is defined by

$$v_{eff} = \frac{\int_0^\infty (r - r_{eff})^2 r^2 n(r)\,dr}{r_{eff}^2 \int_0^\infty r^2 n(r)\,dr} . \tag{1.8}$$

While the log-normal function, especially the bimodal version, is a common representation of the aerosol size distribution for mathematical purposes, in nature aerosol size distributions are not necessarily lognormal or even bimodal lognormal. Sometimes aerosol size distributions are complex superpositions of more than two modes, and modern retrieval methods that allow for irregularly shaped size distributions do retrieve distributions that would not fit a lognormal.

Other mathematical expressions are often used for the size distribution and two of them are given here.

The Junge power law is widely used in old papers because of its simplicity, but it is not often used any more; it gives the number size distribution as

$$n(r) = C\, r^{-\nu-1} \quad \text{in an interval } r_1, r_2 \tag{1.9}$$

$$n(r) = 0 \quad \text{outside this interval;}$$

the interval (r_1, r_2) is often split into a few subintervals.

More useful is the Gamma number size distribution, generally defined with three parameters α, β, γ, as

$$n(r) = C r^\alpha \, \exp(-\beta r^\gamma) , \tag{1.10}$$

where the constant C depends on the total number of particles.

In its modified form, $\gamma=1$ and only two free parameters α, β remain. This modified Gamma size distribution is used in Chapter 2; Equation (2.50) uses as parameters a=r_{eff}, b=v_{eff} defined in Eqs (1.7 and 1.8).

Several standard aerosol models have been proposed (Shettle and Fenn, 1979, Radiation Commission, 1986, d'Almeida et al., 1991). The software package "Optical Properties of Aerosols and Clouds" (OPAC), proposed by Hess et al. (1998) is available at www.meteo.

physik.uni-muenchen.de/strahlung/aerosol/aerosol.html. In addition, increasingly common is the use of the recent results from ground-based remote sensing inversions and networks (Chapters 5 and 6; Dubovik et al., 2002). The climatology aquired by these networks offers aerosol models of total column ambient aerosol conditions without concern for the degree of hygroscopicity and water uptake by the dry particles. Such ambient aerosol models may be preferable for a variety of remote sensing applications (Chapter 8). Aerosol network climatology is available at http://aeronet.gsfc.nasa.gov.

1.2 Why are we interested in aerosols?

There are many reasons to be interested in aerosols. At the Earth's surface, where people live and where the highest concentrations of aerosols are found, these particles present a serious health hazard. Called "particulate matter" by the air quality community, aerosols have been linked to increases in morbidity and mortality rates, and to degradations of environmental quality in terms of acid rain and reduction of visibility. The reduction of visibility and general attenuation of electromagnetic radiation in the infrared spectrum by aerosols is a serious concern for a variety of civilian and military sighting and communication technology. Volcanic ash emissions, another type of aerosol, can cause severe destruction of jet engines and create large-scale disruption of air traffic. In the stratosphere, aerosols play an important role in the heterogeneous atmospheric chemistry in polar stratospheric clouds that depress ozone concentrations at the poles. However, the main reason for the present interest in aerosols and the primary application of aerosol remote sensing is the influence of aerosols on climate.

Aerosols have a "direct effect" on climate, by modifying the Earth's radiation budget and redistributing heating in the atmosphere, and an "indirect effect" by modifying cloudiness, cloud development, precipitation or cloud radiative properties.

The direct aerosol influence on climate depends on the global distribution of these particles and on their optical characteristics, fixed by the size, shape, and refractive indices determined by their chemical composition and listed in Table 1.2. Aerosol effect on climate also depends on their altitude, and on the reflectance or albedo of the underlying surface (see Chapter 3). In most cases, aerosol scattering increases the planetary albedo, and has a cooling influence on climate; that is the case for non-absorbing aerosols above a dark surface, as the ocean. However absorbing aerosols, above a reflecting surface, such as snow, desert or low altitude cloud, have an opposite effect, reducing the planetary albedo, and leading to warming (see Figure 1.1).

The indirect effect is rather complex. Acting as cloud condensation nuclei (CCN) and/or ice nuclei, aerosols increase the number of cloud droplets or crystals, and for a given liquid water content can reduce the droplet size. This in turn increases the cloud reflectance (Twomey, 1977a). On the other hand, absorbing aerosol imbedded either in the cloud droplets or between them can decrease the cloud reflectance (Chylek et al., 1984). Aerosols are also linked to many changes in cloud structure, including cloud coverage, cloud top height and thermodynamic phase, all of which alter the radiation field and may have serious consequences for the hydrological cycle and release of latent heat. The indirect effect is more

Figure 1.1 Aerosols advected over the Atlantic Ocean following wildfires in Portugal in 2008. The dark smoke can be seen on the left side of the image; as it passes over the brighter cloud deck, it darkens the scene, reducing the clouds' ability to reflect sunlight back to space. In contrast, over the darker cloud-free ocean, the smoke appears brighter than the background, increasing the ocean's ability to reflect sunlight back to space. Image: Jacques Descloitres, MODIS Rapid Response Team, NASA/GSFC obtained through NASA's Visible Earth reprinted from Nature Geosciences News & Views.

difficult to analyze than the direct effect because of the complexity of cloud processes and because of the difficulties of measuring cloud microphysics and aerosol properties in the proximity of clouds.

Figure 1.2 (from IPCC, 2007) compares various causes of global radiative forcing; the average impact of aerosol results in cooling, but the uncertainty is still very large, and not all of the processes of aerosols in the atmosphere, general circulation and hydrological cycle are included in that figure.

To narrow the uncertainties represented in Figure 1.2 we must have better understanding of the processes in the atmosphere that aerosols affect so that we can better represent these processes in the models used to predict climate forcing and climate response. We must also have better measurements that quantify aerosol amounts and particle properties, and the distribution of these quantities over the globe and across many different underlying surfaces.

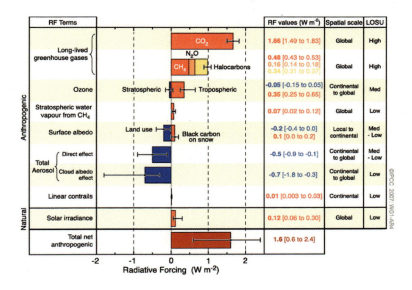

Figure 1.2 Global average radiative forcing estimates and uncertainty ranges in 2005 relative to the pre-industrial climate. Anthropogenic carbon dioxide (CO_2), methane (CH_4), nitrous oxide (N_2O), ozone and aerosols as well as the natural solar irradiance variations are included. Typical geographic extent of the forcing (spatial scale) and the assessed level of scientific understanding (LOSU) are also given. Forcing is expressed in units of watts per square meter (Wm^{-2}). The total anthropogenic radiative forcing and its associated uncertainty are also given. Figure from IPCC (2007).

1.3 How can we observe aerosols?

Aerosol observations can be divided into two broad categories: in situ and remote sensing. In situ measurements are any measurements that are made at the point of interest where the instrument is in direct contact with the aerosol particles it measures. Remote sensing observes these particles from a distance without directly interfering with them.

In situ measurement systems include procedures that collect aerosol particles on filters and analyze the laden filters in a laboratory. There a chemical analysis can be performed, mass concentration determined, and properties such as the refractive index of the bulk material or absorption properties can be directly measured. If the filter is not too heavily loaded, electron microscope analysis can be used to estimate size distribution, and identify particle type and shape. Aerosols can be collected on filters, either at the Earth's surface, or by airborne instruments on balloon or aircraft.

Instead of being collected, aerosols can be directly observed in situ by instruments that measure particles' physical and optical properties. Many of these instruments make use of the particles' scattering and absorption properties by shining a laser on the particles and then measuring the attenuation of the laser light or the amount or direction of the light scattered. Other instruments measure changes to particles in varying humidity fields, temperatures or electrostatic charges in order to characterize hygroscopicity, chemical composition or shape.

Although very useful, in situ methods by their very nature disturb the natural aerosol as they make their measurements. Particles can be lost in the tubing on their way to the measurement chamber, or they can be dried to the point of not resembling their initial ambient size or refractive index. Furthermore, in situ measurements can provide only limited results, both in time and in space. There is a strong need for aerosol observations on a global basis through the entire column of the atmosphere, that is best obtained by remote sensing.

We have seen that aerosols are a very complex component of the atmosphere, and that their influence depends not only on the total loading, but also on several characteristics, as chemical composition, size, shape, and distribution in the atmosphere. These characteristics are determined by the particles' source and transport. Some of these characteristics are more readily measured by remote sensing than others.

Our own eyes provide remote sensing observations. Our eyes can look up at the sky and distinguish between a "clean day" with bright blue skies or a "hazy day" with milky grayish skies even with no clouds present. What we are seeing is the difference in the scattering properties of the gas molecules that make up the bulk of our atmosphere and aerosols. When we look at the sky away from the sun, we are seeing sunlight that has reached either a gas molecule or a suspended aerosol particle, and then has been scattered towards our eye. The gas molecules have a strong spectral signature to their scattering that favors shortwave radiation and scatters blue light to our eyes. Aerosol particles, larger than the gas molecules tend to have less strong spectral dependence. So as the amount of aerosols increases and as these particles become larger, the light reaching our eyes is less blue and more spectrally neutral or white. All aerosol remote sensing techniques make use of the resulting interaction between some component of electromagnetic radiation and the particles of interest, very similar to how our eyes distinguish between clean and hazy conditions from the color of the sky.

1.4 Objective and organization of the book

This book concerns the remote sensing of aerosols, based on analyzing the impact of aerosols on radiation at any wavelength from ultraviolet to thermal infrared. As underlined in the preface, we try to present in a comprehensive and detailed manner the basic physical principles used by remote sensing instruments and retrieval algorithms.

While cloud droplets and crystals are in many ways simply bigger aerosol particles, and the remote sensing of clouds often runs parallel to the remote sensing of aerosols, this book explores only aerosol remote sensing, and not clouds. However, the book offers a comprehensive presentation of aerosol remote sensing from the basic physical principles of how radiation interacts with a small particle and how measurements of the radiation field can be exploited to reveal substantial information about the characteristics of that particle (Chapters 2, 3, 4 and 5). Chapter 6 concerns ground-based shortwave remote sensing instruments and networks, and Chapter 7 covers the history of satellite passive remote sensing. Chapter 8 reviews most of the recent or present satellite-borne instruments.

Longwave remote sensing uses the infrared radiation emitted by the Earth's surface and by the atmosphere; it relies mainly on absorption, re-emission by aerosols, with a contribution of scattering. Chapter 9 presents the principles of longwave remote sensing, with the main instruments.

Active remote sensing uses artificial radiation sources. Presently the principal source used for active remote sensing is the lidar; it can be ground based, airborne, or spaceborne. Lidar remote sensing is complementary to passive remote sensing, as it carries information on the aerosol vertical profile. Chapter 10 presents the principle of lidar, with the lidar equation; the various types of lidar are reviewed, and the major experiments presented.

Finally, in Chapter 11, we present summary tables describing historical and current instruments that use remote sensing techniques to derive aerosol information, discuss the need for validation of remote sensing products and suggest some priorities for future work.

2 Absorption and scattering by molecules and particles

Jacqueline Lenoble, Michael I. Mishchenko, and Maurice Herman

2.1 Introduction

The Earth's atmosphere absorbs, scatters, and emits electromagnetic radiation. Although air molecules are the primary actors in these processes, aerosol particles are also present ubiquitously (see Chapter 1) and modify the radiation field. In fact, this modification constitutes the very physical basis of aerosol remote sensing. Whenever clouds are present, they have a much larger influence on radiation which largely overshadows the aerosol impact. Therefore, in aerosol remote sensing, one often has to limit observations to cloudless conditions and screen cloudy pixels.

In the solar part of the spectrum, molecular absorption is mostly limited to ultraviolet (UV; ozone) and near-infrared (near-IR; carbon dioxide, water vapor) wavelengths and is characterized by strong and narrow oxygen bands. A brief description of atmospheric molecular absorption is presented in Section 2.2. Shortwave aerosol remote sensing is usually performed outside the absorption bands, but some instruments also have channels capturing absorption bands with the objective of quantifying gaseous components.

Absorption in the longwave terrestrial spectrum, both by molecules and aerosols, is accompanied by emission, according to Kirchhoff's law. This subject will be addressed in Chapter 9.

In Sections 2.3 and 2.4, we present some basic definitions concerning the scattering and polarization phenomena; the same formalism will be used throughout this book. Section 2.5 deals with Rayleigh molecular scattering, Section 2.6 summarizes the theory of scattering by spherical particles (the Lorenz–Mie theory), and Section 2.7 addresses the scattering by nonspherical particles. Finally, in Section 2.8, we discuss the main traits of single-scattering and absorption characteristics of spherical and nonspherical aerosols.

2.2 Atmospheric molecular absorption

Solar radiation at very short wavelengths is strongly absorbed in the upper atmosphere, mainly by oxygen in the Schumann–Runge and Herzberg bands (Greenblatt et al., 1990), and, therefore, is not very useful for remote sensing. At wavelengths above 250 nm, the ozone absorption becomes dominant, in the Hartley and Huggins bands, and limits the ability to observe the solar spectrum at the Earth's surface around 300 nm. Figure 2.1 depicts the absorption cross-section of ozone in the UV region. The Hartley and Huggins bands do not yield a line structure, but rather cause small oscillations superposed on the continuum. Numerous measurements of the ozone absorption have been performed over the past decades (Vigroux, 1953; Molina and Molina, 1986; Paur and Bass, 1985; Burrows et al., 1999b), especially in the Huggins bands, and have demonstrated the absorption to be strongly temperature dependent. Remote sensing in the UV can be used to retrieve simultaneously aerosols and ozone, as, e.g., with TOMS and OMI instruments (see Chapters 7 and 8). In the visible, ozone causes the much weaker Chappuis absorption band which is not temperature dependent (Amuroso et al., 1990), as shown in Figure 2.2. This band is used in remote sensing with SAGE and similar occultation instruments (see Chapter 8).

Minor gaseous constituents, such as NO_2, NO_3, OClO, and SO_2, also cause some absorption bands in the UV and visible parts of the spectrum, thereby facilitating remote

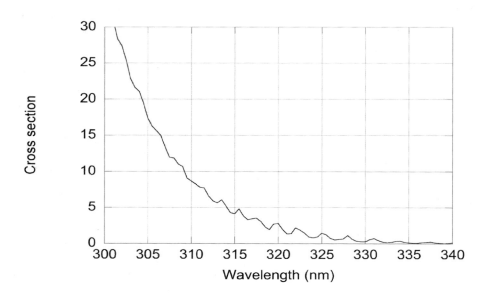

Figure 2.1 Ozone UV absorption cross section $\times 10^{-20}$ (cm^2) at 230 K.

sensing of these gases. The oxygen molecule causes a few narrow bands in the visible and near-IR, the most important bands being centered at 690 and 760 nm. Measurements of the 760-nm band, called the A band, are used in remote sensing of cloud top heights (Fischer and Grassl, 1991).

In the near-IR, besides the strong-absorption bands due to H_2O and CO_2 (see the spectroscopic data banks HITRAN at www.cfa.harvard.edu/hitran, and GEISA at ether.ipsl. jussieu.fr), several species cause absorption bands also used for remote sensing. However, this topic is beyond the scope of our discussion of aerosol remote sensing.

2.3 Definitions: Extinction, scattering, absorption, and phase function

The radiative energy is most generally characterized by the radiant flux density (Wm^{-2}); for simplicity it is called intensity I in this chapter, without considering its distribution in directions; I takes generally the name of irradiance, when received on a surface. In Chapter 3, we will define the radiance L (Wm^{-2}sr^{-1}), used in radiative transfer analysis.

Let us consider a horizontal layer of infinitesimal thickness ds inside the atmosphere or any other medium composed of sparsely distributed scatterers (Figure 2.3a). A paral-

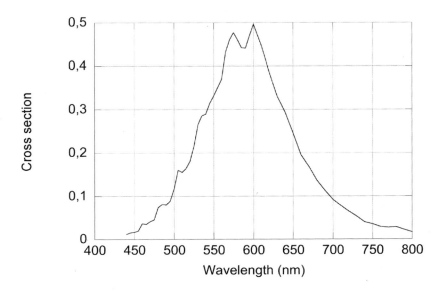

Figure 2.2 Ozone visible absorption cross section $\times 10^{-20}$ (cm^2).

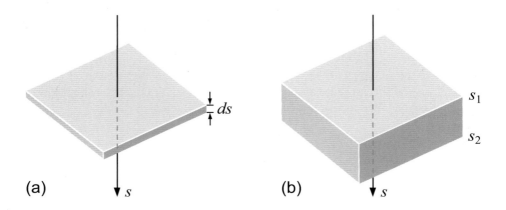

Figure 2.3 Definition of extinction.

lel beam of monochromatic radiation incident from above perpendicularly to this layer is characterized by its intensity I (Wm^{-2}) at the entrance surface of the layer and an intensity $I + dI$ at the exit surface. According to the microphysical theory of radiative transfer (Mishchenko et al., 2006), dI is proportional to the product of I and ds:

$$dI = -\sigma_e I ds, \tag{2.1}$$

where σ_e (m^{-1}) is the local extinction coefficient. Implicit in Eq. (2.1) is the assumption that the scattering medium is composed of spherically symmetric and/or randomly oriented nonspherical particles. Preferential orientation of nonspherical particles requires the introduction of a 4 × 4 extinction-matrix coefficient rather than the scalar extinction coefficient (Mishchenko et al., 2006).

The cumulative extinction is generally due to two different physical phenomena: absorption, wherein the radiative energy is transformed into another form of energy (e.g., via heating and photochemical reactions), and scattering, wherein a part of the incident light changes its direction of propagation and is lost to the incident energy flux. Both molecules and aerosols absorb and scatter radiation. The extinction coefficient is the sum of the absorption coefficient σ_a and the scattering coefficient σ_s:

$$\sigma_e = \sigma_a + \sigma_s. \tag{2.2}$$

To characterize a molecule or a particle, one generally uses the extinction, absorption, and scattering cross-sections (C_e, C_a, and C_s) expressed in m^2. The extinction, absorption, and scattering coefficients defined above are obtained by multiplying the respective cross-sections by the number of molecules or particles per unit volume. The fractional contribution of scattering to the total extinction is given by the single-scattering albedo (SSA):

$$\varpi = \sigma_s / \sigma_e. \tag{2.3}$$

For a finite-thickness layer located between s_1 and s_2 (Figure 2.3b), the integration of Eq. (2.1) yields

$$I(s_2) = I(s_1)\exp(-\tau_e), \tag{2.4}$$

where

$$\tau_e = \int_{s_1}^{s_2} \sigma_e(s)ds \tag{2.5}$$

is the optical thickness of the layer and $\exp(-\tau_e)$ is its optical transmittivity. Of course τ_e can be decomposed into the sum of the absorption, τ_a, and scattering, τ_s, optical thicknesses. Quite often the subscript "e" is omitted when implicit from the context. Equation (2.4) is known as Bouguer-Beer's exponential extinction law and, strictly speaking, applies only to monochromatic radiation. However, it can also be applied to narrow wavelength intervals over which the intensity and the extinction vary slowly. This is the case for scattering (both molecular and aerosol), aerosol absorption, and ozone absorption, but not for gaseous line absorption.

In the atmosphere, the optical depth is traditionally measured along a vertical path and is equal to the optical thickness of the atmospheric layer above a given altitude. The total optical thickness of the atmosphere corresponds to the optical depth at the surface level.

The scattered radiation is lost to the initial parallel energy flux and has a probability to be distributed in any direction depending on the specific type of scatterers. In the case of unpolarized incident light (e.g., sunlight), the phase function $p(\Theta)$ expresses this probability as a function of the scattering angle $\Theta \in [0, \pi]$, i.e., the angle between the incidence and scattering directions (see Figure 2.4 showing the plane of scattering defined by these two directions). The amount of monochromatic radiative power scattered by an elementary volume dV of the scattering medium into a solid angle $d\Omega$ around the direction Θ is given by

$$d\tilde{I} = \sigma_s p(\Theta)IdVd\Omega / 4\pi, \tag{2.6}$$

where $\tilde{I}$ is measured in W. The conservation of energy implies that $p(\Theta)$ is normalized to 4π when integrated over all scattering directions. As before, we assume that the scattering medium is composed of spherically symmetric and/or randomly oriented nonspherical particles.

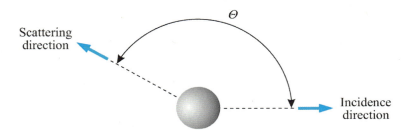

Figure 2.4 Electromagnetic scattering.

It is customary and convenient to expand the phase function in a series of Legendre polynomials P_l:

$$p(\Theta) = \sum_{l=0}^{l=L} \beta_l P_l(\cos\Theta), \qquad (2.7)$$

where

$$\beta_l = \frac{2l+1}{2} \int_{-1}^{+1} p(\Theta) P_l(\cos\Theta) d(\cos\Theta); \qquad (2.8)$$

the upper summation limit L is 2 for molecules and increases rapidly as the particle size exceeds the incident wavelength (Section 2.8). The normalization of the phase function implies that $\beta_0 = 1$.

The asymmetry factor is defined as

$$g = \langle\cos\Theta\rangle = \int_{-1}^{+1} p(\Theta)\cos\Theta d(\cos\Theta) / \int_{-1}^{+1} p(\Theta)d(\cos\Theta) = \beta_1/3. \qquad (2.9)$$

A useful analytical representation of the phase function is the Henyey–Greenstein function (Henyey and Greenstein, 1941)

$$p(\Theta) = \frac{1-g^2}{(1+g^2-2g\cos\Theta)^{3/2}}; \qquad (2.10)$$

it is used quite often and usually gives good results in radiative flux computations.

As stated above, the number of terms in the expansion (2.7) increases rapidly for large particles, when the phase function exhibits a very strong forward peak (see Section 2.8). Therefore, a useful approximation is to write the phase function as the sum of a delta-function term and a much smoother phase-function component:

$$p(\Theta) = 2f\Delta(1 - \cos\Theta) + (1 - f)p'(\Theta), \tag{2.11}$$

where $p'(\Theta)$ can be expanded into series of Legendre polynomials with many fewer terms than $p(\Theta)$. Physically, this approximation implies that the radiation scattered in the forward peak is simply transmitted; the constant f is empirically defined, depending on how the forward peak is truncated (Potter, 1970).

2.4 Polarization and scattering matrix

In general, the scattering process modifies the state of polarization of the radiation incident on a molecule or a particle. In order to account for this phenomenon appropriately, especially when one has to deal with two or more successive scattering events (see Chapter 3), it is first necessary to choose a consistent representation of polarized radiation.

2.4.1 The Stokes parameters

A time-harmonic plane electromagnetic wave propagating in the direction of the wave vector $\mathbf{k}$ is characterized by two orthogonal components $\mathbf{E}_l$ and $\mathbf{E}_r$ of its complex electric vector $\mathbf{E}(\mathbf{r}, t) = \mathbf{E}\exp(i\omega t - i\mathbf{k} \cdot \mathbf{r})$ defined in the wave plane (i.e., the plane normal to $\mathbf{k}$), where t is time, $\mathbf{r}$ is the position vector of the observation point, ω is the angular frequency, and $i = \sqrt{-1}$. The subscripts "l" and "r" denote the components parallel and perpendicular to a reference plane, respectively (Figure 2.5).

Because of high frequency of the time-harmonic oscillations, traditional optical instruments cannot measure the electric and magnetic fields associated with the electromagnetic wave. Instead, optical instruments usually measure quantities that have the dimension of energy flux and are combinations of the four products $E_l E_l^*, E_r E_r^*, E_r E_l^*$, and $E_l E_r^*$, where the asterisk denotes a complex-conjugate value. In particular, the real-valued Stokes parameters I, Q, U, and V of the plane wave (Stokes, 1852; Chandrasekhar, 1950) form a so-called Stokes column vector $\mathbf{I}$ defined as

$$\mathbf{I} = \begin{pmatrix} I \\ Q \\ U \\ V \end{pmatrix} = \frac{\varepsilon_0 c}{2} \begin{pmatrix} E_l E_l^* + E_r E_r^* \\ E_l E_l^* - E_r E_r^* \\ E_l E_r^* + E_r E_l^* \\ iE_l E_r^* - iE_r E_l^* \end{pmatrix} = \frac{\varepsilon_0 c}{2} \begin{pmatrix} E_{l0}^2 + E_{r0}^2 \\ E_{l0}^2 - E_{r0}^2 \\ 2E_{l0}E_{r0}\cos\delta \\ 2E_{l0}E_{r0}\sin\delta \end{pmatrix}, \tag{2.12}$$

where E_{l0} and E_{r0} are the amplitudes of the complex time-harmonic components E_l and E_r,

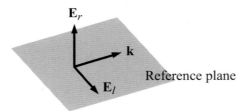

Figure 2.5 Electric field vector components. The reference plane contains the vectors **k** and $\mathbf{E}_l$. The wave plane goes through the vectors $\mathbf{E}_r$ and $\mathbf{E}_l$.

respectively; δ is the retardation in the phase of E_l relative to that of E_r; c is the speed of light in a vacuum; and ε_0 the dielectric constant of a vacuum. All four Stokes parameters have the dimension of the scalar quantity I (Wm^{-2}). If the Stokes vector is associated with radiance, the Stokes parameters are in Wm^{-2}sr^{-1}.

A plane electromagnetic wave is the simplest type of electromagnetic radiation and is well represented by a perfectly monochromatic and perfectly parallel laser beam. Let us consider an arbitrary point within such a beam. It is straightforward to demonstrate (see, e.g., Mishchenko et al., 2002, 2006) that during each time interval $2\pi/\omega$, the end-point of the real electric field vector Re[$\mathbf{E}(\mathbf{r},t)$] describes an ellipse in the wave plane. The sum of the squares of the semi-axes of this ellipse, multiplied by $\varepsilon_0 c/2$, yields the total intensity of the wave I. The ratio of the semi-axes, the orientation of the ellipse, and the sense in which the electric vector rotates (clockwise or counter-clockwise, when looking in the direction of propagation) can be derived from the other three Stokes parameters of the wave, Q, U, and V. Importantly, any plane electromagnetic wave is fully polarized, i.e., satisfies the Stokes identity

$$I^2 \equiv Q^2 + U^2 + V^2. \tag{2.13}$$

Quasi-monochromatic beams of light are encountered much more often than perfectly monochromatic beams and, in general, are described by $\mathbf{E}(\mathbf{r},t) = \mathbf{E}(t)\exp(i\omega t - i\mathbf{k}\cdot\mathbf{r})$, where fluctuations in time of the complex amplitude of the electric field $\mathbf{E}(t)$ around its mean value occur much more slowly than the harmonic oscillations of the time factor $\exp(i\omega t)$, albeit still too fast to be detected by an actual optical detector of electromagnetic energy flux. Therefore, the Stokes parameters measured by the detector are obtained by taking an average of the right-hand side of Eq. (2.12) over a time interval much longer than $2\pi/\omega$:

$$\mathbf{I} = \begin{pmatrix} I \\ Q \\ U \\ V \end{pmatrix} = \frac{\varepsilon_0 c}{2} \begin{pmatrix} \langle E_l E_l^* \rangle + \langle E_r E_r^* \rangle \\ \langle E_l E_l^* \rangle - \langle E_r E_r^* \rangle \\ \langle E_l E_r^* \rangle + \langle E_r E_l^* \rangle \\ i\langle E_l E_r^* \rangle - i\langle E_r E_l^* \rangle \end{pmatrix} = \frac{\varepsilon_0 c}{2} \begin{pmatrix} \langle E_{l0}^2 \rangle + \langle E_{r0}^2 \rangle \\ \langle E_{l0}^2 \rangle - \langle E_{r0}^2 \rangle \\ \langle 2 E_{l0} E_{r0} \cos\delta \rangle \\ \langle 2 E_{l0} E_{r0} \sin\delta \rangle \end{pmatrix}. \tag{2.14}$$

These average Stokes parameters contain all practically available information on the quasi-monochromatic parallel beam (Chandrasekhar, 1950; Mishchenko et al., 2006). I is the total intensity of the beam considered in the previous section. The other three Stokes parameters have the same dimension as I (Wm^{-2}) and satisfy the inequality

$$I^2 \geq Q^2 + U^2 + V^2. \tag{2.15}$$

In general, the end-point of the vector Re$[\mathbf{E}(\mathbf{r}, t)]$ of a quasi-monochromatic beam does not describe a well-defined polarization ellipse. Still, one can think of a "preferential" ellipse with a "preferential orientation", "preferential elongation", and "preferential handedness". Equation (2.15) implies that a quasi-monochromatic beam can be partially polarized and even unpolarized. The latter means that the temporal behavior of Re$[\mathbf{E}(\mathbf{r}, t)]$ is completely "erratic", so that there is no "preferential ellipse". This is, for example, the case for the extraterrestrial solar radiation.

When two or more quasi-monochromatic beams propagating in the same direction are mixed incoherently, which means that there is no permanent phase relation between the separate beams, then the Stokes column vector of the mixture is equal to the sum of the Stokes column vectors of the individual beams:

$$\mathbf{I} = \sum_n \mathbf{I}_n, \tag{2.16}$$

where n numbers the beams. According to Eqs (2.15) and (2.16), it is always possible mathematically to decompose any quasi-monochromatic beam into two incoherent parts, one unpolarized, with a Stokes column vector

$$\begin{bmatrix} I - \sqrt{Q^2 + U^2 + V^2} \\ 0 \\ 0 \\ 0 \end{bmatrix},$$

and one fully polarized, with a Stokes column vector

$$
\begin{bmatrix}
\sqrt{Q^2 + U^2 + V^2} \\
Q \\
U \\
V
\end{bmatrix}.
$$

Thus, the intensity of the fully polarized component is $(Q^2 + U^2 + V^2)^{1/2}$, and so the degree of (elliptical) polarization of the quasi-monochromatic beam can, in general, be defined as

$$
P = \frac{\sqrt{Q^2 + U^2 + V^2}}{I}. \tag{2.17}
$$

The parameters of the "preferential ellipse" of a parallel quasi-monochromatic beam of light can be viewed as quantitative descriptors of the asymmetry in the directional distribution and/or rotation direction distribution of the vector $\mathrm{Re}[\mathbf{E}(\mathbf{r},t)]$ in the wave plane. For example, a common type of polarization encountered in natural conditions is partial or full linear polarization, which implies that the Stokes parameter V is negligibly small. This type of polarization is often described by the total intensity, degree of linear polarization, and angle α defined in the wave plane and specifying the orientation of the plane of preferential oscillations of the real electric field vector with respect to the reference plane (Figure 2.5). The degree of linear polarization is given by

$$
P = \frac{\sqrt{Q^2 + U^2}}{I}, \tag{2.18}
$$

the second and third Stokes parameters are given by

$$
\begin{aligned}
Q &= IP\cos 2\alpha, \\
U &= IP\sin 2\alpha,
\end{aligned} \tag{2.19}
$$

and the orientation angle is given by

$$
\alpha = \tfrac{1}{2}\,\mathrm{arctg}\,(U/V). \tag{2.20}
$$

In general, the parameter V is nonzero and defines the preferential ellipticity of the beam. However, V is usually very small in the atmosphere and is often neglected, the polarization being considered (partially) linear.

It is clear from the above discussion that the Stokes parameters are always defined with respect to a reference plane containing the direction of light propagation (Figure 2.5). If the reference plane is rotated about the direction of propagation then the Stokes parameters are modified according to a rotation transformation rule. Specifically, consider a rotation of the reference plane through an angle $0 \leq \chi < 2\pi$ in the anti-clockwise direction when looking in the direction of propagation. Then the new Stokes column vector is given by

$$\mathbf{I'} = \mathbf{T}(\chi)\mathbf{I}, \tag{2.21}$$

where

$$\mathbf{T}(\chi) = \begin{bmatrix} 1 & 0 & 0 & 0 \\ 0 & \cos 2\chi & \sin 2\chi & 0 \\ 0 & -\sin 2\chi & \cos 2\chi & 0 \\ 0 & 0 & 0 & 1 \end{bmatrix} \tag{2.22}$$

is the Stokes rotation matrix for angle χ.

2.4.2 Scattering matrix

Upon choosing the reference axes parallel and perpendicular to the scattering plane for both the incident and scattered radiation, Eq. (2.6) is replaced by

$$d\tilde{\mathbf{I}} = \sigma_s \mathbf{F}(\Theta)\mathbf{I}dVd\Omega / 4\pi, \tag{2.23}$$

where

$$\mathbf{F}(\Theta) = [F_{ij}(\Theta)], \quad i, j = 1,...,4 \tag{2.24}$$

is the 4×4 so-called normalized Stokes scattering matrix such that $F_{11}(\Theta) = p(\Theta)$. For an ensemble of randomly oriented particles each of which has a plane of symmetry and/or for an ensemble containing an equal number of particles and their mirror counterparts in random orientation, the scattering matrix has the following simplified structure and only six independent elements:

$$\mathbf{F}(\Theta) = \begin{bmatrix} F_{11}(\Theta) & F_{12}(\Theta) & 0 & 0 \\ F_{12}(\Theta) & F_{22}(\Theta) & 0 & 0 \\ 0 & 0 & F_{33}(\Theta) & F_{34}(\Theta) \\ 0 & 0 & -F_{34}(\Theta) & F_{44}(\Theta) \end{bmatrix} \tag{2.25}$$

with

$$F_{12}(0) = F_{12}(\pi) = F_{34}(0) = F_{34}(\pi) = 0, \tag{2.26}$$

$$F_{33}(\pi) = -F_{22}(\pi), \tag{2.27}$$

$$F_{11}(\pi) - 2F_{22}(\pi) - F_{44}(\pi) = 0 \tag{2.28}$$

(van de Hulst, 1957; Mishchenko et al., 2002). Still further simplifications occur for spherical particles:

$$\mathbf{F}(\Theta) = \begin{bmatrix} F_{11}(\Theta) & F_{12}(\Theta) & 0 & 0 \\ F_{12}(\Theta) & F_{11}(\Theta) & 0 & 0 \\ 0 & 0 & F_{33}(\Theta) & F_{34}(\Theta) \\ 0 & 0 & -F_{34}(\Theta) & F_{33}(\Theta) \end{bmatrix}, \tag{2.29}$$

$$F_{33}(0) = F_{11}(0). \tag{2.30}$$

Several recipes can be used to check the physical correctness of the elements of a scattering matrix found as the outcome of laboratory measurements or theoretical computations (Hovenier and van der Mee, 2000).

A useful expansion of the elements of the scattering matrix in so-called generalized spherical functions (Kuščer and Ribarič, 1959; Siewert, 1982; Lenoble et al., 2007) is as follows:

$$F_{11}(\Theta) = p(\Theta) = \sum_{l=0}^{L} \beta_l P_l(\cos\Theta),$$

$$F_{21}(\Theta) = F_{12}(\Theta) = \sum_{l=2}^{L} \gamma_l P_2^l(\cos\Theta),$$

$$F_{22}(\Theta) = \sum_{l=2}^{L} [\alpha_l R_2^l(\cos\Theta) + \zeta_l T_2^l(\cos\Theta)],$$

$$F_{33}(\Theta) = \sum_{l=2}^{L} [\varsigma_l R_2^l(\cos\Theta) + \alpha_l T_2^l(\cos\Theta)], \tag{2.31}$$

$$F_{43}(\Theta) = -F_{34}(\Theta) = \sum_{l=2}^{L} \varepsilon_l P_2^l(\cos\Theta),$$

$$F_{44}(\Theta) = \sum_{l=0}^{L} \delta_l P_l(\cos\Theta),$$

where

$$R_s^l(\cos\Theta) = \frac{1}{2}[P_{s2}^l(\cos\Theta) + P_{s,-2}^l(\cos\Theta)], \tag{2.32a}$$

$$T_s^l(\cos\Theta) = \frac{1}{2}[P_{s2}^l(\cos\Theta) - P_{s,-2}^l(\cos\Theta)]. \tag{2.32b}$$

The $P_{mn}^l(x)$ are the generalized spherical functions introduced by Gel'fand and Shapiro (1956); they reduce to the usual Legendre polynomials P_l for $m = n = 0$. The P_m^l are the standard associated Legendre functions. The first expansion of Eq. (2.31) is, of course, identical to Eq. (2.7). An alternative form of Eq. (2.31) can be found in Mishchenko et al. (2002, 2006), and Hovenier et al. (2004).

2.5 Molecular scattering: Rayleigh theory

The theory of molecular scattering was first developed by Strutt (1871), later known as Lord Rayleigh (1889). He assumed that the incident electromagnetic wave induces an electric dipole moment at the same frequency in the molecule. This dipole emits, according to the classical electromagnetic theory, at the same wavelength. For incident natural radiation, the radiation scattered at $\Theta = 90°$ must be completely polarized, with vibrations of the electric field vector perpendicular to the scattering plane. However, this theoretical prediction has not been confirmed by observations revealing a small depolarization at $90°$. Later on, a correction for the molecular anisotropy was introduced by Cabannes (1929) in order to explain why the degree of polarization is not 100% at $\Theta = 90°$. The depolarization factor d is defined as the ratio of the parallel to the perpendicular components of the electric field vector at $\Theta = 90°$. It has been measured several times, but is still subject to uncertainty, leading to an uncertainty of about 2% in the so-called King factor $(6+3d)/(6-7d)$ appearing in Eq. (2.33) below.

The molecular scattering coefficient for dry air is

$$\sigma_{s,R} = \frac{24\pi^3}{\lambda^4}\left(\frac{m_s^2-1}{m_s^2+2}\right)^2 \frac{6+3d}{6-7d}\frac{N}{N_s^2}, \tag{2.33}$$

where $\lambda = 2\pi/k$ is the wavelength, $k = |\mathbf{k}|$ is the wave number, m_s is the refractive index for standard air (defined as dry air containing 0.03% CO_2 at a pressure of 1013.25 hPa and a temperature of 15°C), N is the molecular density, and N_s is the same quantity in standard conditions. The m_s is also subject to some uncertainty and slightly depends on the wavelength. Bodhaine et al. (1999) carefully analyzed the data available for computing the molecular scattering coefficient and their uncertainties. They proposed the following best-fit equation for the Rayleigh scattering cross section defined, as above, by $C_{s,R} = \sigma_{s,R}/N$:

$$C_{s,R}(\times 10^{-28} \text{cm}^2) = \frac{1.0455996 - 341.29061\lambda^{-2} - 0.90230850\lambda^2}{1 + 0.0027059889\lambda^{-2} - 85.968563\lambda^2}. \qquad (2.34)$$

The corresponding Rayleigh scattering matrix is as follows:

$$\mathbf{F}_R(\Theta) = K \begin{bmatrix} \frac{3}{4}(1 + \cos^2\Theta) & -\frac{3}{4}\sin^2\Theta & 0 & 0 \\ -\frac{3}{4}\sin^2\Theta & \frac{3}{4}(1 + \cos^2\Theta) & 0 & 0 \\ 0 & 0 & \frac{3}{2}\cos\Theta & 0 \\ 0 & 0 & 0 & K'\frac{3}{2}\cos\Theta \end{bmatrix} + (1-K) \begin{bmatrix} 1 & 0 & 0 & 0 \\ 0 & 0 & 0 & 0 \\ 0 & 0 & 0 & 0 \\ 0 & 0 & 0 & 0 \end{bmatrix},$$

$$(2.35)$$

where

$$K = \frac{1 - d}{1 + d/2}, \qquad (2.36)$$

$$K' = \frac{1 - 2d}{1 - d}. \qquad (2.37)$$

2.6 Lorenz–Mie theory

The complete theory of electromagnetic scattering by an individual spherical particle was first presented by Gustav Mie (Mie, 1908; see Mishchenko and Travis, 2008 for a historical perspective on this seminal development). Detailed accounts of the Lorenz–Mie theory can be found in Stratton (1941), van de Hulst (1957), Bohren and Huffman (1983), and Lenoble (1993). The incident electromagnetic field is expressed as a linear combination of elementary solutions of the vector wave equation in spherical coordinates called vector spherical wave functions (VSWFs). The scattered and internal fields are expressed similarly, with unknown coefficients being obtained from the boundary conditions at the particle surface.

Consider an isolated homogeneous spherical particle, having a radius r, illuminated by a plane electromagnetic wave, as defined in Section 2.4.1. Using the scattering plane for reference (Figures 2.4 and 2.5), the scattered field vector components at a large distance ρ from the particle (i.e., in the so-called far-field zone; see Section 3.2 of Mishchenko et al., 2006) are given by

$$E_{sl} = \frac{i}{k\rho} \exp(-ik\rho) S_2(\Theta) E_{0l}, \qquad (2.38a)$$

$$E_{sr} = \frac{i}{k\rho}\exp(-ik\rho)S_1(\Theta)E_{0r},\tag{2.38b}$$

where S_1 and S_2 are complex functions of the scattering angle given by the following series:

$$S_1(\Theta) = \sum_{n=1}^{n_{max}} \frac{2n+1}{n(n+1)}\left(a_n\frac{d}{d\Theta}P_n^1(\cos\Theta) + b_n\frac{1}{\sin\Theta}P_n^1(\cos\Theta)\right),\tag{2.39a}$$

$$S_2(\Theta) = \sum_{n=1}^{n_{max}} \frac{2n+1}{n(n+1)}\left(b_n\frac{d}{d\Theta}P_n^1(\cos\Theta) + a_n\frac{1}{\sin\Theta}P_n^1(\cos\Theta)\right).\tag{2.39b}$$

The so-called Mie coefficients a_n correspond to magnetic oscillations, while the b_n correspond to electric oscillations. They are determined from the boundary conditions at the particle surface and depend on the complex refractive index of the particle relative to the surrounding medium m and on the particle size parameter $x = kr$.

The number of numerically significant terms n_{max} in the series (2.39) is of the order of $2x+3$; the same number is necessary in the expansions (2.31). Therefore, the practical application of the Lorenz–Mie theory for particles larger than the wavelength was delayed until the advent of modern computers. At present, several numerically efficient and highly accurate Lorenz–Mie codes are publicly available on the Internet (see, e.g., ftp://ftp.giss. nasa.gov/pub/crmim/spher.f).

From Eqs (2.38) and the definition of the Stokes parameters, it is straightforward to show that the non-zero elements of the normalized Lorenz–Mie scattering matrix are as follows:

$$F_{11}(\Theta) = F_{22}(\Theta) = p(\Theta) = \frac{\lambda^2}{2\pi C_s}[S_1(\Theta)S_1^*(\Theta) + S_2(\Theta)S_2^*(\Theta)],\tag{2.40a}$$

$$F_{21}(\Theta) = F_{12}(\Theta) = \frac{\lambda^2}{2\pi C_s}[S_2(\Theta)S_2^*(\Theta) - S_1(\Theta)S_1^*(\Theta)],\tag{2.40b}$$

$$F_{33}(\Theta) = F_{44}(\Theta) = \frac{\lambda^2}{2\pi C_s}[S_2(\Theta)S_1^*(\Theta) + S_1(\Theta)S_2^*(\Theta)],\tag{2.40c}$$

$$F_{34}(\Theta) = -F_{43}(\Theta) = \frac{i\lambda^2}{2\pi C_s}[S_1(\Theta)S_2^*(\Theta) - S_2(\Theta)S_1^*(\Theta)],\tag{2.40d}$$

where C_s is the corresponding scattering cross section. The latter is often expressed in terms of the scattering efficiency factor Q_s as follows:

$$C_s = \pi r^2 Q_s, \tag{2.41}$$

with

$$Q_s = \frac{2}{x^2} \sum_{n=1}^{n_{max}} (2n+1)(|a_n|^2 + |b_n|^2). \tag{2.42}$$

Similarly, the extinction cross section is represented as

$$C_e = \pi r^2 Q_e, \tag{2.43}$$

while the extinction efficiency factor is given by

$$Q_e = \frac{2}{x^2} \sum_{n=1}^{n_{max}} (2n+1)\,\mathrm{Re}(a_n + b_n), \tag{2.44}$$

where, as before, Re stands for the real part. If the particle absorbs radiation (i.e., the relative refractive index m has a non-zero imaginary part) then the extinction cross section is greater than the scattering cross section; otherwise they are equal. The extinction and scattering efficiencies are examples of dimensionless scale-invariant quantities depending only on the ratio of the particle size to the wavelength rather than on r and λ separately provided that the relative refractive index m is wavelength-independent (see Section 3.5 of Mishchenko et al., 2006). The normalized Stokes scattering matrix $\mathbf{F}$ defined by Eq. (2.23) is another scale-invariant quantity.

For a polydisperse ensemble of spherical particles, the average scattering and the extinction cross sections are given by

$$C_s = \int_0^\infty \pi r^2 Q_s(x, m)n(r)dr, \tag{2.45a}$$

$$C_e = \int_0^\infty \pi r^2 Q_e(x, m)n(r)dr, \tag{2.45b}$$

where $n(r)dr$ is the fraction of particles with radii between r and $r+dr$ normalized such that

$$\int_0^\infty n(r)dr = 1. \tag{2.46}$$

The corresponding formula for the ensemble-averaged elements of the normalized scattering matrix is as follows:

$$F_{ij}(\Theta) = \frac{1}{C_s} \int_0^{\infty} \pi r^2 Q_s(x, m) n(r) F_{ij}(\Theta; x, m) dr. \tag{2.47}$$

2.7 Nonspherical particles: Theory and measurements

Although spherical (or nearly spherical) aerosols do exist (e.g., Figure 2.6a), many aerosol types exhibit complex particle morphologies (e.g., Figures 2.6b–d), thereby rendering the Lorenz–Mie theory potentially inapplicable. The optical properties of such nonspherical and/or heterogeneous particles must be either computed using an advanced theory of elec-

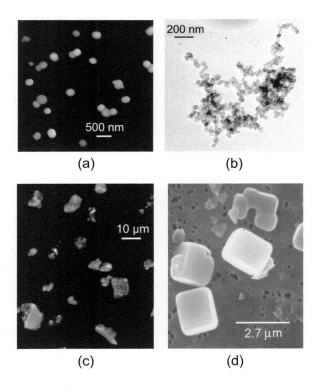

Figure 2.6 Examples of aerosol-particle morphologies. (a) Sub-micrometer-sized quasi-spherical ammonium sulphate and dust aerosols (after Weinzierl et al., 2009). (b) A soot aggregate (after Li et al., 2003). (c) Sahara-desert soil particles (after Weinzierl et al., 2009). (d) Dry sea-salt particles (after Chamaillard et al., 2003).

tromagnetic scattering or measured experimentally, both approaches having their strengths, weaknesses, and limitations. In this section, we provide a brief summary of the existing theoretical and experimental techniques for the determination of single-particle scattering and absorption characteristics. Detailed information and further references can be found in the books by Mishchenko et al. (2000, 2002) and Babenko et al. (2003) as well as in the review by Kahnert (2003).

2.7.1 Numerically-exact and approximate theoretical techniques

The majority of the existing exact theoretical approaches belong to one of two broad categories. Specifically, differential equation methods yield the scattered field via the solution of the Maxwell equations or the vector wave equation in the frequency or in the time domain, whereas integral equation methods are based on the volume or surface integral counterparts of the Maxwell equations.

The classical example of a differential equation method is the Lorenz–Mie theory discussed in the preceding section. By implementing a recursive procedure, one can generalize the Lorenz–Mie solution to deal with concentric multilayer spheres.

Like the Lorenz–Mie theory, the separation of variables method (SVM) for homogeneous or layered spheroids is a frequency-domain technique, wherein all fields and sources are assumed to vary in time according to the same factor $\exp(i\omega t)$. The SVM is based on solving the electromagnetic scattering problem in spheroidal coordinates by means of expanding the incident, internal, and scattered fields in appropriate vector spheroidal wave functions. The expansion coefficients of the incident field can be computed analytically, whereas the unknown expansion coefficients of the internal and scattered fields are determined by applying the appropriate boundary conditions. Unfortunately, the vector spheroidal wave functions are not orthogonal on the surface of a spheroid. Therefore, this procedure yields an infinite set of linear algebraic equations for the unknown coefficients which has to be truncated and solved numerically. The main limitation of the SVM is that it can be applied only to spheroidal scatterers, whereas its primary advantages are the applicability to spheroids with extreme aspect ratios and the ability to produce accurate benchmarks.

Another frequency-domain differential equation technique is the finite element method (FEM) which yields the scattered field via solving numerically the vector Helmholtz equation subject to the standard boundary conditions. The scattering object is intentionally imbedded in a finite computational domain, the latter being discretized into many cells with about 10 to 20 cells per wavelength. The electric field values are specified at the nodes of the cells and are initially unknown. Using the boundary conditions, the differential equation is converted into a matrix equation for the unknown node field values. The latter is solved using the standard Gaussian elimination or one of the preconditioned iterative techniques such as the conjugate gradient method. While scattering in the far-field zone is an open-space problem, the FEM is always implemented in a finite computational domain in order to limit the number of unknowns. Therefore, one has to impose approximate absorbing boundary conditions at the outer boundary of the computational domain, thereby suppressing wave reflections back into the computational domain and allowing the numerical analogs of the outward-propagating wave to exit the domain almost as if the domain were infinite. In principle, the FEM can be applied to arbitrarily shaped and inhomogeneous particles and is simple in terms of its concept and practical implementation. However, FEM computations are spread over the entire computational domain rather than confined to the scatterer itself. This makes the technique rather slow and limits particle size parameters to values less than about 15. The finite spatial discretization and the approximate absorbing boundary conditions limit the accuracy of the method.

Unlike the frequency-domain FEM, the finite difference time domain method (FDTDM) yields the solution of the electromagnetic scattering problem in the time domain by directly solving the Maxwell time-dependent curl equations (e.g., Taflove and Hagness, 2000; P. Yang et al., 2000). The space and time derivatives of the electric and magnetic fields are approximated using a finite difference scheme with space and time discretizations selected so that they constrain computational errors and ensure numerical stability of the algorithm. As before, the scattering object must be imbedded in a finite computational domain, which requires the imposition of absorbing boundary conditions as a model of scattering in the open space. Representing a scattering object with curved boundaries using rectangular grid cells causes a staircasing effect and increases numerical errors, especially for particles with large relative refractive indices. Since the FDTDM yields the near field in the time domain, a special near-zone to far-zone transformation must be implemented in order to compute the requisite scattered far field in the frequency domain. The FDTDM has the same advantages as the FEM and shares its limitations in terms of accuracy and size parameter range.

The interaction of an incident plane electromagnetic wave with an arbitrary particle can also be described fully by the frequency-domain so-called volume integral equation (VIE). The calculation of the scattered field using the VIE would be straightforward except that the internal electric field entering the integrand is unknown beforehand. Therefore, the VIE must first be solved for the internal field. The integral over the particle volume is approximated by partitioning the interior region into a large number N of small cubical cells with about 10 to 20 cells per wavelength and assuming that the electric field and the refractive index within each cell are constant. The resulting system of N linear algebraic equations for the N unknown internal fields is solved numerically. Once the internal fields are found, the scattered field is determined by evaluating the right-hand side of the original VIE. This version of the VIE method is known as the method of moments (MOM). The simple approach to solving the MOM matrix equation for the internal fields by using the standard Gaussian elimination is not practical for particle size parameters exceeding unity. The conjugate gradient method together with the fast Fourier transform can be applied to significantly larger size parameters and reduces computer memory requirements substantially. The traditional drawback of using a preconditioned iterative technique is that computations must be repeated anew for each illumination direction.

Another version of the VIE technique is the so-called discrete dipole approximation (DDA). Whereas the MOM deals with the *actual* electric field in the center of each cell, the DDA exploits the concept of *exciting* fields and is based on discretizing the particle into a number N of elementary polarizable units called dipoles. The form of the electromagnetic response of each dipole to the local exciting electric field is assumed to be known. The field exciting a dipole is a superposition of the external (incident) field and the partial fields scattered by all the other dipoles. This allows one to form a system of N linear equations for the N fields exciting the N dipoles. The numerical solution of the DDA matrix equation is then used to compute the N partial fields scattered by the dipoles and thus the total scattered field. The original formulation of the DDA in the mid-1970s was phenomenological; however, it has been demonstrated since that the DDA can be derived from the VIE and thus is closely related to the MOM.

The main advantages of the MOM and the DDA are that they automatically satisfy the asymptotic radiation condition at infinity and can be applied to inhomogeneous, anisotropic, and optically active scatterers. Furthermore, the actual computation is confined to the scatterer volume, thereby resulting in fewer unknowns than the differential equation methods. However, the numerical accuracy of the MOM and DDA is relatively low and improves rather slowly with increasing N, whereas the computer time grows rapidly with increasing size parameter. Another disadvantage of these techniques is the need to repeat the entire calculation for each new direction of incidence. Further information on the MOM and the DDA and their applications can be found in the recent review by Yurkin and Hoekstra (2007).

The classical Lorenz–Mie solution can be extended to a cluster of non-overlapping spheres by using the translation addition theorem for the participating VSWFs. The total field scattered by the multisphere cluster is represented as a superposition of individual (partial) fields scattered by each sphere. The external electric field illuminating the cluster and the individual fields scattered by the constituent spheres are expanded in VSWFs with origins at the individual sphere centers. The orthogonality of the VSWFs in the sphere boundary conditions is exploited by applying the translation addition theorem wherein a VSWF centered at one sphere origin is re-expanded about another sphere origin. This procedure ultimately results in a matrix equation for the scattered-field expansion coefficients of each sphere. A numerical computer solution of this equation for the specific incident field yields the partial scattered fields and thereby the total scattered field.

Alternatively, the numerical inversion of the cluster matrix equation yields sphere-centered transition matrices (or T matrices) that transform the expansion coefficients of the incident field into the expansion coefficients of the individual scattered fields. In the far-field region of the entire cluster, the individual scattered-field expansions can be transformed into a single expansion centered at a single origin inside the cluster. This procedure yields the cluster T matrix transforming the incident-wave expansion coefficients into the single-origin expansion coefficients of the total scattered field and can be used in the highly efficient semi-analytical averaging of scattering characteristics over cluster orientations (Mackowski and Mishchenko, 1996).

This so-called superposition method (SM) has been extended to spheres with one or more eccentrically positioned spherical inclusions as well as to clusters of spheroids in an arbitrary configuration. Owing to the analyticity of its mathematical formulation, the SM is capable of producing very accurate numerical results.

The T-matrix method (TMM) is based on the expansion of the incident field in VSWFs regular at the origin of the coordinate system and on the expansion of the scattered field outside a circumscribing sphere of the scatterer in VSWFs regular at infinity. The T matrix transforms the expansion coefficients of the incident field into those of the scattered field and, if known, can be used to compute any scattering characteristic of the particle. The TMM was initially developed by Waterman (1971) for single homogeneous objects, but has since been generalized to deal with multilayered scatterers and arbitrary clusters of nonspherical particles. For a homogeneous or concentrically layered sphere, all TMM formulas reduce to

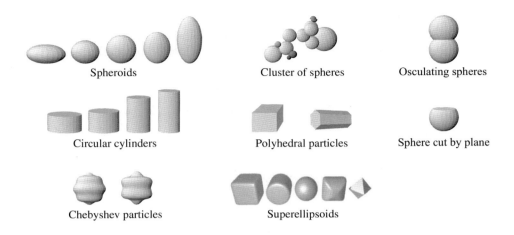

Spheroids   Cluster of spheres   Osculating spheres

Circular cylinders   Polyhedral particles   Sphere cut by plane

Chebyshev particles   Superellipsoids

Figure 2.7 Types of particles that can be treated with the T-matrix method.

those of the Lorenz–Mie theory. In the case of a cluster composed of spherical components, the TMM reduces to the multisphere SM mentioned above.

The T matrix for single homogeneous and multilayered scatterers is usually computed using the so-called extended boundary condition method (EBCM) applicable to any particle shape, although computations become much simpler and much more efficient for bodies of revolution. Special procedures have been developed to improve the numerical stability of EBCM computations for large size parameters and/or extreme aspect ratios. More recent work has demonstrated the practical applicability of the EBCM to particles without axial symmetry, e.g., ellipsoids, cubes, and finite polyhedral cylinders. The computation of the T matrix for a cluster of particles is based on the assumption that the T matrices of all components are known and is based on the use of the translation addition theorem for the VSWFs.

The loss of efficiency for particles with large aspect ratios or with shapes lacking axial symmetry is the main drawback of the TMM. The main advantages of the TMM are high accuracy and speed coupled with applicability to particles with equivalent-sphere size parameters exceeding 200. There are several semi-analytical orientation averaging procedures that make TMM computations for randomly oriented particles as fast as those for a particle in a fixed orientation.

Figure 2.7 shows examples of particles that can be treated using various implementations of the TMM. A representative collection of public-domain T-matrix computer programs has been available at http://www.giss.nasa.gov/staff/mmishchenko/t_matrix.html since 1996 and has been used in more than 780 peer-reviewed publications. These programs have been developed specifically to deal with axially symmetric particles and clusters of spherical monomers. Typical examples are spheroids, finite circular cylinders, Chebyshev particles

(Wiscombe and Mugnai, 1986), osculating spheres, spheres cut by a plane, and clusters of spherical particles with touching or separated components. In all cases, the scattering object in question can be randomly or preferentially oriented. The EBCM-based programs have been thoroughly tested against the SVM for spheroids. The very high numerical accuracy of the T-matrix codes has been used to generate benchmark results with five and more accurate decimals which can be used for testing other numerically-exact and approximate approaches. Extensive timing tests have shown that the numerical efficiency of these T-matrix codes is unparalleled, especially in computations for randomly oriented particles.

The SVM, SM, and TMM are the only methods that can yield very accurate results for particles comparable to and larger than a wavelength. The analytical orientation averaging procedure makes the TMM the most efficient technique for randomly oriented particles with moderate aspect ratios. Particles with larger aspect ratios can be treated with the SVM and an iterative EBCM. Computations for anisotropic objects and homogeneous and inhomogeneous particles lacking rotational symmetry often have to rely on more flexible techniques such as the FEM, FDTDM, MOM, and DDA. All these techniques are conceptually simple, can be easily implemented, and have comparable performance characteristics. However, their simplicity and flexibility are often accompanied by lower efficiency and accuracy and by stricter practical limitations on the range of size parameters and/or refractive indices. A comprehensive collection of computer programs based on various exact numerical techniques is available at http://www.scattport.org/.

The practical importance of approximate theories of electromagnetic scattering diminishes as computers become more efficient while numerically-exact techniques mature and become applicable to a wider range of problems. However, at least one approximation, the geometrical optics method (GOM), is not likely to become obsolete in the near future since its accuracy often improves as the particle size parameter grows, whereas all numerically-exact theoretical techniques for nonspherical particles become inapplicable whenever the size parameter exceeds a certain threshold. The GOM is a phenomenological approach to the computation of electromagnetic scattering by an arbitrarily shaped particle with a size much greater than the wavelength of the incident light. It is based on the assumption that the incident plane wave can be represented as a collection of "independent (or incoherent) parallel rays". The history of each ray impinging on the particle boundary is traced individually using Snell's law and Fresnel's formulas. Each incident ray is partially reflected and partially refracted into the particle. The refracted ray may emerge after an inside–out refraction, possibly preceded by one or more internal reflections, and can be attenuated by absorption inside the particle. Each internal ray is traced until its intensity decreases below a prescribed cutoff value. Varying the polarization state of the incident rays, sampling all emerging rays into predefined narrow angular bins, and adding "incoherently" the respective Stokes parameters of the emerging rays yields a quantitative representation of the particle's scattering properties in terms of the ray-tracing scattering matrix. The ray-tracing extinction cross section does not depend on the polarization state of the incident light and is equal to the geometrical area G of the particle projection on the plane perpendicular to the incidence direction. The presence of the particle modifies the incident plane wave front by eliminating a part that has the shape and size of the geometrical projection of the particle. Therefore, the ray-tracing scattering pattern is artificially supplemented by adding the Fraunhofer pattern caused by diffraction of the incident wave on the particle projection.

Since the particle size is assumed to be much greater than the incident wavelength, the diffraction component of the scattering matrix is confined to a narrow angular cone centered at the exact forward-scattering direction.

The main advantage of the GOM is its applicability to essentially any particle shape. However, this technique is approximate by definition, which implies that its range of applicability in terms of the smallest size parameter must be examined thoroughly by comparing GOM results with numerically-exact solutions of the Maxwell equations. Such comparisons with Lorenz–Mie and T-matrix results have demonstrated that although the main geometrical optics features can be reproduced qualitatively by particles with size parameters less than 100, obtaining good quantitative accuracy in GOM computations of the scattering matrix requires size parameters exceeding a few hundred. Even then, the GOM fails to reproduce scattering features caused by various interference effects.

The so-called physical optics or Kirchhoff approximation (KA) has been developed with the purpose of improving the GOM performance (see, e.g., P. Yang and Liou, 2006). This technique is based on expressing the scattered field in terms of the electric and magnetic fields on the exterior side of the particle boundary. The latter are computed approximately using Fresnel's formulas and the standard ray-tracing procedure. The KA partially preserves the phase information and reproduces some physical optics effects ignored completely by the simple GOM.

2.7.2 Measurement techniques

The majority of existing laboratory measurement techniques fall into two basic categories:

- scattering of visible or infrared light by particles with sizes from several hundredths of a micron to several hundred microns;

- scattering of microwaves by millimeter- and centimeter-sized objects.

Measurements in the visible and infrared parts of the spectrum benefit from the availability of sensitive detectors of electromagnetic energy, diverse sources of radiation, and high-quality optical elements. They usually involve less expensive and more portable instrumentation and can be performed in the field as well as in the laboratory. However, they may be more difficult to interpret due to lack of independent information on sample microphysical characteristics and composition. Microwave scattering experiments often require more cumbersome and expensive instrumentation and large stationary measurement facilities, but allow almost full control over the scattering object.

Traditional detectors of electromagnetic energy in the visible and near-infrared spectral regions are polarization-insensitive, which means that the detector response is determined only by the first Stokes parameter of the radiation impinging on the detector. This implies that in order to measure all 16 elements of the scattering matrix, one must use optical elements that can vary the polarization state of light before and after scattering in a specific and controllable way. Figure 2.8 depicts the scheme of a modern laboratory setup used to measure the scattering matrix for a small group of natural or artificial particles. The laser beam first passes through a linear polarizer and a polarization modulator and then illuminates particles contained in the scattering chamber. Light scattered by the particles at an

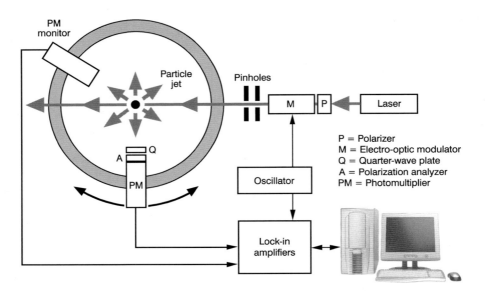

Figure 2.8 Schematic view of a laboratory scattering setup based on measurements of visible or near-infrared light (after Hovenier, 2000).

angle Θ relative to the incidence direction passes a quarter-wave plate and a polarization analyzer, after which its intensity is measured by a detector. The transformation Mueller matrices of the polarizer, modulator, quarter-wave plate, and analyzer depend on their orientation with respect to the scattering plane and can be varied precisely. Because the detector measures only the first element of the resulting Stokes column vector, several measurements with different orientations of the optical components with respect to the scattering plane are necessary for the full determination of the scattering matrix. This procedure must be repeated at different scattering angles in order to determine the full angular profile of the scattering matrix, perhaps with the exception of near-forward and/or near-backward directions. This laboratory technique has been used to accumulate a large and representative set of scattering-matrix data for samples of natural and artificial aerosols (see, e.g., the review by Muñoz and Volten, 2006 and references therein)

In accordance with the above-mentioned electromagnetic scale invariance rule (see Section 3.5 of Mishchenko et al., 2006), the main idea of the microwave analog technique is to manufacture a centimeter-sized scattering object with desired shape and refractive index, measure the scattering of a microwave beam by this object, and finally extrapolate the result to visible or near-infrared wavelengths by keeping the ratio of the object size to the wavelength fixed. In a modern microwave scattering setup (e.g., Gustafson, 2009), radiation from a transmitting conical horn antenna passes through a collimating lens and a

polarizer. The lens produces a nearly flat wave front which is scattered by an analog particle model. The scattered wave passes through another polarizer and lens and is measured
by a receiving horn antenna. Positioning the receiver end of the setup at different scattering
angles yields information on the angular distribution of the scattered radiation. By varying
the orientations of the two polarizers, one can measure all 16 elements of the scattering
matrix.

2.8 Illustrative theoretical and laboratory results

The review by Hansen and Travis (1974) and monographs by Mishchenko et al. (2000,
2002) provide a detailed discussion of extinction, scattering, and absorption properties of
aerosol particles having diverse morphologies and compositions. Plentiful information on
light scattering by nonspherical and morphologically complex particles can also be found
in several special issues of the Journal of Quantitative Spectroscopy and Radiative Transfer
(Hovenier, 1996; Lumme, 1998; Mishchenko et al., 1999b, 2008; Videen et al., 2001, 2004;
Kolokolova et al., 2003; Wriedt, 2004; Moreno et al., 2006; Voshchinnikov and Videen,
2007; Horvath, 2009; Hough, 2009). Therefore, the limited purpose of the several illustra-
tive examples given in this section is to highlight the most typical traits of the single-scattering patterns caused by small particles.

Figure 2.9 shows the extinction efficiency factor defined by Eq. (2.43) versus size parameter x for monodisperse spheres with three real-valued relative refractive indices. Each
curve exhibits a succession of major low-frequency maxima and minima with a super-
imposed high-frequency ripple consisting of sharp, irregularly spaced extrema many of
which are super-narrow spike-like features. The major maxima and minima are called the
"interference structure" since they are usually interpreted as being the result of interference of light diffracted and transmitted by the particle. Unlike the interference structure,
the so-called morphology-dependent resonances (MDRs) forming the ripple are caused by
the resonance behavior of the Lorenz–Mie coefficients a_n and b_n at specific size-parameter
values. The interference structure and the MDRs are typical attributes of all scattering
characteristics of nonabsorbing monodisperse spheres.

Irrespective of m, the extinction efficiency rapidly vanishes in the Rayleigh domain of
size parameters (i.e., as x approaches zero). Indeed, it is well known that for nonabsorbing
particles much smaller than the wavelength,

$$Q_e \equiv Q_s \underset{x \to 0}{\propto} \frac{1}{\lambda^4}, \tag{2.48}$$

as first demonstrated by Lord Rayleigh and hence called Rayleigh scattering. As the particle size becomes much greater than the wavelength, Q_e tends to the asymptotic geometri-
cal-optics value 2, with equal contributions from the rays striking the particle surface and
the light diffracted by the particle projection. Figure 2.9 also demonstrates that for nonabsorbing particles with size parameters of order one, the extinction cross section can exceed

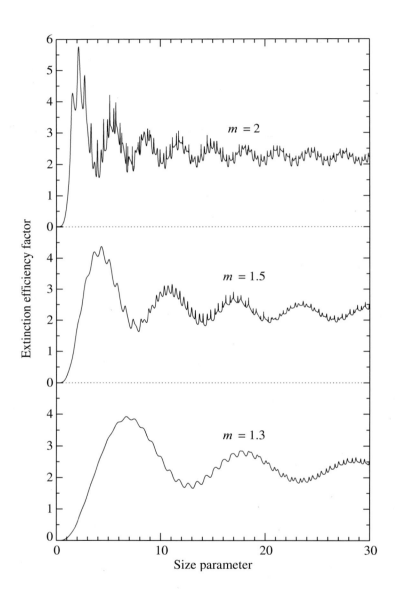

Figure 2.9 Extinction efficiency factor Q_e versus size parameter x for monodisperse, homogeneous spherical particles with relative refractive indices $m = 1.3, 1.5$, and 2.

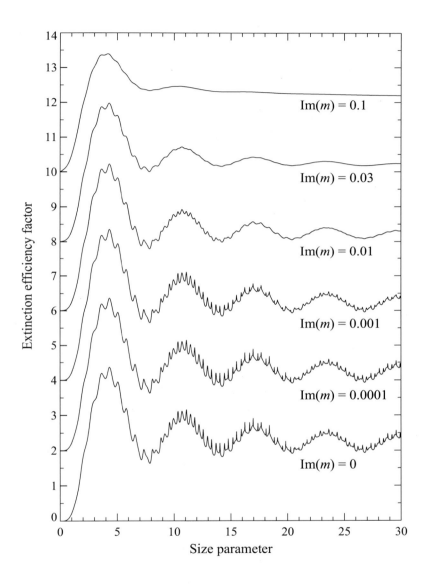

Figure 2.10 The effect of increasing absorption on the interference and ripple structures of the extinction efficiency factor for monodisperse spherical particles with the real part of the relative refractive index $Re(m) = 1.5$. The vertical axis scale applies to the curve with $Im(m) = 0$, the other curves being successively displaced upward by 2.

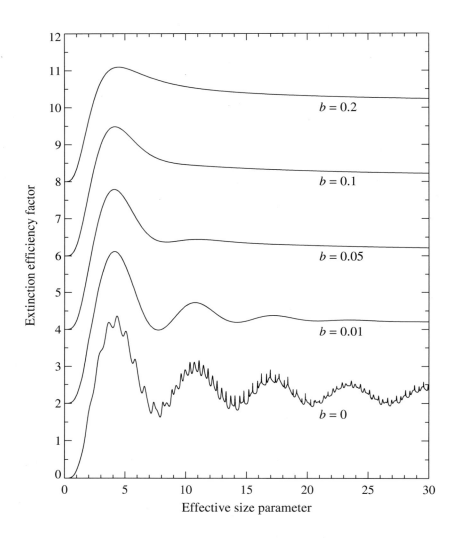

Figure 2.11 The effect of increasing width of the size distribution on the interference structure and ripple for nonabsorbing spherical particles with the real-valued relative refractive index $m = 1.5$ and effective size parameters x_{eff} ranging from 0 to 30. The vertical axis scale applies to the curve with $b = 0$, the other curves being successively displaced upward by 2.

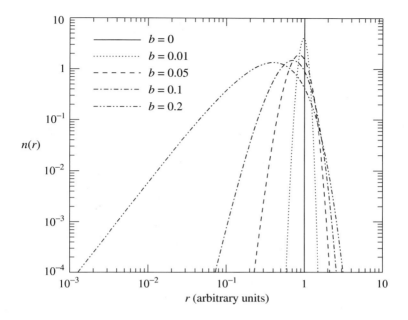

Figure 2.12 Gamma size distributions (2.50) with $a = 1$ (in arbitrary units of length) and $b = 0, 0.01$, $0.05, 0.1$, and 0.2. The value $b = 0$ corresponds to monodisperse particles.

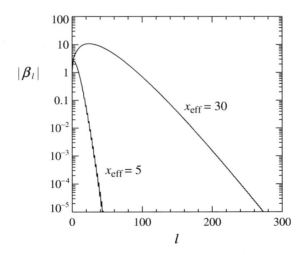

Figure 2.13 Expansion coefficients for two models of polydisperse spherical particles with effective size parameters $x_{\text{eff}} = 5$ and 30.

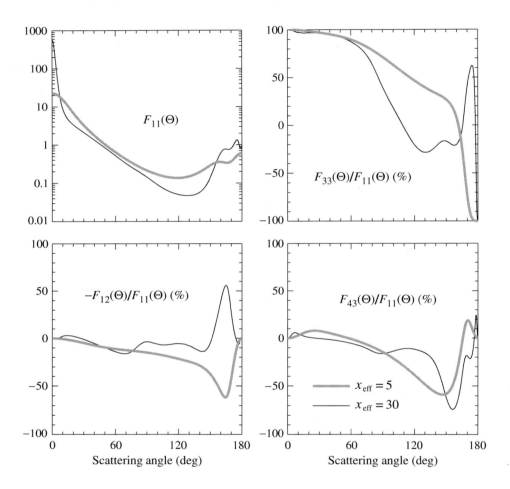

Figure 2.14 Elements of the normalized Stokes scattering matrix for two models of polydisperse spherical particles.

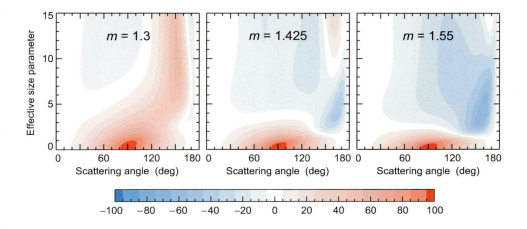

Figure 2.15 Degree of linear polarization for unpolarized incident light (in %) versus scattering angle Θ and effective size parameter ka for polydisperse spherical particles with relative refractive indices 1.3, 1.425, and 1.55. The effective variance of the gamma size distribution (2.50) is fixed at $b = 0.2$.

the particle geometrical cross section by more than a factor of 5.

For absorbing particles, extinction in the Rayleigh limit of size parameters is dominated by absorption and varies as the inverse wavelength:

$$Q_e \approx Q_a \underset{x \to 0}{\propto} \frac{1}{\lambda}. \tag{2.49}$$

The MDRs rapidly weaken and then vanish with increasing absorption, as Figure 2.10 demonstrates. Increasing the imaginary part of the relative refractive index Im(m) beyond 0.001 starts to affect and eventually suppresses most of the interference structure as well. However, the first interference maximum at $x \approx 4$ survives, although becomes significantly less pronounced, even at Im(m) = 0.1.

A very similar smoothing effect on the interference structure and MDRs is caused by particle polydispersity. Figure 2.11 shows the results of Lorenz–Mie computations for the gamma distribution of particle radii

$$n(r) = \frac{1}{(ab)^{(1-2b)/b}\Gamma[(1-2b)/b]} r^{(1-3b)/b} \exp\left(-\frac{r}{ab}\right), \quad r \in (0, \infty), \quad b \in [0, 5) \tag{2.50}$$

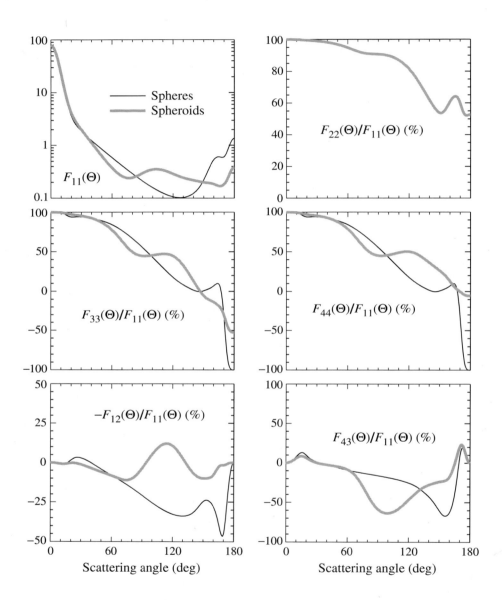

Figure 2.16 Elements of the normalized Stokes scattering matrix for the gamma distribution of spheres and surface-area-equivalent, randomly oriented oblate spheroids with $ka = 10$, $b = 0.1$, and $m = 1.53 + i0.008$. The ratio of the larger to the smaller spheroid axes is 2.

with the effective variance (width) b increasing from 0 (which corresponds to monodisperse particles) to 0.2 (which corresponds to a moderately wide size distribution). The efficiency factor is now defined as the average extinction cross section divided by the average area of the particle geometrical projection and is plotted against the effective size parameter defined as $x_{eff} = ka$. It is seen that increasing the width of the size distribution (see Figure 2.12) initially suppresses the MDRs and then eliminates the interference structure. Interestingly, as narrow a dispersion of sizes as that corresponding to $b = 0.01$ washes the MDRs out completely. The first major maximum of the interference structure persists to much larger values of b, but eventually fades away too.

To illustrate the dependence of the coefficients β_l entering the expansion (2.7) on particle physical characteristics, Figure 2.13 depicts them versus l for two polydispersions of spherical particles, each described by the gamma distribution (2.50). For both polydispersions, the relative refractive index is $m = 1.5$ and the effective variance is $b = 0.2$. The effective size parameter x_{eff} is equal to 5 for the first model and to 30 for the second model. Figure 2.14 visualizes the four non-zero independent elements of the normalized Stokes scattering matrix (2.29) for each polydispersion. Figure 2.13 reveals the typical behavior of the expansion coefficients β_l with increasing index l: they first grow in magnitude and then decay to absolute values below a meaningful numerical threshold. The greater the particle size parameter, the larger the maximal absolute value of the expansion coefficients and the slower their decay. This trend is largely explained by the rapid growth of the height of the forward-scattering peak in the phase function $p(\Theta) = F_{11}(\Theta)$ (as well as in the element $F_{33}(\Theta)$) with increasing x_{eff} (see Figure 2.14).

Besides the magnitude of the forward-scattering peak, the angular profiles of the two phase functions in Figure 2.14 are qualitatively similar. The same is largely true of the ratio $F_{43}(\Theta)/F_{11}(\Theta)$. On the other hand, the angular profiles of the ratios $F_{33}(\Theta)/F_{11}(\Theta)$ and $-F_{12}(\Theta)/F_{11}(\Theta)$ (the latter represents the degree of linear polarization of singly scattered light for unpolarized incident light) reveal a strong dependence on the effective size parameter. Indeed, at certain scattering angles these ratios can differ not just in magnitude but even in sign. Figure 2.15 demonstrates the equally strong dependence of the ratio $-F_{12}(\Theta)/F_{11}(\Theta)$ on the relative refractive index. These results illustrate well why measurements of polarization contain much more information on particle microphysics than measurements of intensity only (e.g., Mishchenko et al., 2010 and references therein).

The dependence of all scattering and absorption characteristics on particle microphysical properties can become much more intricate for nonspherical and/or morphologically complex particles, particularly those having a preferred orientation. This is especially true of the interference structure and MDRs, which now strongly depend on the particle orientation with respect to the incidence and scattering directions and on the polarization state of the incident light. However, averaging over orientations reinforces the effect of averaging over sizes and extinguishes many resonance features, thereby making scattering patterns for randomly oriented, polydisperse nonspherical particles even smoother than those for surface- or volume-equivalent polydisperse spheres. In fact, it is often difficult to distinguish spherical and randomly oriented nonspherical particles based on *qualitative* differences in their scattering patterns.

However, there can be significant *quantitative* differences in specific scattering patterns. As an example, Figure 2.16 contrasts the elements of the normalized Stokes scattering matrix for polydisperse spheres and surface-equivalent, randomly oriented spheroids with a relative refractive index $m = 1.53 + i0.008$. The left-hand top diagram shows the corresponding phase functions and reveals the following five distinct scattering-angle ranges:

$$
\begin{aligned}
\text{nonsphere} &\approx \text{sphere} && \text{from } \Theta = 0° \text{ to } \Theta \sim 15°-20°; \\
\text{nonsphere} &> \text{sphere} && \text{from } \Theta \sim 15°-20° \text{ to } \Theta \sim 35°; \\
\text{nonsphere} &< \text{sphere} && \text{from } \Theta \sim 35° \text{ to } \Theta \sim 85°; \\
\text{nonsphere} &\gg \text{sphere} && \text{from } \Theta \sim 85° \text{ to } \Theta \sim 150°; \\
\text{nonsphere} &\ll \text{sphere} && \text{from } \Theta \sim 150° \text{ to } \Theta = 180°.
\end{aligned}
\tag{2.51}
$$

Although the specific boundaries of these regions can be expected to shift with changing particle shape and relative refractive index, the enhanced side-scattering and suppressed backscattering appear to be rather typical characteristics of nonspherical particles.

The degree of linear polarization for unpolarized incident light, $-F_{12}(\Theta)/F_{11}(\Theta)$, tends to be positive at scattering angles around 120° for the spheroids, but is negative at most scattering angles for the surface-equivalent spheres. Whereas $F_{22}(\Theta)/F_{11}(\Theta) \equiv 1$ for spherical particles, the $F_{22}(\Theta)/F_{11}(\Theta)$ curve for the spheroids significantly deviates from 100% and causes a non-zero value of the linear depolarization ratio defined by

$$
\delta_{\mathrm{L}} = \frac{F_{11}(180°) - F_{22}(180°)}{F_{11}(180°) + F_{22}(180°)}.
\tag{2.52}
$$

Similarly, $F_{33}(\Theta)/F_{11}(\Theta) \equiv F_{44}(\Theta)/F_{11}(\Theta)$ for spherically symmetric particles, whereas the $F_{44}(\Theta)/F_{11}(\Theta)$ for the spheroids tends to be greater than the $F_{33}(\Theta)/F_{11}(\Theta)$ at most scattering angles, especially in the backscattering direction. The violation of the Lorenz–Mie equality $F_{44}(180°) = -F_{11}(180°)$ by the spheroids yields a non-zero value of the circular depolarization ratio defined by

$$
\delta_{\mathrm{C}} = \frac{F_{11}(180°) + F_{44}(180°)}{F_{11}(180°) - F_{44}(180°)} \equiv \frac{2\delta_{\mathrm{L}}}{1 - \delta_{\mathrm{L}}}
\tag{2.53}
$$

(cf. Eq. (2.28)). The ratios $F_{43}(\Theta)/F_{11}(\Theta)$ for the spheres and the spheroids also reveal significant quantitative differences at scattering angles exceeding 60°. On the other hand, the nonspherical/spherical differences in the integral scattering and absorption characteristics are not nearly as significant as those in the scattering matrix elements.

Despite the significant progress in our ability to model scattering by nonspherical and morphologically complex particles, direct theoretical computations for many types of natural and artificial particles with sizes comparable to and greater than the wavelength (Figure 2.6) remain highly problematic. Therefore, there have been several attempts to simulate the scattering and absorption properties of actual particles using simple model shapes. These attempts have been based on the realization that in addition to size and orientation averaging, averaging over shapes can also be necessary in many cases. Indeed, usually ensembles of natural and artificial particles exhibit a vast variety of shapes and morphologies, thereby making quite questionable the utility of a single model shape (however "irregular" it may look to the human eye) in the representation of scattering properties of an ensemble.

As an illustration, Figure 2.17 shows the phase functions computed for polydisperse, randomly oriented prolate spheroids with varying aspect ratios (Mishchenko et al., 1997). It is seen indeed that even after size and orientation averaging, each spheroidal shape produces a unique, aspect-ratio-specific angular scattering pattern, whereas laboratory and in situ measurements for real nonspherical particles usually show smooth and mostly featureless phase functions. On the other hand, the grey curves in Figure 2.18 (Dubovik et al., 2006) show that *shape mixtures* of polydisperse, randomly oriented prolate and oblate spheroids can provide a good quantitative fit to the results of accurate laboratory measurements of the scattering matrix for natural irregular particles. On the other hand, the Lorenz–Mie results depicted by black curves disagree with the laboratory data quite substantially.

These examples lead to two important conclusions. First of all, they provide evidence that the often observed smooth scattering-angle dependence of the elements of the scattering matrix for natural and artificial ensembles of nonspherical aerosols (e.g., Muñoz and Volten, 2006) is largely caused by the diversity of particle shapes in the ensemble. Secondly, they suggest that at least some scattering properties of ensembles of irregularly and randomly shaped aerosols can be modeled adequately using polydisperse shape mixtures of simple particles such as spheroids. It goes without saying that forming representative mixtures of less regular particles than spheroids should be expected to eventually provide an even better model of electromagnetic scattering by many natural and artificial aerosols (e.g., Bi et al., 2009; Zubko et al., 2009).

In most cases nonspherical–spherical differences in the optical cross-sections and the single-scattering albedo are much less significant than those in the scattering matrix elements. This does not mean, however, that the effects of nonsphericity and morphology on the integral scattering and absorption characteristics are always negligible or unimportant. An important type of particle characterized by integral radiometric properties substantially different from those of volume-equivalent spheres are clusters composed of large numbers of small monomers such as soot aggregates shown in Figure 2.6b (see, e.g., the reviews by Sorensen, 2001 and Moosmüller et al., 2009). The overall morphology of a dry soot aerosol is usually described by the following statistical scaling law:

$$N_S = k_0 \left(\frac{R_g}{a} \right)^{D_f}, \tag{2.54}$$

where a is the monomer mean radius, k_0 is the prefactor, D_f is the fractal dimension, N_S is the number of spherical monomers in the cluster, and R_g, called the radius of gyration, is a measure of the overall cluster radius. The fractal dimension is especially important for the quantitative characterization of the aggregate morphology. Densely packed aggregates have D_f values approaching 3, whereas the fractal dimension of chain-like branched clusters can be significantly smaller.

Detailed computations for fractal soot clusters based on the DDA and the superposition T-matrix method have been reported by Klusek et al. (2003), Liu and Mishchenko (2005,

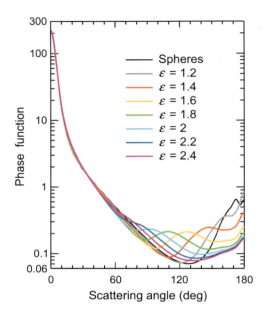

Figure 2.17 *T*-matrix computations of the phase function for micrometer-sized polydisperse spheres and randomly oriented surface-equivalent prolate spheroids with aspect ratios ε ranging from 1.2 to 2.4 at a wavelength of 443 nm. The relative refractive index is fixed at 1.53 + i0.008.

2007), Liu et al. (2008), and Kahnert (2010). These numerically exact results have demonstrated that the integral radiometric properties of the clusters can often be profoundly different from those of the volume-equivalent spheres. This is especially true of the scattering cross-section, single-scattering albedo, and the asymmetry parameter defined by Eq. (2.9).

Figure 2.19 depicts the results of *T*-matrix computations of the scattering matrix elements averaged over 20 soot-cluster realizations randomly computer-generated for the same values of the fractal parameters using the procedure developed by Mackowski (2006). In a rather peculiar way (cf. West, 1991), the angular scattering properties of the clusters appear to be a mix of those of wavelength-sized compact particles (the nearly isotropic Rayleigh phase function of the small individual spherules evolves into a forward scattering phase function) and Rayleigh scatterers (i.e., the ratio $-F_{12}(\Theta)/F_{11}(\Theta)$ is systematically positive, almost symmetric with respect to the scattering angle $\Theta = 90°$, and reaches a nearly 100% maximum at $\Theta \approx 90°$, while the ratio $F_{34}(\Theta)/F_{11}(\Theta)$ is very close to zero). The deviation of the ratio $F_{22}(\Theta)/F_{11}(\Theta)$ from 100% is the only unequivocal manifestation of the nonsphericity of the soot-cluster shape.

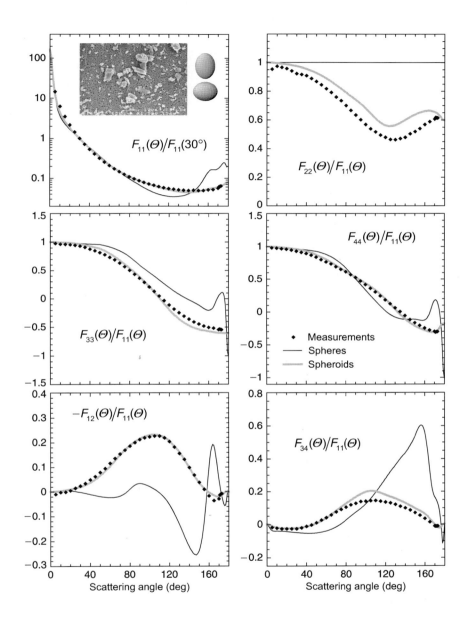

Figure 2.18 The diamonds depict the results of laboratory measurements of the ensemble-averaged Stokes scattering matrix for micrometer-sized feldspar particles at a wavelength of 633 nm (Volten et al., 2001). The grey curves show the result of fitting the laboratory data with T-matrix results computed for a shape mixture of polydisperse, randomly oriented prolate and oblate spheroids. The real and model particle shapes are contrasted in the inset. The black curves show the corresponding results for volume-equivalent polydisperse spherical particles.

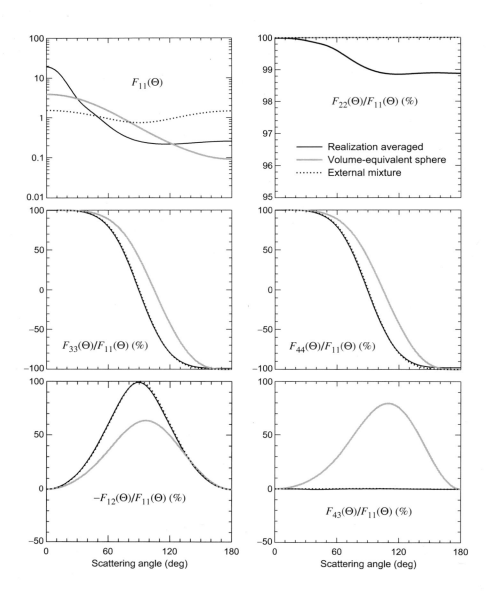

Figure 2.19 Realization-averaged scattering matrix elements for randomly oriented fractal clusters with $D_f = 1.82$, $k_0 = 1.19$, $N_s = 400$, and a = $0.02\,\mu$m. The soot refractive index is $1.75 + i0.435$ and the wavelength of the incident light is 628 nm. Also shown are Lorenz–Mie results for the corresponding homogeneous volume-equivalent sphere and the "equivalent" external mixture of soot monomers.

Also depicted in Figure. 2.19 are two sets of approximate computations. The first one is the result of applying the single-scattering approximation to the corresponding external mixture of the constituent monomers (i.e., by assuming that all monomers are widely separated and randomly positioned rather than form a cluster with touching components). The second set was computed by applying the Lorenz–Mie theory to a homogeneous sphere with a volume equal to the cumulative volume of the cluster monomers. Clearly, the external-mixture model provides a poor representation of the cluster phase function, whereas the performance of the equal-volume-sphere model is inadequate with respect to all scattering matrix elements.

3 Radiative transfer in the Earth's atmosphere

Jacqueline Lenoble, and Maurice Herman

3.1 Introduction

In Chapter 2, we analyzed the scattering and absorption properties of air molecules and aerosol particles. These properties create a specific signature of aerosol impact on radiation and this signature provides the basis for aerosol remote sensing. We have seen that this signature depends on the shape, size, and chemical nature of the particle, so that by measuring the result of an incident beam's interaction with an aerosol particle, we can derive information about that particle. The problem becomes challenging because generally we have to deal with an ensemble of different particles, mixed within the air molecules, not just a single particle.

Moreover, the photons are submitted to several successive encounters with molecules and particles, between the source and the detector, which makes the problem of recognizing the aerosol signature even more complex. The general radiative transfer theory analyzes the interaction of the radiation field with a scattering-absorbing-emitting medium, where the photons are submitted to multiple scatterings. This is the subject of the present chapter.

For the Earth's atmosphere, the natural sources of radiation are the solar beam received at the top of the atmosphere, and the longwave emission of the surface.

Remote sensing methods using these natural sources are called passive remote sensing, either shortwave (solar) or longwave (thermal). The problem of emission by the surface and by the atmosphere itself (longwave remote sensing) is considered in Chapter 9. In the present chapter, we will limit ourselves to radiative transfer of the solar radiation in a scattering-absorbing atmosphere (shortwave remote sensing).

An artificial light source, such as a laser as part of a lidar system, can be used instead of the solar beam for aerosol remote sensing; this method is called active remote sensing, and it is presented in Chapter 10.

Section 3.2 establishes the equation of radiative transfer without polarization (scalar form) and with polarization (vector form), and Section 3.3 presents the boundary conditions at the Earth's surface. In Section 3.4, we present a few of the most popular methods of solving this equation of transfer, and in Section 3.5, we discuss the choice of radiation codes. Finally, Section 3.6 shows some examples of simulated aerosol impact on radiation, which are or can be used in remote sensing. We insist on the specific information contained in polarization observations.

3.2 The scalar and the vector equations of transfer

3.2.1 The scalar equation of radiative transfer

The term "scalar" refers to the radiative transfer problem concerned only with the problem of propagation of radiative energy without taking into account its state of polarization. The usual (scalar) radiative transfer equation (RTE) governs the radiance $L(M,\mathbf{k})$, of a diffuse radiation field at any point M, and for any direction of propagation $\mathbf{k}$, in a scattering, absorbing and emitting medium. The following developments concern monochromatic radiation; the subscript λ is omitted for simplicity; as explained in Section 2.3 for Beer's law, the formalism presented below can be extended to finite spectral intervals, where the air properties vary only slightly. $L(M,\mathbf{k})$, in $Wm^{-2}sr^{-1}$, is such that the monochromatic radiative power within a solid angle element $d\omega$, through a surface element $d\Sigma$ perpendicular to $\mathbf{k}$, is $L(M,\mathbf{k})\,d\Sigma d\omega$. If we go back to Eq. (2.1), defining the extinction coefficient and written for the radiance L, the loss of $L(M,\mathbf{k})$, due to absorption and scattering processes within an atmosphere element in M (upper Figure 3.1) has to be completed, in order to include the gain of radiation in the element ds along the propagation direction $\mathbf{k}$ (lower Figure 3.1). The radiance variation now writes as

$$dL = -\sigma_e(M)L(M,\mathbf{k})ds + \sigma_e(M)J(M,\mathbf{k})ds. \qquad (3.1)$$

The gain in radiation is expressed by the so-called source function $J(M,\mathbf{k})$. It is due to two different causes: local emission and scattering of radiation coming from other directions

$$J(M,\mathbf{k}) = J_{em}(M,\mathbf{k}) + J_{sc}(M,\mathbf{k}). \qquad (3.2)$$

In the simple conditions of local thermodynamic equilibrium, emission occurs according to Kirchhoff's law; it is proportional to the blackbody radiance $B(T)$ at the local temperature $T(M)$ and to the absorption coefficient σ_a; with the single scattering albedo (SSA or ϖ) defined by Eqs (2.2 and 2.3)

$$J_{em}(M,\mathbf{k}) = (1-\varpi(M))B(T(M)). \tag{3.3}$$

At the temperatures of the Earth's atmosphere, emission is negligible in ultraviolet, visible, and near infrared. The analysis of the transfer equation with an emission source function is therefore postponed to Chapter 9, which concerns remote sensing in the thermal infrared domain.

In this chapter, we limit ourselves to the RTE with a scattering source function. According to Eq. (2.6), in Figure 3.1b, the total radiative power scattered within $d\omega$ by the element $dV = d\Sigma ds$, illuminated by the irradiances $L(M,\mathbf{k}')d\omega'$ of the radiation field leads to

$$J_{sc}(M,\mathbf{k}) = \frac{\varpi(M)}{4\pi} \iint p(M,\mathbf{k},\mathbf{k}')L(M,\mathbf{k}')d\omega' , \tag{3.4}$$

where $d\omega'$ is the solid angle around the direction $\mathbf{k}'$, and the integration is performed over all directions; $p(M,\mathbf{k},\mathbf{k}')$ is the phase function defined in Chapter 2, at point M, for the scattering angle between directions $\mathbf{k}$ and $\mathbf{k}'$.

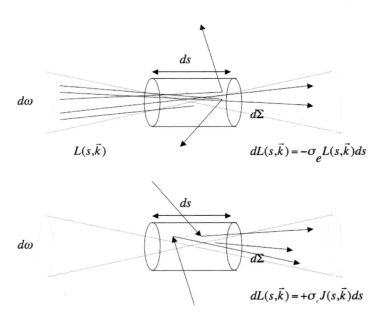

Figure 3.1 The scalar equation of transfer (see text for explanation).

The RTE has been simply established here on energetical considerations; this approach is used in many textbooks (Chandrasekhar, 1950; Van de Hulst, 1980; Liou, 1992; Lenoble, 1993).

A more correct method based on basic electromagnetism leads to the same result (Mishchenko et al., 2006).

3.2.2 Formalism for a plane parallel atmosphere

For the simplicity of formalism, we will consider a plane parallel atmosphere, i.e. an atmosphere limited by two parallel horizontal planes, and with properties invariant in a horizontal plane.

The sphericity of the Earth has to be taken into account in the following cases:

• for observations very close to the horizon;

• when the sun is very low above the horizon, or even below the horizon (twilight observations);

• for limb observations from space.

The assumption of invariant properties in a horizontal plane is not valid in the presence of clouds; 3D radiative transfer has to be used in this case, and it is presently the subject of active research (Cahalan et al., 2005). For aerosols, horizontal inhomogeneities may also lead to uncertainties.

In a plane parallel atmosphere, a position is uniquely defined by its altitude z above the ground level; the top of the atmosphere can be defined only approximately by the altitude above which the number of air molecules becomes negligible. This difficulty is generally overcome by, instead of z, using as the altitude variable the optical depth τ, defined in Chapter 2

$$\tau = \int_z^\infty \sigma_e(z)dz , \qquad (3.5)$$

where $\sigma_e(z)$ is the local extinction coefficient.

The optical depth of the atmosphere at the surface of the Earth, called total optical depth (or thickness), is noted τ^*.

A direction $\mathbf{k}$ is defined by two angles (see Figure 3.2): the first is the angle with the upward vertical, or zenith angle θ, counted between 0 and π, or preferably by $\mu = \cos\theta$; note that μ is positive for upward directions, and negative for downward directions. Sometimes, a somewhat different convention is used, with $-\mu$ (μ positive) for downward directions. The second direction coordinate is an azimuth angle φ, between 0 and 2π, counted from an arbitrary origin; the sun meridian is generally taken as origin for the azimuths. The sun direction is $\mathbf{k}_0$ (μ_0, φ_0).

With these coordinates, the RTE can be expressed as

$$\mu\frac{dL(\tau,\mu,\varphi)}{d\tau} = L(\tau,\mu,\varphi) - J(\tau,\mu,\varphi), \qquad (3.6)$$

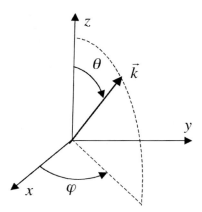

Figure 3.2 The coordinates in a plane-parallel atmosphere.

where J stands for the scattering part of the source function,

$$J(\tau,\mu,\varphi) = \frac{\varpi(\tau)}{4\pi} \int_0^{2\pi} d\varphi' \int_{-1}^{+1} p(\tau,\mu,\varphi,\mu',\varphi') L(\tau,\mu',\varphi') d\mu' \qquad (3.7)$$

The RTE needs to be completed by boundary conditions, at the surface and at the top of the atmosphere. At the ground, the simplest case is a nonscattering, nonreflecting, non emitting, fully absorbing (black) surface, i.e.,

$$L(\tau = \tau^*, \mu, \varphi) = 0 \text{ for } \mu > 0; \qquad (3.8)$$

the case of a reflecting surface will be presented in Section 3.3.

The boundary condition at the top of the atmosphere represents the incoming solar beam in the particular direction (μ_0, φ_0); this can be expressed as

$$L(\tau = 0, \mu, \varphi) = \Delta(\mu - \mu_0)\Delta(\varphi - \varphi_0)E_0, \qquad (3.9)$$

where E_0 is the extraterrestrial solar irradiance, and $\Delta(x)$ stands for the usual Dirac function ($\Delta(0)=1$, otherwise $\Delta(x)=0$).

The solar beam is partly directly transmitted into the atmosphere, introducing a discontinuity in the downward radiance, as

$$L(\tau,\mu,\varphi) = L_{dif}(\tau,\mu,\varphi) + \Delta(\mu - \mu_0)\Delta(\varphi - \varphi_0)E_0 \exp(\tau/\mu_0); \tag{3.10}$$

the second term on the right-hand side of Eq. (3.10) represents the directly transmitted radiance in the sun direction, and L_{dif} stands for the sky diffuse radiance.
Introducing Eq. (3.10) into Eqs (3.6 and 3.7), it is easy to find that the diffuse radiance satisfies Eq. (3.6), with

$$J(\tau,\mu,\varphi) = \frac{\varpi(\tau)}{4\pi} p(\tau,\mu,\varphi,\mu_0,\varphi_0)E_0 \exp(\tau/\mu_0)$$

$$+ \frac{\varpi(\tau)}{4\pi} \int_0^{2\pi} d\varphi' \int_{-1}^{+1} p(\tau,\mu,\varphi,\mu',\varphi')L(\tau,\mu',\varphi')d\mu' \tag{3.11}$$

where we have dropped the subscript "*dif*" for simplicity in L.

3.2.3 The vector equation of transfer

As we have seen in Chapter 2, a complete description of the radiation field must include polarization, and this is easily achieved by using the Stokes parameters. The usual scalar radiance is replaced by a radiance vector (Eq. 2.14), and the effect of scattering is expressed by a 4x4 scattering matrix (Eqs 2.23 and 2.24) instead of the phase function. The Stokes parameters are additive when several independent streams of radiation are combined; therefore a vector equation of transfer can be established for the vector radiance $\mathbf{L}$, exactly as has been done for the scalar radiance L. The only condition is that the same axes of reference $(\mathbf{l},\mathbf{r})$ in the wave plane are used for all the added streams. The most obvious choice is to always use $\mathbf{l}$ and $\mathbf{r}$ respectively parallel and perpendicular to the vertical plane containing the direction of propagation. A difficulty appears because the scattering matrix is defined with axes parallel and perpendicular to the scattering plane (see Chapter 2). Therefore, it is necessary to perform two rotations (see Figure 3.3), first by an angle χ' in order to bring the axes $\mathbf{l'}$, $\mathbf{r'}$ associated to the incident radiance matrix $\mathbf{L}(\tau,\mu',\varphi')$ to be parallel and perpendicular to the scattering plane. After multiplication by the scattering matrix, a second rotation by an angle $-\chi$ is necessary to return to the axes $\mathbf{l}$, $\mathbf{r}$ for the scattered radiance. The vector RTE is written in a form similar to Eq. (3.6)

$$\mu \frac{d\mathbf{L}(\tau,\mu,\varphi)}{d\tau} = \mathbf{L}(\tau,\mu,\varphi) - \mathbf{J}(\tau,\mu,\varphi), \tag{3.12}$$

where the source matrix is

$$\mathbf{J}(\tau,\mu,\varphi) = \frac{\varpi(\tau)}{4\pi}\mathbf{P}(\tau,\mu,\varphi,\mu_0,\varphi_0)\mathbf{E_0} \exp(\tau/\mu_0)$$

$$+ \frac{\varpi(\tau)}{4\pi} \int_0^{2\pi} d\varphi' \int_{-1}^{+1} \mathbf{P}(\tau,\mu,\varphi,\mu',\varphi')\mathbf{L}(\tau,\mu',\varphi')d\mu'. \tag{3.13}$$

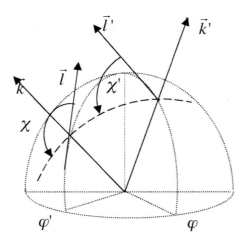

Figure 3.3 The rotation angles for the phase matrix (see text for explanation).

Equation (3.13) is similar to Eq. (3.11) in the scalar case, with the major difference that

$$\mathbf{P}(\tau,\mu,\varphi,\mu',\varphi') = \mathbf{T}(-\chi)\mathbf{F}(\tau,\Theta)\mathbf{T}(\chi'); \qquad (3.14)$$

$\mathbf{T}(\chi)$ is the transformation matrix in a rotation by an angle χ, defined in Eq. (2.22), and $\mathbf{F}(\tau,\Theta)$ is the scattering matrix defined in Chapter 2, for a scattering angle Θ, between the incidence direction (μ',φ') and the scattering direction (μ,φ). We call $\mathbf{P}(\tau,\mu,\varphi,\mu',\varphi')$ the phase matrix.

$\mathbf{E_0}=(E_0,0,0,0)^{\mathrm{T}}$ is the extraterrestrial non polarized solar irradiance vector.

3.2.4 Expansion into Fourier series of azimuth

For scalar problems, expanding the various quantities into Fourier series of the azimuth is a standard and easy procedure, which strongly simplifies the resolution of the RTE. Because the radiance is symmetric about the vertical plane containing the solar incident beam (μ_0,φ_0) (unless this symmetry is broken by the boundary conditions), and the phase function is symmetric about the vertical plane containing the incident direction (μ',φ'), we can write

$$L(\tau,\mu,\varphi) = \sum_{s=0}^{S}(2-\delta_{0s})\cos(s(\varphi-\varphi_0))L^s(\tau,\mu), \qquad (3.15)$$

$$p(\tau,\mu,\varphi,\mu',\varphi') = \sum_{s=0}^{S}(2-\delta_{0s})\cos(s(\varphi-\varphi'))p^s(\tau,\mu,\mu'),\tag{3.16}$$

where δ_{0s} (=1 for s=0, =0 otherwise) is introduced for convenience.

Using the development of the phase function into series of Legendre polynomials (Eq. 2.7) and the additivity theorem of these polynomials, it is easy to find

$$p^s(\tau,\mu,\mu') = \sum_{l=s}^{L}\beta_l(\tau)P_s^l(\mu)P_s^l(\mu'),\tag{3.17}$$

where $P_s^l(\mu)$ stands for the associated Legendre function, normalized as

$$\int_{-1}^{+1}P_s^l(\mu)P_s^{l'}(\mu)d\mu = \frac{2}{2l+1}\ ,\text{ for }l'=l,\tag{3.18}$$

$$= 0,\text{ for }l'{\neq}l.$$

In Eqs (3.15) and (3.16), the order S of the development is clearly limited to the value L used in Eq. (2.7). The useful numerical value for S can generally be much smaller. Introducing Eqs (3.16) and (3.17) into the RTE (Eq. 3.6 with 3.11), and integrating over φ', one easily gets the classical separation into azimuth of the scalar RTE

$$\mu\frac{dL^s(\tau,\mu)}{d\tau} = L^s(\tau,\mu) - \frac{\varpi(\tau)}{4\pi}p^s(\tau,\mu,\mu_0)E_0\exp(\tau/\mu_0)$$

$$-\frac{\varpi(\tau)}{2}\int_{-1}^{+1}p^s(\tau,\mu,\mu')L^s(\tau,\mu')d\mu'\qquad,\tag{3.19}$$

for s=0,1,.....L

The problem is much more complex (Siewert, 1982, Deuzé et al., 1989, Van der Mee and Hovenier, 1990) for the vector RTE.

Because of the added complexity, we restrict the derivation to scattering laws with axial symmetry, i.e. with a phase matrix in the form $\mathbf{P}(\tau,\mu,\mu',\varphi-\varphi')$ and a similar symmetry for the reflectance (see Section 3.3), so that the radiance field is symmetric about the solar incident plane with

$$I(\tau,\mu,\varphi) = \sum_{s=0}^{S}(2-\delta_{0s})\cos(s(\varphi-\varphi_o))I^s(\tau,\mu)\quad,$$

$$Q(\tau,\mu,\varphi) = \sum_{s=0}^{S}(2-\delta_{0s})\cos(s(\varphi-\varphi_o))Q^s(\tau,\mu)\ ,$$

$$U(\tau,\mu,\varphi) = \sum_{s=1}^{S}2\sin(s(\varphi-\varphi_o))U^s(\tau,\mu)\qquad,\tag{3.20}$$

$$V(\tau,\mu,\varphi) = \sum_{s=1}^{S}2\sin(s(\varphi-\varphi_o))V^s(\tau,\mu)\qquad,$$

i.e

$$L(\tau,\mu,,\varphi) = \sum_{s=0}^{S}(2-\delta_{0s})[\cos(s(\varphi-\varphi_{o}))L_{\cos}^{s}(\tau,\mu) + \sin(s(\varphi-\varphi_{o}))L_{\sin}^{s}(\tau,\mu)] , \quad (3.21)$$

with

$$L_{\cos}^{s}(\tau,\mu) = (I^{s},Q^{s},0,0)^{T} \text{ and } L_{\sin}^{s}(\tau,\mu) = (0,0,U^{s},V^{s})^{T}. \quad (3.22))$$

The parity of U and V corresponds to polarized vibrations with opposite inclinations and opposite turning directions when refered to the $(\mathbf{l},\mathbf{r},\mathbf{k})$ axes on either side of the incident plane.

Given the matrix multiplication in Eq. (3.14), $\mathbf{P}(\mu,\varphi,\mu',\varphi')$ is in the form

$$\mathbf{P}(\mu,\varphi,\mu',\varphi') = \begin{pmatrix} F_{11} & c'F_{12} & s'F_{12} & 0 \\ cF_{12} & cc'F_{22}+ss'F_{33} & cs'F_{22}-sc'F_{33} & sF_{43} \\ sF_{12} & sc'F_{22}-cs'F_{33} & ss'F_{22}+cc'F_{33} & -cF_{43} \\ 0 & -s'F_{43} & c'F_{43} & F_{44} \end{pmatrix}, \quad (3.23)$$

where F_{ij} stand for the elements $F_{ij}(\Theta)$ of the scattering matrix, Eq. (2.25), and c stands for $\cos(2\chi)$, s for $\sin(2\chi)$, and similarly for c' and s' with χ'. The variable τ has been omitted, but the elements $F_{ij}(\Theta)$ vary with τ in inhomogeneous atmospheres.

Since scattering directions symmetric about the incident plane correspond to similar values for the $F_{ij}(\Theta)$ elements but opposite numerical values for χ and χ' (see Figure 3.3), $\mathbf{P}(\mu,\varphi,\mu',\varphi')$ clearly separates into a symmetric and an anti-symmetric matrices with respect to $(\varphi-\varphi')$,

$$\mathbf{P}_{\cos}(\tau,\mu,\varphi,\mu',\varphi') = \begin{pmatrix} P_{11} & P_{12} & 0 & 0 \\ P_{21} & P_{22} & 0 & 0 \\ 0 & 0 & P_{33} & P_{34} \\ 0 & 0 & P_{43} & P_{44} \end{pmatrix} \quad \text{and} \quad \mathbf{P}_{\sin}(\tau,\mu,\varphi,\mu',\varphi') = \begin{pmatrix} 0 & 0 & P_{13} & P_{14} \\ 0 & 0 & P_{23} & P_{24} \\ P_{31} & P_{32} & 0 & 0 \\ P_{41} & P_{42} & 0 & 0 \end{pmatrix}, \quad (3.24)$$

which may be expanded, respectively, into series of $\cos(s(\varphi-\varphi'))$ and $\sin(s(\varphi-\varphi'))$. Then $\mathbf{P}$ is in the form

$$\mathbf{P}(\mu,\varphi,\mu',\varphi') = \sum_{s=0}^{L}(2-\delta_{0s})[\cos[s(\varphi-\varphi')]\mathbf{P}_{\cos}^{s}(\mu,\mu') + \sin[s(\varphi-\varphi')]\mathbf{P}_{\sin}^{s}(\mu,\mu')]. \quad (3.25)$$

Inserting Eqs (3.21) and (3.25) into Eq. (3.13) and integrating over φ' leads immediately to the development of the source function into a Fourier series, with

$$
\begin{aligned}
\mathbf{J}(\tau,\mu,\varphi) =& \frac{\varpi(\tau)}{4\pi} \sum_{s=0}^{S} (2-\delta_{0s})\cos(s(\varphi-\varphi_0))\mathbf{P}_{\cos}^s(\tau,\mu,\mu_0)\mathbf{E}_0 \exp(\tau/\mu_0) \\
& \frac{\varpi(\tau)}{2} \sum_{s=0}^{S} (2-\delta_{0s})\cos(s(\varphi-\varphi_0))\int_{-1}^{+1}(\mathbf{P}_{\cos}^s\mathbf{L}_{\cos}^s - \mathbf{P}_{\sin}^s\mathbf{L}_{\sin}^s)d\mu' \\
& + \frac{\varpi(\tau)}{2} \sum_{s=1}^{S} 2\sin(s(\varphi-\varphi_0))\int_{-1}^{+1}(\mathbf{P}_{\sin}^s\mathbf{L}_{\cos}^s - \mathbf{P}_{s\cos}^s\mathbf{L}_{\sin}^s)d\mu'
\end{aligned}
\tag{3.26}
$$

The variables τ ,μ, μ' have been omitted in the integral terms. The problem is to derive the coefficients of the matrices $\mathbf{P}_{\cos}^s(\mu,\mu')$ and $\mathbf{P}_{\sin}^s(\mu,\mu')$. Kuščer and Ribarič (1959) first suggested an answer to this problem. It consists of developing the $F_{ij}(\Theta)$ elements in Eq. (3.23) into series of convenient generalized Legendre functions, $\mathbf{P}_{m,n}^l(\Theta)$ according to Eqs (2.31, 2.32). The additivity theorem of these functions is such that the products $\cos(-m\chi + n\chi')\,\mathbf{P}_{m,n}^l(\Theta)$ and $\sin(-m\chi + n\chi')\,\mathbf{P}_{m,n}^l(\Theta)$ may be expanded into Fourier series of $\cos(s(\varphi-\varphi'))$ and $\sin(s(\varphi-\varphi'))$, respectively, thus providing the required coefficients. The development is somewhat intricate. Details are given in (Siewert, 1982; Van der Mee and Hovenier, 1990; Deuzé et al., 1989; Lenoble et al., 2007).

The results are in the form

$$
\mathbf{P}_{\cos}^s(\mu,\mu') =
\begin{pmatrix}
\sum_{l=s}^{L}\beta_l P_s^l P_s^{l'} & \sum_{l=s}^{L}\gamma_l P_s^l R_s^{l'} & 0 & 0 \\
\sum_{l=s}^{L}\gamma_l R_s^l P_s^{l'} & \sum_{l=s}^{L}\alpha_l R_s^l R_s^{l'} + \zeta_l T_s^l T_s^{l'} & 0 & 0 \\
0 & 0 & \sum_{l=s}^{L}\alpha_l T_s^l T_s^{l'} + \zeta_l R_s^l R_s^{l'} & \sum_{l=s}^{L}-\varepsilon_l R_s^l P_s^{l'} \\
0 & 0 & \sum_{l=s}^{L}\varepsilon_l P_s^l R_s^{l'} & \sum_{l=s}^{L}\delta_l P_s^l P_s^{l'}
\end{pmatrix},
\tag{3.27}
$$

$$
\mathbf{P}_{\sin}^s(\mu,\mu') =
\begin{pmatrix}
0 & 0 & \sum_{l=s}^{L}\gamma_l P_s^l T_s^{l'} & 0 \\
0 & 0 & \sum_{l=s}^{L}\alpha_l R_s^l T_s^{l'} + \zeta_l T_s^l R_s^{l'} & \sum_{l=s}^{L}-\varepsilon_l T_s^l P_s^{l'} \\
\sum_{l=s}^{L}-\gamma_l T_s^l P_s^{l'} & \sum_{l=s}^{L}-\alpha_l T_s^l R_s^{l'} - \zeta_l R_s^l T_s^{l'} & 0 & 0 \\
0 & \sum_{l=s}^{L}-\varepsilon_l P_s^l T_s^{l'} & 0 & 0
\end{pmatrix},
\tag{3.28}
$$

where terms X_s^l stand for $X_s^l(\mu)$ and terms $X_s^{l'}$ stand for $X_s^l(\mu')$.

Inserting Eqs (3.27) and (3.28) into Eq. (3.26), leads directly after rearrangement to

$$\mu \frac{d\mathbf{L}^s(\tau,\mu)}{d\tau} = \mathbf{L}^s(\tau,\mu) - \frac{\varpi(\tau)}{4\pi} \mathbf{P}^s(\tau,\mu,\mu_0)\mathbf{E}_0 \exp(\tau/\mu_0) \qquad (3.29)$$

$$\frac{\varpi(\tau)}{2} \int_{-1}^{+1} \mathbf{P}^s(\tau,\mu,\mu')\mathbf{L}^s(\tau,\mu')d\mu' \qquad ,$$

where

$$\mathbf{L}^s(\tau,\mu) = \left(I^s(\tau,\mu), Q^s(\tau,\mu), U^s(\tau,\mu), V^s(\tau,\mu) \right)^T, \qquad (3.30)$$

and

$$\mathbf{P}^s(\tau,\mu,\mu') = \begin{pmatrix} \sum_{l=s}^{L} \beta_l P_s^l P_s^{l'} & \sum_{l=s}^{L} \gamma_l P_s^l R_s^{l'} & \sum_{l=s}^{L} -\gamma_l P_s^l T_s^{l'} & 0 \\ \sum_{l=s}^{L} \gamma_l R_s^l P_s^{l'} & \sum_{l=s}^{L} \alpha_l R_s^l R_s^{l'} + \zeta_l T_s^l T_s^{l'} & \sum_{l=s}^{L} -\alpha_l R_s^l T_s^{l'} - \zeta_l T_s^l R_s^{l'} & \sum_{l=s}^{L} \varepsilon_l T_s^l P_s^{l'} \\ \sum_{l=s}^{L} -\gamma_l T_s^l P_s^{l'} & \sum_{l=s}^{L} -\alpha_l T_s^l R_s^{l'} - \zeta_l R_s^l T_s^{l'} & \sum_{l=s}^{L} \alpha_l T_s^l T_s^{l'} + \zeta_l R_s^l R_s^{l'} & \sum_{l=s}^{L} -\varepsilon_l R_s^l P_s^{l'} \\ 0 & \sum_{l=s}^{L} -\varepsilon_l P_s^l T_s^{l'} & \sum_{l=s}^{L} \varepsilon_l P_s^l R_s^{l'} & \sum_{l=s}^{L} \delta_l P_s^l P_s^{l'} \end{pmatrix}, \qquad (3.31)$$

which generalizes Eq. (3.19) to polarized light.

3.3 Boundary conditions

The top of the atmosphere receives only the parallel solar beam, and no diffuse radiation. At the lower boundary, the condition is fixed by the ground surface reflection. Note, in remote sensing applications the lower boundary condition is much more important for a sensor in space looking down through the atmosphere than for a ground-based sensor looking up.

The simplest case for the lower boundary condition is a black surface, that is, completely absorbing with no reflectance. Alternatively, a commonly-used model is a Lambertian ground reflectance, in which reflectance from the surface is isotropic. A Lambertian surface may be a good approximation for land surfaces and it requires a few more calculations than assuming a black ground. Because the reflected radiance is isotropic and depends only on the incident flux, only the azimut independent term, $L^{s=0}(\tau,\mu)$, is modified, with the boundary condition (3.8) replaced by

$$L(\tau = \tau^*,\mu,\varphi) = R F^-(\tau^*)/\pi \quad for \ \mu > 0, \qquad (3.32)$$

where R is the reflectance and $F^-(\tau^*)$ is the downward flux at ground level.

Moreover, the radiances $L^+_R(0,\mu,\varphi)$ reflected, or $L^-_R(\tau^*,\mu,\varphi)$ transmitted by the atmosphere with a surface of Lambertian reflectance R (Figure 3.4) may be derived from radiances calculated for the same atmosphere but with a black surface, L^+_0 and L^-_0 according to

$$L^+_R(0,\mu,\varphi) = L^+_0(0,\mu,\varphi)) + \frac{RF^-_0(\tau^*)T(\tau^*,\mu)}{(1-RS)} ; \tag{3.33.a}$$

$$L^-_R(\tau^*,\mu,\varphi) = L^-_0(\tau^*,\mu,\varphi)) + \frac{RF^-_0(\tau^*)S(\tau^*,\mu)}{(1-RS)} ; \tag{3.33.b}$$

$F^-_0(\tau^*)$ is the downward flux at ground level for a black surface. $T(\tau^*,\mu)$, and $S(\tau^*,\mu)$ are respectively the atmosphere flux total (direct+diffuse) transmittance and the atmosphere flux reflectance, S is the spherical albedo of the atmosphere. Figure 3.4 explains the meaning of the different terms in Eqs (3.33); the factor $1/1-RS$ corresponds to the series of successive reflections between the surface and the atmosphere. More details can be found in (Tanré et al., 1979, Lenoble, 1993).

A more exact simulation of the atmospheric radiance must take into account the bidirectionnal surface reflectance or polarized reflectance.

For water surfaces, a major effect is the anisotropic polarized light resulting from the Fresnel's reflection on the agitated sea surface. This term, nearly independent on the wave-

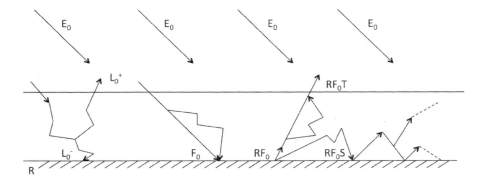

Figure 3.4 Lambertian reflecting boundary.

length and mainly localized around the specular direction, is generally accounted for by the model of Cox and Munk (1954) which proves to be quite realistic (Bréon and Henriot, 2006). A second term, sensitive in all directions, is the water leaving radiance at short wavelengths, whose spectral dependance leads to the ocean color effect. Modeling of the bidirectional and polarized properties of the water leaving radiance needs much involved calculations of the RTE for the coupled ocean–atmosphere system (e.g. Plass and Kattawar, 1972, Fischer and Grassl, 1984, Chami et al., 2001, Chowdhary et al., 2006, Ota et al., 2010). These studies show that, in a first approximation, for the purpose of aerosol retrieval, the water leaving radiance may be accounted for by a Lambertian unpolarized reflectance. A last contribution comes from the white caps reflectance. It is also accounted for by a Lambertian term with the reflectance derived from the wind speed at sea level according to the model of Koepke (1984).

Land surface reflectances also mix diffuse and specular components, with spectral and bidirectional dependances much more variable than water surfaces. The bidirectional reflectance ranges from nearly diffuse Lambertian to nearly specular reflection, with shadowing effect leading to specific behaviors like the hotspot. Various analytical models (e.g. Walthall et al., 1985, Roujean et al., 1992, Rahman et al., 1993) tend to account by the way of a few parameters for this variability of the total reflected light, including the hotspot feature ((Bréon et al., 2002, Maignan et al., 2004). Concerning the polarized component of the reflected light, it is generally admitted that it emanates from single specular reflection on surface elements of the ground (Vanderbilt et al., 1990, Rondeaux and Herman 1991, Bréon et al. 1995, Maignan et al. 2009). This model explains the main polarization features of land surfaces. Minor polarization features in backscattering directions, which are not relevant from Fresnel's reflection and the explanation of which is not clearly established (Mishchenko et al., 2000, Litvinov et al., 2010) would necessitate specific modeling.

Let us first consider the scalar problem. The bidirectional reflectance distribution function (BRDF), $R(\mu,\varphi,\mu_0,\varphi_0)$, generalizes the Lambertian reflectance with

$$L(\tau^*,\mu > 0,\varphi) = R(\mu,\varphi,\mu_0,\varphi_0)(-\mu_0)E_0 \exp(\tau^*/\mu_0)/\pi$$
$$+ \int_0^{2\pi} d\varphi' \int_{-1}^0 d\mu' R(\mu,\varphi,\mu',\varphi')(-\mu')L(\tau^*,\mu',\varphi')/\pi. \tag{3.34}$$

Given some BRDF, symmetric about the vertical plane containing the incident direction (μ',φ'), it may be expanded in the form

$$R(\mu,\varphi,\mu',\varphi') = \sum_{t=0}^{T} (2-\delta_{0t})\cos(t(\varphi-\varphi'))R^t(\mu,\mu'). \tag{3.35}$$

Introducing Eqs (3.15) and (3.35) into Eq. (3.34) leads to

$$L^s(\tau^*,\mu > 0) = -\mu_0 R^s(\mu,\mu_0)E_0 \exp(\tau^*/\mu_0)/\pi + 2\int_{-1}^0 -\mu' R^s(\mu,\mu')L^s(\tau^*,\mu')d\mu', \tag{3.36}$$

which preserves separation into azimuth of the scalar RTE.

Generalization to polarized light is obtained by introducing the (4×4) reflectance matrix $\mathbf{R}(\mu,\varphi,\mu',\varphi')$ such that

$$L(\tau^*,\mu > 0,\varphi) = \mathbf{R}(\mu,\varphi,\mu_0,\varphi_0)(-\mu_0)\mathbf{E}_0 \exp(\tau^*/\mu_0)/\pi$$
$$+ \int_0^{2\pi} d\varphi' \int_{-1}^0 d\mu' \mathbf{R}(\mu,\varphi,\mu',\varphi')(-\mu')\mathbf{L}(\tau^*,\mu',\varphi')/\pi,$$

(3.37)

where separating the diffuse and specular components of the reflectance leads to

$$\mathbf{R}(\mu,\varphi,\mu',\varphi') = \mathbf{R}_{dif}(\mu,\varphi,\mu',\varphi') + \mathbf{R}_{sp}(\mu,\varphi,\mu',\varphi').$$

(3.38)

Obviously, $\mathbf{R}_{dif}(\mu,\varphi,\mu',\varphi')$ is characterized by the scalar BRDF (Eq. 3.35) with

$$\mathbf{R}_{dif}(\mu,\varphi,\mu',\varphi') = \begin{pmatrix} R(\mu,\varphi,\mu',\varphi') & 0 & 0 & 0 \\ 0 & 0 & 0 & 0 \\ 0 & 0 & 0 & 0 \\ 0 & 0 & 0 & 0 \end{pmatrix}.$$

(3.39)

Let us consider $\mathbf{R}_{sp}(\mu,\varphi,\mu',\varphi')$. Given incident and reflected directions, respectively $(\mu'<0,\varphi')$ and $(\mu>0,\varphi)$, we assume that some fraction of the surface is occupied by unshadowed elements whose normal, (μ_n,φ_n), bisects the directions $(-\mu',\varphi'-\pi)$ and (μ,φ) and which act as Fresnel's specular reflectors for the incidence angle i (Figure 3.5). Clearly the reflected radiance is proportional to (i) the fraction of the surface occupied by elements with their normal oriented conveniently, which must be given by some structure function of the surface, say $S(\mu,\varphi,\mu',\varphi')$, and (ii) the Fresnel reflection coefficient

$$r_F(i) = \frac{1}{2}(r_l(i)^2 + r_r(i)^2),$$

(3.40)

with

$$r_r(i) = \frac{\cos i - \sqrt{m^2 - \sin^2 i}}{\cos i + \sqrt{m^2 - \sin^2 i}}, \text{ and } r_l(i) = \frac{m^2\cos i - \sqrt{m^2 - \sin^2 i}}{m^2\cos i + \sqrt{m^2 - \sin^2 i}},$$

(3.41)

assuming for simplicity real refractive index, m, of the reflectors. Therefore, referring the Stokes parameters to directions parallel and perpendicular to the reflection plane (which contains the normal to the elements and the incident and reflected directions, see (Figure 3.5), the polarized reflected radiance is proportional to $S(\mu,\varphi,\mu',\varphi')$ and to the Fresnel's reflectance matrix, $\mathbf{r}_F(i)$, (Deuzé et al., 1989) referred to these axes:

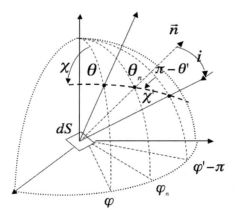

Figure 3.5 Specular reflectors (see text for explanation).

$$
\mathbf{r}_F(i) = \frac{1}{2}
\begin{pmatrix}
r_i^2(i) + r_r^2(i) & r_i^2(i) - r_r^2(i) & 0 & 0 \\
r_i^2(i) - r_r^2(i) & r_i^2(i) + r_r^2(i) & 0 & 0 \\
0 & 0 & 2r_i(i)r_r(i) & 0 \\
0 & 0 & 0 & 2r_i(i)r_r(i)
\end{pmatrix}.
\tag{3.42}
$$

Finally, as for the case of the scattering matrix, taking into account the rotations of the reference axes (Figure 3.5) needed to come back to axes parallel and perpendicular to the vertical planes which the radiances are referred to, the required $\mathbf{R}_{sp}(\mu,\varphi,\mu',\varphi')$ matrix for reflected polarized radiance is

$$
\mathbf{R}_{sp}(\mu,\varphi,\mu',\varphi') = S(\mu,\varphi,\mu',\varphi')\mathbf{T}(-\chi)\,\mathbf{r}_F(i)\mathbf{T}(\chi').
\tag{3.43}
$$

Let $\mathbf{M}_F(\mu,\varphi,\mu',\varphi')$ stand for the matrix $\mathbf{T}(-\chi)\mathbf{r}_F(i)\mathbf{T}(\chi')$. The expansion of $\mathbf{M}_F(\mu,\varphi,\mu',\varphi')$ into Fourier series of the azimuth is straightforward when noticing the analogy of the reflection and scattering processes: the reflection plane is similar to the scattering plane, and $(\pi -2i)$ is similar to the scattering angle Θ (Figure 3.5). Developing the elements of $\mathbf{r}_F(i)$ into the same form as the corresponding elements of $\mathbf{F}(\Theta)$, that is

$$(r_l^2(i) + r_r^2(i))/2 = \sum_{l=0}^{P} b_l P_l(\cos \Theta)$$

$$(r_l^2(i) - r_r^2(i))/2 = \sum_{l=2}^{P} g_l P_2^l(\cos \Theta)$$

$$(r_l^2(i) + r_r^2(i))/2 = \sum_{l=2}^{P} [a_l R_2^l(\cos \Theta) + z_l T_2^l(\cos \Theta)] , \qquad (3.44)$$

$$\eta(i) r_r(i) = \sum_{l=2}^{P} [z_l R_2^l(\cos \Theta) + a_l T_2^l(\cos \Theta)]$$

$$\eta(i) r_r(i) = \sum_{l=0}^{P} d_l P_l(\cos \Theta)$$

where $\Theta = \pi - 2i$, leads to the separation of $\mathbf{M}_F(\mu, \phi, \mu', \phi')$ into symmetric and anti-symmetric matrices with respect to the azimuth, and thus to

$$\mathbf{M}_F(\mu, \varphi, \mu', \varphi') = \sum_{p=0}^{P} (2 - \delta_{0s})[\cos[p(\varphi - \varphi')]\mathbf{M}_{F\cos}^{p}(\mu, \mu') + \sin[p(\varphi - \varphi')]\mathbf{M}_{F\sin}^{p}(\mu, \mu')] \quad (3.45)$$

with $\mathbf{M}_{F\cos}^{p}(\mu, \mu')$ and $\mathbf{M}_{F\sin}^{p}(\mu, \mu')$ derived from Eqs (3.27, 3.28) by substituting a_l, b_l, c_l, d_l, e_l, respectively, to $\alpha_l, \beta_l, \gamma_l, \delta_l, \varepsilon_l$.

A last difficulty comes from the azimuthal dependance of the distribution function.

According to Cox and Munk (1954), restricting to their approximate isotropic model which depends only on the wind speed, w, and preserves the symmetry of the reflectance law, the appropriate function for the agitated sea surface is

$$S(\mu, \varphi, \mu', \varphi') = \frac{1}{4\sigma^2 \mu(-\mu')\mu_n^4} \exp\left(-\frac{1 - \mu_n^2}{\sigma^2 \mu_n^2}\right), \qquad (3.48)$$

where σ^2 is about $0.003 + 0.00512w$, and $\mu_n = \cos\theta_n$.

According to Maignan et al. (2009), an appropriate function for the polarized reflectance of land surfaces measured by the POLDER/PARASOL instrument (Deschamps et al., 1994) is

$$S(\mu, \varphi, \mu', \varphi') = \frac{C \exp(-\tan i)}{4(\mu - \mu')}, \qquad (3.49)$$

where C depends principally on the Earth biome, according to the IGBP classification, and on the vegetation index (Maignan et al., 2009).

Both functions depend on the azimuth and may be developed in the form

$$S(\mu, \varphi, \mu', \varphi') = \sum_{m=0}^{M} (2 - \delta_{0m}) \cos(m(\varphi - \varphi')) S^m(\mu, \mu') . \qquad (3.50)$$

Multiplication by the symmetric scalar function $S(\mu,\varphi,\mu',\varphi')$ does not change the parity of $\mathbf{M}_F(\mu,\varphi,\mu',\varphi')$, and the final reflectance matrix

$$\mathbf{R}_{sp}(\mu,\varphi,\mu',\varphi') = S(\mu,\varphi,\mu',\varphi')\mathbf{M}_F(\mu,\varphi,\mu',\varphi') \tag{3.51}$$

separates, as $\mathbf{M}_F(\mu,\varphi,\mu',\varphi')$, into

$$\mathbf{R}_{sp}(\mu,\varphi,\mu',\varphi') = \sum_{s=0}^{M}(2-\delta_{0s})[\cos[s(\varphi-\varphi')]\mathbf{R}^s_{sp\,\cos}(\mu,\mu') + \sin[s(\varphi-\varphi')]\mathbf{R}^s_{sp\,\sin}(\mu,\mu')], \tag{3.52}$$

and rearrangement of the products $\cos(p\varphi).\cos(m\varphi)$ or $\sin(p\varphi).\cos(m\varphi)$ into series of $\cos(s\varphi)$ or $\sin(s\varphi)$ (Lenoble et al., 2007) leads to matrices $\mathbf{R}^s_{sp\,\cos}(\mu,\mu')$ and $\mathbf{R}^s_{sp\,\sin}(\mu,\mu')$ given by

$$\mathbf{R}^s_{sp\,\cos}(\mu,\mu') = \mathbf{M}^0_{F\,\cos}(\mu,\mu') + \sum_{k=1}^{M+P}\mathbf{M}^k_{F\,\cos}(\mu,\mu')(S^{k+s}(\mu,\mu') + S^{|k-s|}(\mu,\mu')) \tag{3.53}$$

and

$$\mathbf{R}^s_{sp\,\sin}(\mu,\mu') = -\sum_{k=1}^{M+P}\mathbf{M}^k_{F\,\sin}(\mu,\mu')(S^{k+s}(\mu,\mu') - S^{|k-s|}(\mu,\mu')). \tag{3.54}$$

Inserting Eqs (3.52) and (3.45) in Eq. (3.43) leads to the boundary condition

$$\mathbf{L}^s(\tau^*,\mu > 0) = \mathbf{R}^s_{sp}(\mu,\mu_0)(-\mu_0)\mathbf{E}_0 e^{\tau^*/\mu_0}/\pi + 2\int_{-1}^{0}d\mu'\mathbf{R}^s_{sp}(\mu,\mu')(-\mu')\mathbf{L}^s(\tau^*,\mu'), \tag{3.55}$$

where

$$\mathbf{R}^s_{sp}(\mu,\mu') = \begin{pmatrix} R^s_{sp\,\cos11}(\mu,\mu') & R^s_{sp\,\cos12}(\mu,\mu') & -R^s_{sp\,\sin13}(\mu,\mu') & -R^s_{sp\,\sin14}(\mu,\mu') \\ R^s_{sp\,\cos21}(\mu,\mu') & R^s_{sp\,\cos22}(\mu,\mu') & -R^s_{sp\,\sin23}(\mu,\mu') & -R^s_{sp\,\sin24}(\mu,\mu') \\ R^s_{sp\,\sin31}(\mu,\mu') & R^s_{sp\,\sin32}(\mu,\mu') & R^s_{sp\,\cos33}(\mu,\mu') & R^s_{sp\,\cos34}(\mu,\mu') \\ R^s_{sp\,\sin41}(\mu,\mu') & R^s_{sp\,\sin42}(\mu,\mu') & R^s_{sp\,\cos43}(\mu,\mu') & R^s_{sp\,\cos44}(\mu,\mu') \end{pmatrix}, \tag{3.56}$$

which preserves the separation into azimuth of the matricial RTE.

3.4 Solution of the radiative transfer problem

Several methods have been proposed for solving the scalar radiative transfer problem, and several numerical codes are available. Some of the methods have been directly extended to the vector problem. We are not trying here to give an exhaustive presentation of these methods, but will limit ourselves to the description of some very popular methods. A fairly large bibliography concerning the early works can be found in Lenoble (1985).

The most direct approach consists in solving the scalar or the vector RTE given above (Eqs 3.6 or 3.12), generally starting with the expansion into Fourier series of azimuth presented in Section 3.2.4. A further difficulty is due to the vertical inhomogeneity of the atmosphere, which is approximated by a division into homogeneous layers. Two methods for solving Eq. (3.6), and further extended to Eq. (3.12), are presented respectively in Sections 3.4.1 and 3.4.2.

A somewhat different approach consists of defining a reflection and a transmission function computed by solving the RTE for a thin layer, and then superposing such thin layers; this is the basis for the Adding-Doubling method presented in Section 3.4.3.

Monte Carlo methods can be applied to the problem of radiative transfer (see Section 3.4.4); they are useful for non plane-parallel cases.

Finally, fast approximate methods are available, but have not had widespread use in remote sensing applications.

3.4.1 Successive orders of scattering (OS)

The idea underlying the method of successive orders of scattering is to separate the radiance into the contributions of photons scattered once, twice, ..., n times,

$$L(\tau,\mu,\varphi) = \sum_{n=1}^{N} L_n(\tau,\mu,\varphi). \tag{3.57}$$

The method was suggested very early, first limited to two terms, by Hammad and Chapman (1939), it has been used by several authors, in somewhat different forms (Dave and Furukawa,1966), and recently by Min and Duan (2004) and Lenoble et al. (2007).

Integrating Eq. (3.12), with Eq. (3.57) leads to

$$L_n(\tau,\mu < 0,\varphi) = -\int_0^\tau e^{-(\tau'-\tau)/\mu} J_n(\tau',\mu,\varphi)d\tau'/\mu, \tag{3.58.a}$$

$$L_n(\tau,\mu > 0,\varphi) = L_n^{up}(\tau^*,\mu > 0,\varphi)e^{-(\tau^*-\tau)/\mu} + \int_\tau^{\tau^*} e^{-(\tau'-\tau)/\mu} J_n(\tau',\mu,\varphi)d\tau'/\mu, \tag{3.58.b}$$

where

$$J_1(\tau,\mu,\varphi) = \frac{\varpi(\tau)}{4\pi} P(\tau,\mu,\varphi,\mu_0,\varphi_0) E_0 e^{\tau/\mu_0}, \tag{3.58.c}$$

$$J_{n>1}(\tau,\mu,\varphi) = \frac{\varpi(\tau)}{4\pi} \int_0^{2\pi}\int_{-1}^{+1} P(\tau,\mu,\varphi,\mu',\varphi')L_{n-1}(\tau,\mu',\varphi')d\mu'd\varphi', \qquad (3.58.d)$$

and, considering that one reflection is similar to one scattering,

$$L_1^{up}(\tau^*,\mu > 0,\varphi) = (-\mu_o)R(\mu,\varphi,\mu_o,\varphi_o)E_0 e^{\tau^*/\mu_0}/\pi, \qquad (3.58.e)$$

$$L_{n>1}^{up}(\tau^*,\mu > 0,\varphi) = \int_0^{2\pi}\int_{-1}^{0}(-\mu')R(\mu,\varphi,\mu',\varphi')L_{n-1}(\tau^*,\mu',\varphi')d\mu'd\varphi'/\pi, \qquad (3.58.f)$$

where R stands for the reflectance matrix defined in Section 3.3.

Similar equations apply after the expansion in azimuth. The integrals over τ' are replaced by finite sums, when the atmosphere is divided into homogeneous layers.

The method of successive orders of scattering is very easy to implement and very efficient when the scattering layers are not too thick (approximately $\tau^* \leq 2$).

3.4.2 Discrete ordinates (DISORT)

The method was first developed for the scalar equation (Chandrasekhar, 1950; Stamnes et al., 1988); its vector form can be found in Siewert (2000) and in Rozanov et al. (2005). We will present briefly the method for the scalar RTE, after separation into azimuth terms, as seen in Section 3.2.4.

The idea is to approximate the integral in Eq. (3.19), by a finite sum as

$$\int_{-1}^{+1} f(\mu')d\mu' = \sum_{j=-n}^{j=+n} a_j f(\mu_j); \qquad (3.59)$$

the best approximation of the integral in the interval (-1+1) is given by the Gauss quadrature. The μ_j 's are the zeros of the Legendre polynomial $P_{2n}(\mu)$, and the a_j's the corresponding weights; they satisfy the relations $\mu_{-j}=-\mu_j$ and $a_{-j}=a_j$.

If we write now Eq.(3.19), with this approximation, and for each of the μ_i , we get

$$\mu\frac{dL(\tau,\mu_i)}{d\tau} - L(\tau,\mu_i) + \frac{\varpi(\tau)}{2}\sum_{j=-n}^{j=+n} a_j p(\tau,\mu_i,\mu_j)L(\tau,\mu_j) = -\frac{\varpi(\tau)}{4\pi}p(\tau,\mu,\mu_0)E_0\exp(\tau/\mu_0),$$

$$(3.60)$$

for i=-n,....-1,+1,.....+n; order s is omitted .

Equation (3.60) represent a system of 2n differential equations for the 2n unknown functions $L(\tau,\mu_i)$. In an homogeneous sublayer, the coefficients are constant, and the solution of the system is easily obtained as

$$L(\tau,\mu_i) = \sum_{l=-n}^{l=+n} k_l g_i(v_l) \exp(v_l \tau) + h_i \exp(\tau/\mu_0),$$ (3.61)

where the second term on the right handside of Eq. (3.61) is a particular solution of Eq. (3.60); the functions $g_i(v_l)exp(v_l\tau)$ are solutions of the homogeneous system associated with Eq. (3.60); the 2n values v_l are found as eigenvalues of a matrix; the k_l are constants to be found from boundary conditions and continuity conditions between the sublayers.

DISORT is more complicated to use than the OS method, but it has the advantage that it works well for thick scattering layers, such as clouds.

3.4.3 Adding–Doubling method. (de Haan et al.,1987; Stammes et al.,1989; Stammes, 2001)[1]

Let us define the reflectance function of a layer of optical depth τ above a black ground surface, receiving the solar irradiance from the direction $(-\mu_0,\varphi_0)$ as

$$S(\tau^*,\mu,\varphi,\mu_0,\varphi_0) = \frac{\pi L^+(0,\mu,\varphi)}{\mu_0 E_0},$$ (3.62)

where L^+ is the upward radiance.

Similarly, the diffuse transmittance function is

$$T(\tau^*,\mu,\varphi,\mu_0,\varphi_0) = \frac{\pi L^-(\tau^*,-\mu,\varphi)}{\mu_0 E_0},$$ (3.63)

where L^- is the downward radiance; in Eq. (3.63), we use the convention $-\mu$, with $\mu>0$, for downward directions. The total transmittance is

$$T_{tot}(\tau^*,\mu,\varphi,\mu_0,\varphi_0) = T(\tau^*,\mu,\varphi,\mu_0,\varphi_0) + \frac{\pi}{\mu_0}\Delta(\mu-\mu_0)\Delta(\varphi-\varphi_0)\exp(-\tau^*/\mu_0).$$ (3.64)

The atmosphere is divided into layers thin enough to have a very rapid solution of the RTE in each layer, giving the reflectance and transmittance of the layer. Then two layers are combined, taking into account the successive reflections between them. This is the principle of the adding method; it can be simplified to doubling for an homogeneous atmosphere.

_____

1   μ and μ_0 are taken as positive in this section.

3.4.4 Monte Carlo method

Monte Carlo methods are very general probabilistic methods that can be applied to different kinds of problems, including radiative transfer in the atmosphere (Marchuk et al.,1980).
In this case, one photon at a time is followed on its three dimensional path in the scattering
medium. The various events are defined by suitable probability distributions; a set of random numbers is then used to make a particular choice for the result of each event. The principle is simple, and a Monte Carlo code is relatively easy to write; but a large number of
photons is necessary for achieving a good precision. Therefore, the method needs a rather
large amount of computer time, and is not recommended for plane parallel problems. Its
major advantage is applicability to any geometry of the scattering medium, like spherical
atmospheres, or horizontal inhomogeneity of the surface or of the atmosphere.

3.4.5 Approximate methods

Some very rapid methods allow to obtain approximate solutions of the transfer problem.
They are generally of the "two-stream" type, considering only one upward and one downward direction (Meador and Weaver, 1980). They are mainly used when repetitive calculations are necessary, as in climate models. Although they give reasonably good results for
fluxes, they fail to reproduce directional distribution of radiation; therefore they cannot be
used for analysis of remote sensing data.

A better approximate solution consists in limiting the solution of the RTE to single scattering; it is a particular case of the OS method, and gives good results for thin scattering
layers. It is often applied to stratospheric studies, and it permits to solve spherical problems
as well as plane parallel cases.

We present it here for the scalar RTE, and a nonreflecting ground surface.

Equations (3.58.a and b) write in this case as,

$$L(\tau,\mu < 0,\varphi) = -\int_0^\tau e^{-(\tau'-\tau)/\mu} J_1(\tau',\mu,\varphi)d\tau'/\mu, \tag{3.65}$$

for downward directions, and

$$L(\tau,\mu > 0,\varphi) = \int_\tau^{\tau^*} e^{-(\tau'-\tau)/\mu} J_1(\tau',\mu,\varphi)d\tau'/\mu, \tag{3.66}$$

for upward directions, with

$$J_1(\tau,\mu,\varphi) = \frac{\varpi(\tau)}{4\pi} p(\tau,\mu,\varphi,\mu_0,\varphi_0)E_0 e^{\tau/\mu_0}. \tag{3.67}$$

A simple integration gives for downward radiance

$$L(\tau,\mu < 0,\varphi) = \frac{\varpi\mu_0}{4\pi(\mu - \mu_0)} p(\mu,\varphi,\mu_0,\varphi_0)E_0[e^{\tau/\mu} - e^{\tau/\mu_0}] \quad \text{for } \mu \neq \mu_0, \tag{3.68}$$

$$L(\tau,\mu_0,\varphi) = -\frac{\varpi\tau}{4\pi\mu_0}\, p(\mu_0,\varphi,\mu_0,\varphi_0) E_0 e^{\tau/\mu_0} \qquad \text{for } \mu=\mu_0, \tag{3.69}$$

and for upward radiance

$$L(\tau,\mu>0,\varphi) = \frac{\varpi\mu_0}{4\pi(\mu-\mu_0)}\, p(\mu,\varphi,\mu_0,\varphi_0) E_0 e^{\tau*/\mu_0} [e^{-(\tau*-\tau)/\mu} - e^{-(\tau*-\tau)/\mu_0}]. \tag{3.70}$$

A further simplification uses the expansion of the exponential functions for τ small. Sometimes an empirical and approximate correction G is added to Eqs (3.68, 3.69, 3.70) to take into account multiple scatterings and surface reflection.

Equations (3.68) and (3.67) are used in Chapter 5 as Eq. (5.12).(NOTE that μ for downward directions and μ_0 are negative here, according to the general usage in Chapter 3.)

3.5 Radiative transfer codes

There are now several radiative transfer codes (RT codes) freely available, either from the internet or by contacting the authors directly.

We must clarify here what an RT code is, as it is not simply the computer version of a method for solving the RTE, but includes a complete environment describing the physical problem. The description of the problem includes:

- the atmospheric characteristics, i.e.:
  - i) the molecular optical depth, which depends on the site altitude, by the surface pressure,
  - ii) the aerosol scattering and absorption optical depths, the aerosol profile, and the aerosol phase matrix, characteristics that depend on the aerosol composition, shape, and size distribution,
  - iii) eventually the molecular gaseous absorption;

- the wavelengths, either discrete corresponding to an instrument channel, or with a given step over a fixed interval;

- the conditions of illumination, solar zenith angle, extraterrestrial solar irradiance; for remote sensing, results are often expressed as reflectances or normalized radiances, and not absolute radiances, which avoids the problem of uncertainties on the extraterrestrial flux.

The results sought include for all chosen wavelengths:

- the radiances for specific directions, at specific levels, more generally at the surface and at the top of the atmosphere (TOA); the irradiance at the surface and the exitance at the TOA are by-products easy to obtain;

- the state of polarization for the same directions and levels.

Then it is necessary to discretize the atmosphere into thin, assumed homogeneous, layers, before applying the chosen method for solving the RTE. An angular discretization is also necessary in most methods.

Generally, an RT code has been first written with the objective of solving a specific physical problem, using one of the resolution methods mentioned above. Most codes have first been concerned with the scalar problem, and thereafter have been extended to the polarization problem. Depending on the wavelength interval of interest, some codes have included a treatment of gaseous line absorption; also some codes are extended to the case of longwave transfer. The most advanced codes are modular, and permit the users to apply the code to very different problems, by changing only the input instructions.

A good example of such a complete and sophisticated code is the libRadtran software package (Mayer and Kylling, 2005, http://www.libradtran.org). It derives from the uvspec code, originally designed for computing the ultraviolet and visible spectral irradiance at the Earth's surface. It can be used now for computing the complete radiation field, including polarization, and covering the solar and thermal spectral ranges. It also proposes to the user the choice between several methods for solving the RTE (solvers), including Monte Carlo.

The DISORT code uses the Discrete Ordinate method; it is also fairly flexible, and allows various problems to be tackled without modifying the program; it is available from ftp://climate.gsfc.nasa.gov/pub/wiscombe/Multiple_Scatt/. It was widely used for the scalar case and is now also used for the vector case.

The OS and the Adding–Doubling codes are respectively built around the Successive Orders of Scattering and the Doubling–Adding methods described above. They are freely available from the authors who may help with their adaptation to the user's specific problem. Both include polarization.

6S stands for "Second Simulation of the Satellite Signal in the Solar Spectrum". It is rather simple and rapid (Tanré et al., 1990, Vermote et al., 1997). The name clearly tells the specific objective of the code; it uses the Successive orders of scattering method for solving the RTE. A vector version 6Sv also exists.

3.6 Examples of aerosol impact

Some numerical simulations of the aerosol impact on the atmospheric radiance and polarization in spaceborne or ground-based measurements are shown below; the computations are done with the OS code developed at the Laboratoire d'Optique Atmosphérique (LOA) of the University of Lille, France. For the sake of simplicity, the calculations are conducted for a few predefined models of particles corresponding to the main types of terrestrial aerosols (Remer and Kaufman, 1998; Dubovik et al., 2002): small particles corresponding to the accumulation mode, large maritime aerosols and large nonspherical particles. The phase functions ($p(\Theta)$ or $F_{11}(\Theta)$ in Eq. (2.25)) and the polarized phase functions ($q(\Theta) = -F_{12}(\Theta)$ defined in Eq. (2.25)) corresponding to these models are shown in Figures 3.6a and 3.6b, for 4 wavelengths. The radiance and the polarized radiance are normalized by choosing $E_0 = \pi$ in the calculations

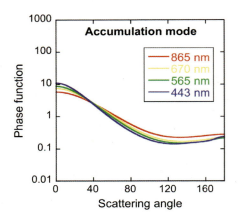 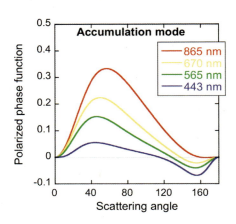

Figure 3.6a Small particles, accumulation mode, LND size distribution with r_m=0.10μm, lnσ=0.46, real refractive index=1.45; left: phase function, right: polarized phase function.

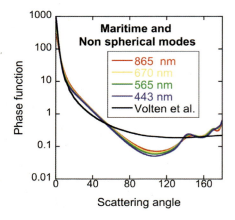 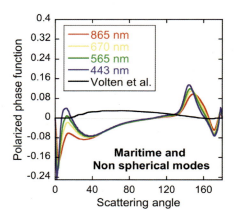

Figure 3.6b Large spherical particles, maritime, LND size distribution with r_m=1.0μm, lnσ=0.70, real refractive index=1.33, and non-spherical mineral particles with average characteristics from Volten et al. (2001); left: phase function, right: polarized phase function.

As the scattering properties of small particles are not overly sensitive to shape (Mishchenko and Travis, 1994a and b), Figure 3.6a corresponds to spherical particles with a lognormal size distribution (Eq. (1.1)) (mode radius on a logarithmic scale of the number distribution $r_m=0.10\mu$m, standard deviation $\sigma=1.584$ ($\ln\sigma= 0.46$), i.e. effective radius reff$=0.17\mu$m, real refractive index $m=1.45$). Figure 3.6b corresponds to models of large particles. The maritime model consists of large spherical particles ($r_m=1.0\mu$m, $\sigma=2.014$, $\ln\sigma=0.70$, $r_{eff}=3.40\mu$m, real refractive index $m=1.33$). Large nonspherical particles are illustrated by the average model of Volten et al. (2001) for mineral particles. Elaborated modeling of nonspherical aerosol scattering, based on spheroid models (see Chapter 2, Mishchenko et al., 1997, Dubovik et al., 2006) indicate that the spectral dependence is probably small, similar to that of the maritime model.

3.6.1 Ground-based observations

The best opportunity for detailed remote sensing of aerosols is provided by ground-based measurements of the sky's brightness in conjunction with measurements of the direct solar beam (Nakajima et al., 1983; Kaufman et al., 1994). Multi-spectral multiangle ground-based measurements of direct sun and sky radiance permit retrieval of both the aerosol optical thickness, from the solar beam transmission, and aerosol phase functions, especially in the forward scattering directions, from analysis of the diffuse sky light. Differences between large and small particles are apparent, both from the characteristic forward diffraction peak of large particles, in Figure 3.6b left, and from the very different spectral behavior of their optical thickness. According to Figures 2.10 and 2.11 (Chapter 2), with mean Mie parameters, $x_{eff}= 2\pi r_{eff}/\lambda$, respectively, about 40 and 2 in the visible range, the scattering coefficient of large particles is nearly independent of the wavelength, while that of small particles increases largely from large to short wavelengths; moreover, this increase itself, often characterized by the Ångström coefficient α with the spectral dependence of τ approximated into the form $\tau(\lambda) = \tau(\lambda_0)(\lambda_0/\lambda)^\alpha$, is sensitive to the particle size. The polarized phase function reported in Figure 3.6a right shows that the large polarization exhibited by small particles provides a third way to characterize particle properties (Vermeulen et al. 2000). Moreover, the measurement analysis is simplified since ground-based measurements are nearly insensitive to the aerosol vertical profile and much less sensitive to the ground reflectance than spaceborne observations.

Ground-based observations of the diffuse sky light are exemplified in Figures 3.7 to 3.12 by simulated measurements of the sky light in the almucantar (i.e with viewing azimuth φ ranging from 0 to 2π and viewing zenith angle $\theta = \pi- \theta_0$ constant). All calculations are for a large solar zenith angle, $\theta_0 = 70°$ (which enlarges the range of scattering angles), at wavelength $\lambda = 865nm$ (which reduces the molecular scattering screening effect; the assumed surface pressure is 1013 mb.).

In Figure 3.7, the results are shown for an atmosphere free from aerosols ($\tau=0.0$), and for each aerosol type with $\tau=0.20$; the surface is black. Figures 3.7 left and right show, respectively, the radiance and the polarized radiance; in Figure 3.7 right, positive and negative figures correspond to directions of the linearly polarized radiance, respectively, nearly perpendicular and nearly parallel to the scattering plane.

At near infrared wavelengths, the aerosol contribution to the diffuse sky light is quite clear because the molecular scattering component is negligible. As explained in Chapter 5, retrieval of the size distribution and refractive index of the aerosols is mainly based on joint analysis of the spectral dependence of $\tau(\lambda)$ and the angular dependence of their phase functions. The large contribution of singly scattered light entails good correlation between the measured sky radiance as a function of angle and aerosol type, and the derived phase functions, as can be seen by comparing Figures 3.6 and 3.7. Neither the surface reflectance nor multiple scattering, in a large range of optical thicknesses, modify this result severely.

For the case of the same small mode particles used in previous examples, Figure 3.8 compares simulations of the downward radiance and polarized radiance at $\lambda = 865nm$, for a black surface and for representations of surface reflectance of specific surface types. Overall, we see a lack of sensitivity to surface type in the modeled downward sky radiance. For oceanic surfaces, the influence comes essentially from the sunglint contribution and is not dependent on the assumed wind speed. The major change here is for the polarized light, due to the large polarization of the sunglint for solar zenith angles near to the Brewster point. The influence of the land surface reflection is shown for a Lambertian reflectance $R=0.20$. It is nearly unchanged when the surface BRDF is largely modified, provided the surface albedo is the same. For typical polarization of land surfaces, the change in the sky light polarization is indiscernable.

Lastly, when moving the mean altitude of the aerosol layer from 1 to 10km, the change in the radiance is indiscernable, even at $\lambda = 443nm$ where molecular scattering is maximum. The polarized radiance depends slightly on the aerosol altitude at short wavelengths, but not at all in the near infrared channels, at $670nm$ and $865nm$.

For 3 different modes of small particles in the range of Mie scattering: $r_m = 0.05, 0.10,$ $0.15\ \mu m$, with $\sigma = 1.6$ and $m = 1.45$, Figure 3.9 shows that the downward radiance and polarized radiance in the almucantar, at $\lambda = 865nm$, show large sensitivity to particle size which allows us to discriminate the characteristic angular behaviors of the particles up until optical thicknesses of about $\tau = 0.50$.

In the same conditions, for the nonspherical model, Figure 3.10 compares the downward radiance with the one computed using the single scattering approximation for different aerosol optical thicknesses. Single scattering is a good approximation until about $\tau = 1.0$. In this low aerosol optical thickness regime the forward scattering peak clearly emerges and correction of the measurements from the multiple scattering contribution is tractable. For lower solar zenith angles, the forward peak detectivity may be extended to aerosol optical thicknesses as large as $\tau = 4.0$.

Finally, as the total scattered light may be compared with the optical thickness for extinction, retrieval of the aerosol absorption is possible provided both measurements are carefully calibrated.

To illustrate the sensitivity of the measurements to the single scattering albedo (SSA) of the aerosols, for small mode particles with AOT $\tau = 0.40$, the radiance in the almucantar at $\lambda = 865nm$ is reported in Figure 3.11 for various SSA (1.00, 0.90, 0.80). Figure 3.12 shows the same comparison for the case of the nonspherical model, but for a larger measured optical thickness for extinction, $\tau = 1.00$. The results are similar at other wavelengths and polarization measurements exhibit similar behavior. Increasing absorption approximately entails a mere translation of the downward radiance or polarized radiance, about $dL = -1.5Ld\varpi$ whatever the particle dimension.

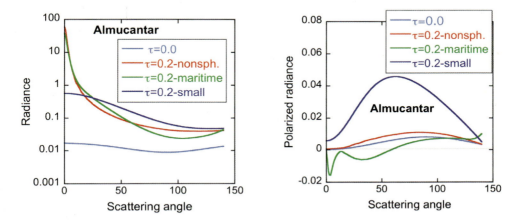

Figure 3.7 Simulation of measurements in the almucantar for the 3 models and with no aerosols (τ=0.00), θ_0=70°, λ=865nm; left: radiance; right: polarized radiance.

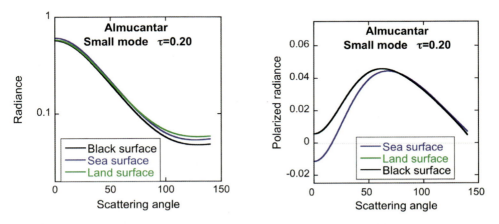

Figure 3.8 Influence of the surface reflectance for the small mode, θ_0=70°, λ=865nm; left: radiance, right: polarized radiance.

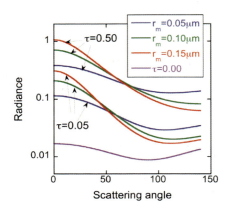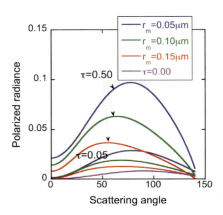

Figure 3.9 Influence of the mode radius and optical depth for small particles; left: radiance, right: polarized radiance.

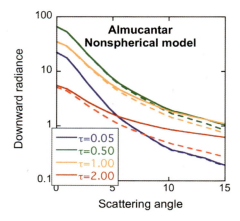

Figure 3.10 Non-spherical particles, comparison of exact radiance (solid lines) with single scattering approximation (dashed lines) for various optical depths.

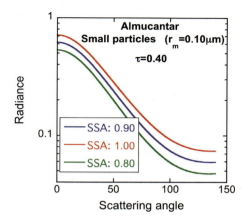

Figure 3.11 Influence of absorption (fixed by the SSA) on radiance for small particles.

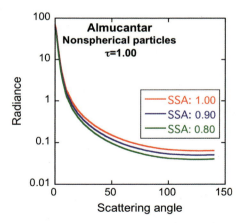

Figure 3.12 Same as Figure 3.11, for large non-spherical particles.

3.6.2 Spaceborne measurements

Remote sensing of the aerosols on a global scale requires spaceborne observations. We limit the calculations here to observations over the ocean which are much more tractable than over land surfaces where the highly variable reflectance prevents easy detection of the aerosol contribution in the upward radiance.

Over the ocean, the major constraint is to avoid too large contamination by the sunglint. For $\theta_0 = 40°$, Figure 3.13 is a polar plot of viewing directions (θ, φ) corresponding to scattering angles ranging to 80° to 180° ($\varphi = 0°$ for the solar incident plane) and to sunglint radiance $L_G = 0.001$ for wind speeds ranging from 2 to 11 m/s.

The most complete information about the aerosols will be obtained from multispectral measurements of the radiance and polarized radiance at different angles, by scanning a given ground target through some convenient plane. In Figure 3.13, black and green points correspond, respectively, to scans along the satellite subtrace (scanning plane vertical, including nadir view), and along a parallel to the subtrace (scanning plane tilted, here by 30°; Earth curvature is neglected). The assumed azimuth of the satellite subtrace with respect to the solar incident plane here is 50°.

The following figures are for observations with the tilted plane in Figure 3.13, which provides useful data until 80° scattering angle, typical of POLDER data (Deschamps et al., 1994) at mid and high latitudes over about one third of the swath. In these conditions, the influence of the sea surface reflection is nearly insensitive to the wind speed. Observations are over case I waters (Morel, 1983) in near infrared channels where the sea water leaving radiance is negligible (Lambertian reflectances $R_w(865nm) = 0.0$ and $R_w(670nm) = 0.001$) The main uncertainty concerns the foam reflectance; the assumed wind speed in the Koepke (1984) model here is 7m/s; best observations are for low wind speeds (w<5m/s), with the foam reflectivity negligible. Rayleigh scattering corresponds to a standard atmosphere (1013 mb at sea level).

The main point in spaceborne measurements is to retrieve the AOT, which needs some estimate of the phase function. Given predefined models of large and small particles, this may be derived from the spectral dependence of the radiance, as used in the very first aerosol global characterizations over the ocean from the AVHRR data (Husar et al.,1997; Nakajima and Higurashi, 1998). Improved estimates of the aerosol properties may be derived from measurements of the directionnal and polarized radiances (Deuze et al., 2000; Chowdhary et al., 2001; Herman et al., 2005).

As an example, for conservative aerosols of the maritime model (SSA= 1.00) with AOT 0.05, simulations of the upward radiance, $L_\lambda(\Theta)$, and polarized radiance, $L_{\lambda,pol}(\Theta)$, at 865 and 670nm, are reported in Figure 3.14 left as a function of the scattering angle Θ. Because molecular scattering varies largely from 865 to 670nm, the same results but for an aerosol free atmosphere, say $L_\lambda^R(\Theta)$ and $L_{\lambda pol}^R(\Theta)$, have been calculated, and for the sake of clarity Figure 3.14 right shows the corrected data: $L_\lambda(\Theta) - L_\lambda^R(\Theta)$ and $L_{\lambda,pol}(\Theta) - L_{\lambda,pol}^R(\Theta)$. Figure 3.15 shows the same calculations for the nonspherical model in place of the maritime model. The main characteristics of large spherical and nonspherical particles appear more clearly in Figures 3.14 right and 3.15 right and permit identification of maritime and desertic dust particles (Herman et al., 2005).

For small conservative aerosols, the sensivity of the measurements to the particle dimension is exemplified in Figure 3.16 by the directional radiance and polarized radiance, corrected as previously, corresponding to the two extreme small modes of Figures 3.9. Figure 3.16 left shows results for the 0.05 μm mode and Figure 3.16 right shows results for the 0.15 μm mode. The optical thicknesses have been adjusted to provide radiances of the same order of magnitude.

Indication of the particle dimension is firstly provided by the rate of scattering increasing from 865 to 670nm (linked to the Ångström coefficient of the particles). Figure 3.16 left, for the smaller particles shows larger differences between 865nm and 670nm than does Figure 3.16 right, for larger particles. Another indication is the increasing asymmetry of the directional radiance as the particle size increases, in Figure 3.16. Finally, in the same figures, the large change in the polarization provides a third very useful independent information on the particle dimension. More generally, as large nonspherical and small particles exhibit quite similar directional radiances in spaceborne observations (cf. Figures 3.15 and 3.16), one see that polarization is especially interesting in cases of aerosols mixing small and large particles (Dubovik et al., 2002), with polarized light nearly independent on the large mode contribution (Herman et al., 2005).

Moreover, for large enough aerosol contents, polarization may provide information about the particle refractive index, as suggested in Figure 3.17. For three small modes of conservative particles, with respective refractive index 1.35, 1.45, and 1.60, the median radius (respectively $r_m = 0.102, 0.100. 0.092\mu m$; $\sigma = 1.6$) and the optical thickness (respectively $\tau(865nm) = 0.42, 0.40, 0.35$) have been adjusted in order that the corrected reflectances be nearly the same here, in this range of scattering angles. These same parameters will yield large differences in the forward directions that are attainable by ground-based measurements but are not attainable from satellite observations. Figure 3.17 right shows that multidirectionnal polarization measurements permit some retrieval of the particle refractive index.

Previous figures are for conservative particles. The phase matrix of aerosols and the spectral dependence of their scattering coefficient do not change largely when the particle absorption (i.e. the imaginary refractive index of the particles) increases, but the impact of the absorption on radiative transfer becomes noticeable for large enough aerosol contents (say AOT larger than 0.5). Absorption tends to diminish the contribution of multiple scatterings. As this contribution increases with the AOT, changing the absorption changes the spectral variation of the radiance, which is very useful for the aerosol characterization as outlined previously. Retrieval of the aerosol absorption, therefore, is more difficult than in the case of ground-based observations and has to be conducted in parallel with the retrieval of the aerosol size distribution and refractive index.

On-going studies show that polarization measurements could bring more information on aerosol absorption, but they still need to be confirmed.

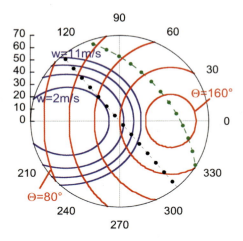

Figure 3.13 Polar plot of viewing angles θ,φ ($\varphi=0$ for the solar incident plane). Red lines correspond to scattering angles ranging from 80° to 160° (step 20°). Blue lines correspond to sunglint radiance $L_G=0.001$ for wind speeds ranging from 2, 5, 7 and 11ms⁻¹. Black and green dots correspond to two satellite scans (see text).

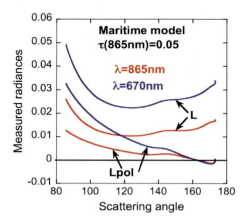
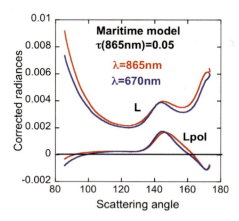

Figure 3.14 left: Simulated measured radiance and polarized radiance for large maritime particles; right: same after subtracting the Rayleigh molecular contribution.

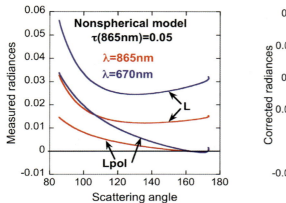
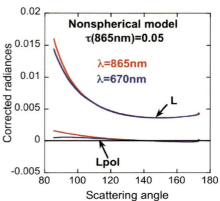

Figure 3.15 Same as Figure 3.14 but for non-spherical particles.

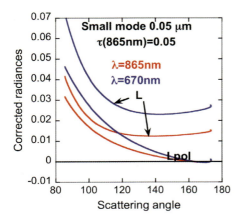
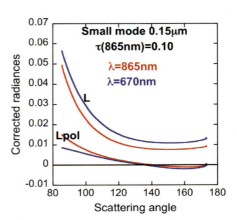

Figure 3.16 Influence of particle size on corrected radiance for the small mode; left: r_m=0.05μm; right: r_m=0.15μm.

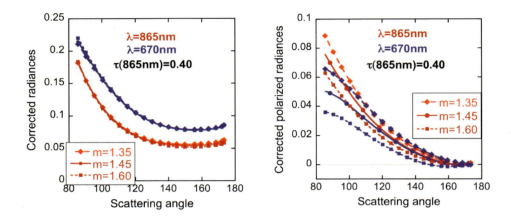

Figure 3.17 left: Comparison of corrected radiances for 3 models of small particles; right: same for corrected polarized radiance. See text for discussion.

3.7 Summary

This chapter provides the methods for solving the so-called "direct problem", i.e. the problem of computing the radiative field, radiance and polarization, knowing the atmosphere composition, and the ground surface reflectance. Examples are shown of the calculated impact of the presence of aerosols and of their characteristics. The "inverse problem", retrieving aerosol amount and characteristics from radiation measurements, will be considered in Chapter 5.

4 Direct observation of the sun for aerosol retrieval

Colette Brogniez, Jacqueline Lenoble, Glenn Shaw

4.1 Introduction

The simplest remote sensing methods rely on the observation of the extinction of radiation, as it is defined in Chapter 2, in Eqs 2.4 and 2.5 (Beer exponential extinction law), which we recall here as Eqs 4.1 and 4.2:

$$I(s_2) = I(s_1) \exp(-\tau_e),$$ (4.1)

where

$$\tau_e = \int_{s_1}^{s_2} \sigma_e(s)ds.$$ (4.2)

If the source radiance $I(s_1)$ is known, the simple measure of $I(s_2)$ at a distance (s_2-s_1) from the source directly gives access to the total extinction optical thickness of the atmospheric layer between the points s_1 and s_2 along the radiation path length.

This procedure has been used with artificial sources, especially at the Earth's surface, in order to observe fog or pollution episodes. Among natural sources (passive remote sensing), the moon and stars can be used, but of course the sun is the most important and widely-used source.

Ground-based observation of the direct solar beam, known as sunphotometry, has been used for decades. Sunphotometers are often associated with instruments measuring the sky radiance (and eventually the sky polarization) in selected directions and at selected wavelengths; this allows a more complete retrieval of the aerosol characteristics. A history of ground-based aerosol remote sensing and a description of present-day networks that use sunphotometry with sky observations is presented in Chapter 6. In this chapter, we will address the basic principles of sunphotometry and the problems linked to calibration.

From space, it seems less obvious as to how we might observe the sun directly attenuated by atmospheric layers. However, this is achieved by the so-called "occultation methods", allowing an efficient sounding of the stratosphere and higher layers of the troposphere. Section 4.3 presents the basic principles of satellite occultation methods as used in the Stratospheric Aerosol and Gas Experiment (SAGE), the Polar Ozone and Aerosol Measurement (POAM) instruments, and some others.

Section 4.4. presents a few conclusions.

4.2 Ground-based sunphotometry

At the Earth's surface, the transmitted solar radiation can easily be observed by a sunphotometer pointing to the sun. The monochromatic irradiance measured on a plane perpendicular to the solar beam is

$$E_\lambda = E_{0\lambda} \exp(-M\tau_\lambda), \tag{4.3}$$

where $E_{0\lambda}$ is the extraterrestrial solar irradiance, τ_λ the vertical total atmospheric optical depth defined in Chapter 2, and M the air mass depending on the solar zenith angle. For the plane parallel approximation, M is the secant of the solar zenith angle. Note that atmospheric air mass deviates from the plane parallel approximation for low solar elevation angles. Calculations of M for spherical atmospheres can and have been made and must be used for solar elevations less than about 10 degrees. Useful tables and convenient fitted closed form mathematical approximations for the air mass function are given in Kasten (1965) and Young (1994). Eq. (4.3) allows the determination of τ_λ, if the extraterrestrial irradiance $E_{0\lambda}$ is known. Subtracting the gaseous optical depth (gas absorption and molecular scattering) from the total optical depth provides the aerosol optical depth.

Translating a ground-based measurement of the solar radiation into aerosol optical depth requires knowing the $E_{0\lambda}$ and the optical depths of water vapor, ozone and Rayleigh scattering for the bandpass of every wavelength of the instrument. Spectral transmission over the solar spectrum for the non-aerosol components is well known. All that is needed is the total column loading of these constituents. Ozone amounts can be estimated from climatology or, for more precise measurements, from open data sources. The magnitude of the Rayleigh scattering is a function of atmospheric pressure, which can be approximated by altitude above sea level or auxilliary information. Total column water vapor amounts can be obtained by auxilliary measurements, or derived from the sunphotometer measurements themselves using observations in highly absorbing bands. The best option in terms of correcting for water vapor absorption in the aerosol measurement is to choose spectral bands for the instrument that measure in the atmospheric "windows", avoiding water vapor absorption altogether. Assuming that the corrections for gas absorption and molecular scattering are known precisely, then the accuracy of the aerosol optical depth measurement is dependent on the accuracy of determining $E_{0\lambda}$.

Because the solution of Eq. 4.3 for τ_λ involves the ratio $E_\lambda / E_{0\lambda}$ the units of the value of the extraterrestrial irradiance are immaterial and only need to match the units measured by the sunphotometer. Often values are left in instrument-measured "voltage". The point is that the extraterrestrial irradiance needed in Eq. 4.3 to solve for τ_λ is the irradiance that would be measured by the individual instrument in question if that instrument were to be transported to the top of the atmosphere. Calibrating a sunphotometer involves determining $E_{0\lambda}$ for a particular instrument and for every wavelength band in that instrument. Calibration is a continuing necessity during the life span of the instrument, as band pass filters can degrade and other instrument characteristics can change over time (Ichoku et al., 2002a).

The typical approach to calibration is the so-called Langley (or Bouguer-Langley) method (Shaw et al., 1973), measuring the irradiance for a large range of solar zenith angles, and plotting $\ln(E_\lambda)$ versus M. The extrapolation of the plotted line to $M=0$ provides $E_{0\lambda}$, and the slope is $-\tau_\lambda$ (see Figure 4.1). This method was used extensively at the Smithsonian Institution between 1920 and 1950, with the objective of obtaining the solar constant, i.e.

$$E_0 = \int_0^\infty E_{0\lambda}\, d\lambda \qquad (4.4)$$

The Langley plots are now used mainly for measuring the aerosol optical depth.

Langley plot calibration of sunphotometers, though simple in principle, has a number of subtle pitfalls. One must exercise great care in the Langley calibration because it is here that the largest and most difficult-to-define systematic errors occur. The Langley method assumes that, during the measurements at different elevation angles, the atmosphere is temporally invariant and horizontally homogeneous (within about 50 km of the observer). The latter problem may be reduced by choosing the observation location to ensure that power plant plumes or other interfering aerosol plumes are not passing through the field of view, but the problem of temporal stability of the atmosphere is much more difficult to satisfy. Almost all calibrations of sunphotometers conducted at continental locations have the possibility (and high probability) of being seriously marred because of time-changing drifts in atmospheric transmission. The only exceptions are those measurements made from high-altitude mountain observatories, but even those locations often have problems because of complex mountain meteorology and upslope thermally-driven currents. For example, though the air at the Mauna Loa Observatory (MLO) can be exceedingly transparent, it is affected in late morning and afternoon hours by marine aerosol that reaches the observatory as the marine inversion layer breaks down under solar heating. Langley plots from Mauna Loa give calibration constants varying slightly from day to day, by about one percent when data are taken all day long. However, when data acquired only before about 10 or 11 am local time are analyzed (before up-flowing air had reached the observatory) the calibration constants repeat to a few parts in a thousand day after day.

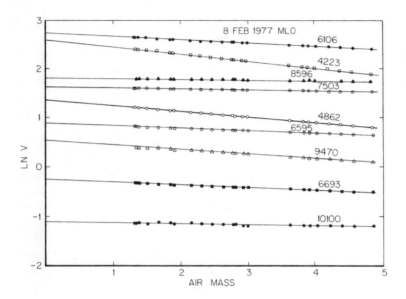

Figure 4.1 Example of a Langley-Bouguer plot for multiwavelength sun photometer data taken at Mauna Loa. The ordinate is instrument voltage and the abscissa is the atmospheric air mass (secant of the solar zenith angle for plane parallel approximation). Theory suggests a straight line or linear relationship, but such a relationship can also occur when some of the assumptions in the theory formulation break down (see text). The individual line fits are for data taken at different narrow (10 nm) wavelength bands through well-blocked interference filters. Numbers indicate wavelengths in Ångströms.

At locations not as favorably endowed as Mauna Loa, the problem of time-varying atmospheric transmission, as it relates to the accuracy of calibration by the Langley method, is more severe and challenging to handle. What can be particularly annoying is that time variations in transparencies often have systematic trends of a type that provide a nearly linear Langley plot. If such time trends occur and if they persist day after day because of systematic diurnal trends in local meteorological conditions, then inaccurate calibration constants for the sunphotometer will be derived (Shaw, 1976) and – if cross-comparisons of instruments are made – propagate errors through an entire sunphotometer network.

To illustrate the insidious nature of the problem of local time drifts in atmospheric transparency, a modelled Langley plot is shown in Figure 4.2, for a situation when the atmospheric optical depth varies parabolically about the noon point; a case that actually occurs, to some approximation, quite frequently in polluted areas due to trapping and build up of aerosols under a temperature inversion and break up and dispersal of the aerosol in the afternoon hours. Photometers calibrated by the Langley method at Tucson, Arizona, for example, consistently show lower calibration voltage by 1 to 3 percent when compared to calibration conducted during pristine conditions in the morning hours at Mauna Loa because of this reason. Even though the Langley plots are consistently providing low values of extrapolated zero air mass intercepts at Tucson, the squared sum of the residuals about the least square fit to the exponential Bouguer law is often very low, because of the semi-parabolic variation of optical depth around solar noon. Notice that the modelled Langley plots are very close to straight lines, even though the optical depth is strongly varying (Shaw, 1976).

The only practical way to guard against this sort of calibration systematic errors is to conduct Langley plot calibrations from high-altitude observatories during stable conditions in the mornings and, simultaneously, constantly measure the aerosol particle concentration to ensure both that aerosols below the station remain below it during the calibration period and that the calibration repeats on several days. Even this is insufficient, because it may be possible that new particles are produced by sunlight, a kind of natural photochemical smog, which could conceivably maximize around the solar noon point. If this occurs, as Figure 4.2 illustrates, the calibration constants could still be in error. This error, however, would be very low if the aerosol optical depth is small, as indeed it typically is from an excellent station such as Mauna Loa.

Sunphotometers typically are manufactured with baffling and collimators to maintain a relatively small field of view (FOV) (several degrees) to minimize the scattered sky radiation. Though a correction for this forward-scattered radiation is performed, the photometers have to be accurately pointed at the sun. With handheld instruments, "operator error", the inability of an operator to keep the instrument steadily pointed at the solar disk, is a common source of inaccuracy.

Care must also be taken when performing accurate determinations of optical depth to ensure that the temperature of the detector/filter combination remains within fairly narrow bounds. It is common for optical detectors to have a rather large and usually positive temperature coefficient at the redder wavelengths (perhaps 1% per 10 degrees C), and a smaller and sometimes negative coefficient at bluer wavelengths, for example in the 350 to 450 nm region.

Sunphotometers are calibrated frequently either by referencing to a standard lamp or by comparing with a "master" instrument which has undergone accurate and extensive Langley plot calibration from a high-altitude mountain station with known excellent optical conditions, such as the Mauna Loa Observatory. It is well to use a great deal of caution when performing such intercomparisons, for a number of reasons. First, the spectral distribution of sunlight may be quite different in a pristine mountain environment compared with that in a polluted environment, and thus small amounts of signal due to out-of-band light leakage can be rather different for the two situations. Second, to be reliable, the pass bands and detector sensitivities in the two instruments (master and instrument undergoing calibration) must be either known or carefully included in the comparison, or must be identical, which is rarely the case. Even filters ordered in batch quantities often have spurious slight differences in pass band characteristics. It is quite possible that one of the two pass bands might contain a slight gaseous absorption feature not detected in the other, for instance. There must be other, perhaps not well understood, systematic error sources.

(a)

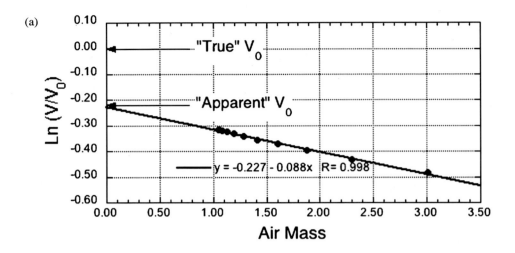

(b)

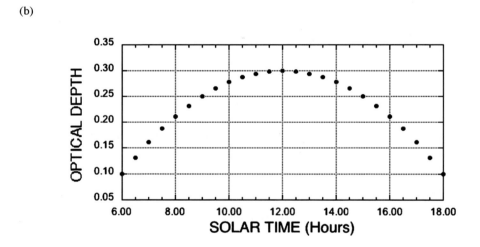

Figure 4.2 (a) Modelled Langley-Bouguer plot for a hypothetical situation where the optical depth varies parabolically around solar noon. Note the goodness of the linear fit, in spite of massive breakdown in the assumption of optical depth remaining constant during the observation period. The "correct" zero air mass intercept in this modelled exercise is 1.0. Note the significant error in extrapolation.

(b) Assumed parabolically shaped variation of optical depth with time, varying around the solar noon point for the Langley model illustrated in Figure 4.2(a).

One of us (GS) frequently has seen unexpected differences of several percentage points in the extrapolated calibration constant when comparing two instruments, both referenced against master instruments that had been calibrated at Mauna Loa using the direct Langley method.

Extrapolated zero air mass intercepts can be performed to an accuracy of about one half percent, or possibly slightly better, if calibrations are very carefully performed under only ideal conditions, rejecting data when there is interfering aerosol plume as detected by a scattering nephelometer, or very thin cirrus or other visible sky contamination. A data set taken over a period of one year at MLO (Shaw, 1982) provided extrapolated values of zero air mass intercepts identical to one another within a half percent for most wavelength bands. Figure 4.3 shows a histogram for Langley plot intercepts taken during clean conditions at Mauna Loa. This, incidentally, indicates that extraterrestrial solar spectral irradiance was constant to at least half of one percent over a one-year period!

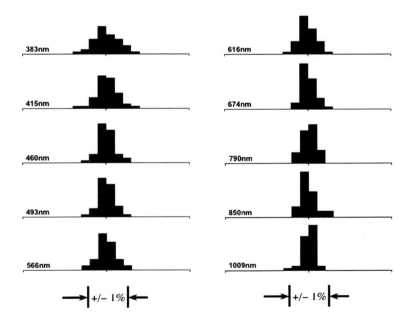

Figure 4.3 Histograms of the frequency of occurrence for about 200 especially clean days of the zero air mass intercept of the extraterrestrial solar spectral irradiance from Langley plots taken at Mauna Loa Observatory, for several wavelengths indicated in nm. The rms variation in day-to-day extrapolated values is approximately 0.3 percent for most wavelengths.

Sometimes investigators calibrate sunphotometers by viewing a standard lamp; for example, type FEL 1000-watt lamps are frequently used in conjunction with stable electric current sources. One of us (GS) has had quite a lot of experience with intercomparisons of instruments calibrated in this manner, but with somewhat disappointing results. For example, a sunphotometer calibrated with a standard lamp, referencing an instrument accurately calibrated at MLO, might, when the secondary instrument is taken to MLO, provide zero air mass intercepts disagreeing by a few percentage points. The reasons probably have to do with the very different spectral distribution of the lamp's light (approximating a black body source at 1600 degrees C) with that of the sun (approximating a black body source at 5500 degrees C).

In summary, one should exercise a great deal of scepticism in calibrating sunphotometers. It must always be kept firmly in mind that knowing the zero air mass calibration value to only 2 percent error, a small calibration error for most instrumental calibration purposes, can bring inadmissible errors into sun photometry, especially in situations where the optical depths are quite small. For performing successful science or when assessing slow changes in atmospheric turbidity, one must keep an open mind and always stay alert to possible subtle errors that can and do creep into sun photometry. Calibration errors are often very subtle and difficult to detect. It is recommended that instruments at mountaintop observatories are calibrated regularly.

Another technique for measuring atmospheric turbidity, simpler, but less accurate, consists of measuring alternately global and diffuse irradiances on a horizontal plane with a movable shading disc or shading ring that occasionally cuts off the direct solar component. The direct sun signal is deduced by subtracting the "diffuse" sky radiation signal from the "global" (diffuse sky plus direct sun). This can be done using spectrometers (Lenoble et al., 2008, Brogniez et al., 2008), or filter radiometers (Bigelow et al., 1998). Such instruments need to be carefully calibrated and a good stability of the atmosphere during the two measures is necessary; this last condition is especially compelling for instruments that need quite a long time (a few minutes) for recording a large spectrum. In addition, a bias is due to the size of the shadower being larger than the solar disc; this bias is approximately corrected by computing the contribution of the circumsolar sky radiance, which needs an assumption about the atmospheric scattering properties (including aerosols).

A few sunphotometers, instead of being ground-based, operate from airborne platforms (Russell et al., 2005), enabling vertical profiles of aerosol optical thickness to be retrieved. A paragraph on airborne sunphotometers with ample references for further review can be found in Chapter 6, Section 6.7.

4.3 Occultation methods

From space, observing the direct solar radiation is performed in occultation experiments. The first observations of aerosols by solar occultation were performed by the Stratospheric Aerosol Measurement (SAM) instrument flown on the Apollo-Soyuz mission launched in 1975 by NASA (Pepin and McCormick, 1976). Though the mission lasted only few days it proved the capability of this technique of observation for stratospheric studies and opened

the way to other missions. History of space-borne instruments using the occultation technique and the long-term data record of stratospheric aerosols obtained by this technique is
presented in Chapter 7.

As the satellite orbits the Earth, the solar occultation instrument points towards the sun
and measures its irradiance. The instrument observes sunsets as the spacecraft moves from
the sunlit toward the dark side of the Earth. Before each sunset starts, the line of sight
(LOS) between the instrument and the sun is well above the atmosphere so that the sun's
irradiance as measured by the instrument is unattenuated. As the spacecraft moves toward
the dark side, the LOS passes through a portion of the atmosphere, and the sun's irradiance is attenuated due to scattering and absorption by atmospheric constituents. A set of
measurements for a series of LOS constitutes an event. During sunrise events, the measurement sequence is just the reverse of that during sunset. As the spacecraft moves from the
dark towards the sunlit side of Earth, the sun is first viewed through the atmosphere, and
then along an unobstructed path when the spacecraft moves toward the sunlit side. During its rise and set the sun is thus observed through the atmospheric layers and each LOS
corresponds to a tangent altitude Z_t, which is the point on the LOS closest to the Earth's
surface (cf. Figure 4.4). The precise determination of the tangent altitude is a difficult step,
critical for deriving vertical profiles of all species, gas and spectral aerosol extinctions and
is achieved differently for different instruments (for example, Chu et al.,1989, Glaccum
et al.,1996). It should be noted that spacecraft sunsets do not correspond systematically to
astronomical ones, it depends on spacecraft orbital parameters.

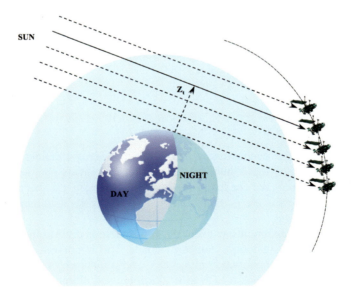

Figure 4.4 Solar occultation measurement geometry (without refraction). Z_t denotes the tangent
altitude.

The sequence of LOS irradiance measurements during an event enables the characterization of the composition of the atmosphere. In the visible and near-infrared, the emission term in the source function of the transfer equation (Eqs 3.1 to 3.4) is negligible compared to the directly transmitted radiance. Moreover, provided that the FOV is small, the contribution of the scattering source term for single and multi-scattered radiations is negligible as well. The slant path transmission along the LOS at a tangent altitude Z_t is obtained by dividing the corresponding irradiance measurement by the unattenuated irradiance measured when the LOS does not intersect the atmosphere, at Z_t^∞ tangent altitude

$$T(Z_t) = E(Z_t)/E(Z_t^\infty) \qquad (4.5)$$

This is analogous to solving for τ_λ in Eq. 4.3 for ground-based sunphotometry. In the occultation technique there is no need for on-board calibration facilities because, due to the unattenuated light measurement, the instrument is recalibrated just after or just before an event so that there is no sensitivity to any drift in the instrument performances. This provides a more accurate calibration than can be attained by ground-based sunphotometers because irradiance at the top of atmosphere is measured directly rather than being extrapolated from a set of measurements in a Langley plot. Therefore this technique is well suited for long-term monitoring of atmospheric species.

The attenuation is caused by the combination of molecular scattering, gaseous absorption and aerosol extinction. Thus, to enable characterization of each contributor, more information is needed than can be provided by a single wavelength, and measurements are generally performed at several wavelengths, as are made for ground-based sunphotometers. In the UV-Visible-NIR wavelength domain the absorbing gases are ozone, nitrogen dioxide, water vapor, oxygen and carbon dioxide. In the far IR several other gases such as methane, nitrous oxide, CO, HCl, contribute also to the absorption.

The total slant path optical depth is derived from the transmission $T(Z_t, \lambda)$ at each tangent altitude Z_t and at wavelength λ as

$$\tau_{total}^{SP}(Z_t, \lambda) = -\ln\left(T(Z_t, \lambda)\right) \qquad (4.6)$$

Vertical profiles of gas concentration and of spectral aerosol extinction coefficients can be retrieved combining these multi-wavelength measurements obtained at all tangent altitudes. Thus two steps are needed in the retrieval processing: a spectral inversion and a spatial inversion (details are given in Chapter 8). Below we summarize some important features of the retrieval.

For the spatial inversion of one event the atmosphere is divided into thin spherical layers and assumption of spherical symmetry is made. This hypothesis is not always fulfilled since it would require horizontal homogeneity over a distance of several hundreds of kilometers. For example, along a LOS at a 10 km tangent altitude, a 1-km thickness layer located at 40 km should be homogeneous over a distance of about 1200 km (Hamill et al., 2006). Heterogeneity along the LOS can lead to significant errors in the retrieved products (Swartz et al., 2006, Berthet et al., 2007).

(a)

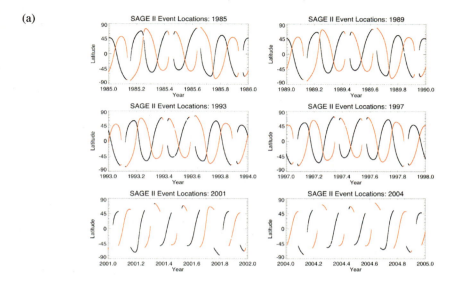

(b)

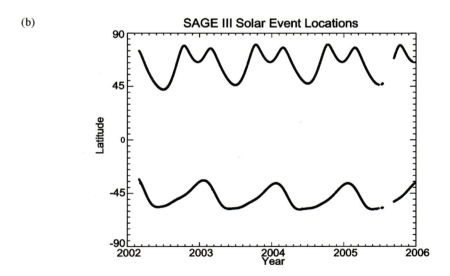

Figure 4.5 Time-latitude measurement coverage for (a) SAGE II/ERBS for few years between 1984 and 2005. Sunrises, as seen from the spacecraft, are in black; sunsets in red. (b) SAGE III/Meteor-3M for all years between 2002–2005. Sunrises, as seen from the spacecraft, occur in the Southern hemisphere, sunsets in the Northern hemisphere. (Courtesy of L. Thomason, NASA LaRC).

Due to the long path through the atmosphere, the occultation technique is able to detect minor species in the stratosphere or in the troposphere that shorter path lengths, for example from a nadir-viewing satellite, would be lost in the noise. The only caveat is that the optical depth cannot be large enough to overwhelm the signal. Indeed, for retrieval of any species using occultation methods the optical thickness cannot be so large as to obscure the source (the sun). In general this is the case outside strong absorption bands and outside the wavelength domain where the molecular scattering is large, i.e. in the UV range. Of course, the presence of clouds limits the altitude coverage.

An advantage of the space occultation technique compared with ground-based sunphotometry, is that it enables retrieval of information on the vertical distribution of aerosols while ground-based measurements only give access to an integrated column. Compared with airborne sunphotometry, which is limited to altitudes lower than about 10 km, space-based occultation provides information on higher altitudes.

The same method can also be applied to sources other than the sun, such as the moon (for example, with the SAGE III lunar mode) or another star (for example, with Global Ozone Monitoring by Occultation of Stars (GOMOS)).

Space occultation measurements provide better vertical resolution (though it depends on the field-of-view) than passive space nadir measurements, but the geographical and temporal coverage at any location are worse, depending on the spacecraft orbit. Typically, the spacecraft orbits the Earth in approximately 90 minutes or 16 times per day, depending on orbital parameters. Since each orbit provides two measurement opportunities (one sunrise and one sunset), the instrument can acquire 32 separate solar occultation measurements during a 24-hour period at different geographical locations over the Earth depending on the spacecraft orbit. The geographical and temporal coverage of the measurements depend on the satellite orbital parameters. For example, Figure 4.5 shows the time-latitude measurement coverage of SAGE II (inclined orbit) and SAGE III (polar orbit). SAGE II was able to observe sunrises and sunsets in both hemispheres, between 90°S and 90°N, whereas SAGE III observed sunsets in the Northern hemisphere between about 45°N and 80°N, and sunrises in the Southern hemisphere between about 30°S and 60°S. For SAGE II, one can notice the slight drift over the duration of the experiment. During the last years of SAGE II, coverage was not full due to instrumental problems. Similarly, SAGE III encountered a failure in mid-2005.

The number of measurement opportunities and the geographical coverage can be increased if measurements are made using both lunar and solar occultation events, or if stellar occultation events are used.

A few instruments have operated from balloon-borne platforms rather than from spaceborne ones (Renard et al., 1996, 2000), and a few others have pointed at the Earth's limb rather than at the sun (SAGE III, Optical Spectograph and InfraRed Imaging System – OSIRIS, on the Odin satellite). In this latter case, the aerosol characterization is much more difficult to achieve but the geographical coverage is extended. More details on the retrievals are available for SAGE III in Rault and Loughman (2007) and for OSIRIS in Boussara et al. (2007).

4.4 Conclusion

The direct observation of the sun with ground-based sunphotometers allows retrieval of the total atmospheric optical depth, at one or several wavelengths. Although very simple in principle, the method needs great care in the instrument calibration and utilization for obtaining accurate results. Subtracting from the total optical depth the other components (molecular scattering and gas absorption) leads to the aerosol optical depth. The molecular scattering contribution estimate is provided by ancillary measurements of pressure and temperature. Concerning gas absorption, a correction is often needed even if measurements generally avoid strong gas absorption bands. Spectroscopic data banks are thus necessary, as well as gas concentration data coming, for example from climatology, as is the case to correct for the ozone contribution. These aerosol optical depth data enable the estimation of the columnar aerosol content. An estimate of the aerosol size representative of the column atmosphere can be achieved via spectral variations of the aerosol optical depth (see Chapter 5), the simultaneity of the spectral observations is thus critical. Some instruments employ a filter wheel to measure different wavelengths introducing a time delay between wavelengths. Some others perform measurements simultaneously in several channels, for example, Microtops instruments (Ichoku et al., 2002a). Nevertheless, the time delay is often short (a few seconds for Cimel-AERONET instruments), therefore the precision is similar for both techniques. For additonal aerosol characterization, such as detailed size distribution and information on real and imaginary refractive index, additional measurements are required. Ground-based radiometers linked to sunphotometry that measure sky radiance distribution and/or polarization can produce these additional data, and are described in Chapter 6.

Solar occultation measurements with space-borne instruments also use the extinction of the direct solar beam through the atmosphere during sunset or sunrise, and allow the retrieval of the spectral slant optical depth. Generally, one aims at retrieving not only aerosol, but also a few important gas components, such as ozone. Species separation leads to spectral aerosol slant optical depths and is followed by retrieval of vertical profiles of spectral aerosol extinction coefficients. Species separation and spatial inversion are explained in Chapter 8.

5 Determination of aerosol optical properties from inverse methods

Michael D. King and Oleg Dubovik

5.1 Introduction

Light scattering by atmospheric aerosols modifies the diffuse and direct solar radiation observed at the Earth's surface as well as from spaceborne radiometers. When the atmospheric characteristics are known, including the vertical distribution of aerosols and their optical and microphysical properties, the diffuse radiation field can be computed using the methods outlined in Chapter 3. Ground-based measurements of diffuse and direct solar radiation have been used to infer the size distribution of aerosol particles in the intervening atmosphere as well as their optical properties. This is often accomplished by comparing measurements with computations for a wide range of aerosol parameters. This method is referred to as look-up table procedures whereby a solution is obtained by comparing measurements with theoretical calculations. The look-up table solution is stable and generally fast in implementation, though it is limited to a set of potentially admissible solutions included in the look-up table. A rigorous, but more complex, technique has also been invoked that consists of "inverting" a set of measurements as a function of wavelength and/or scattering angle to infer the size distribution and optical properties of the radiation that produces the measurements. This method is not limited to a predefined set of aerosol classes, but instead searches for the best fitting set of aerosol parameters through the continuous space of all possible solutions. However, in practice, often several different combinations of aerosol parameters produce the same, or nearly the same, radiation distribution. Therefore, the general solution is fundamentally nonunique or becomes nonunique in the presence of minor measurement noise. Though the inversion result is not fundamentally unique, it's known to be satisfactory once minor *a priori* information is added that constrains the solution while reproducing the measurement field within the error bars established for the measurements.

This method has been applied to measurements from ground-based sunphotometers as well as from ground-based sun/sky radiometers, invoking *a priori* assumptions on smoothness of the size distribution or spectral smoothness of the optical constants. In this chapter, these inversion methods and their sensitivity to ancillary assumptions will be described and applied to real measurements.

5.2 Diffuse and direct solar radiation

Solar radiation transmitted through the atmosphere and scattered by the intervening medium varies as a function of direction, wavelength, and polarization, and depends on the scattering and absorption properties of the atmosphere as well as the reflectance properties of the underlying surface. In the absence of clouds, the light scattering field can be computed to a high degree of accuracy.

5.2.1 Spectral aerosol optical thickness

The attenuation of sunlight passing through the atmosphere containing aerosol particles, molecules, and absorbing gases is given by the well-known Lambert-Beer Law as

$$E(\lambda) = E_0(\lambda)\, e^{-\tau^*(\lambda)M(\theta_0)}, \tag{5.1}$$

where $E(\lambda)$ is the solar flux density (irradiance) reaching the detector at wavelength λ, $E_0(\lambda)$ the solar irradiance incident on the top of the atmosphere ($\tau = 0$ level), $M(\theta_0)$ the atmospheric air mass, a function of solar zenith angle θ_0, and τ^* the total optical thickness. From a measurement of the transmitted solar radiation as a function of air mass one can readily determine the total optical thickness of the atmosphere (cf. Chapter 4).

From the values of total optical thickness one can determine corresponding values of the aerosol optical thickness τ_{aer} by subtracting contributions due to Rayleigh scattering τ_{Ray} and contributions due to ozone absorption τ_{O_3}.

Under the assumption that atmospheric aerosol particles can be modeled as a polydisperse collection of spherical particles with a single refractive index m, the integral equation that relates the aerosol optical thickness to an aerosol size distribution can be written in the form

$$\tau_{aer}(\lambda) = \int_0^\infty \int_0^\infty \pi r^2 Q_e(r,\lambda,m) n(r,z)\,dz dr, \tag{5.2}$$

where $n(r, z)$ is the height-dependent aerosol number density in the radius range r to $r + dr$, and $Q_e(r, \lambda, m)$ the extinction efficiency factor from Mie theory. Note that Eq. (5.2) follows from Eq. (2.5) using the relationship between extinction efficiency factor and extinction coefficient $\sigma_e(r, \lambda, m) = \pi r^2 Q_e(r, \lambda, m)$.

Performing the height integration, Eq. (5.2) can be rewritten as

$$\tau_{aer}(\lambda) = \int_0^\infty \pi r^2 Q_e(r,\lambda,m) n_c(r) dr, \tag{5.3}$$

where $n_c(r)$ is the unknown columnar aerosol size distribution, i.e. the number of particles per unit area per unit radius interval in a vertical column through the atmosphere, and defined by $dN_c(r)/dr$, following Eqs (1.1) and (1.2). Equation (5.3) is a Fredholm equation (because the limits of integration are fixed, not variable) of the first kind (because the unknown function $n_c(r)$ appears only in the integrand). Making the limits of integration finite with $r_1 = r_{min}$ and $r_2 = r_{max}$, this equation becomes

$$\tau_{aer}(\lambda) = \int_{r_{min}}^{r_{max}} K_\tau(r,\lambda,m) n_c(r) dr, \tag{5.4}$$

where the kernel function of optical thickness $K_\tau(r,\lambda,m) = \pi r^2 Q_e(r,\lambda,m)$ can be computed from Mie theory for homogeneous spherical particles of known refractive index. Hence, if the columnar aerosol size distribution is known and the refractive index estimated with reasonable accuracy, the aerosol optical thickness can readily be computed. This is the *forward problem* and generally poses no difficulty using modern computers.

Figure 5.1a illustrates the extinction efficiency factor as a function of radius for four representative wavelengths in the visible wavelength region. In fact $Q_e(r,\lambda,m)$ is invariant in the ratio r/λ for a fixed value of m, and thus the extinction efficiency factor for various wavelengths is simply shifted to the right or left in Figure 5.1a. We have assumed the complex refractive index of the aerosol particles $m = 1.45 + 0.0i$ in all calculations presented in Figure 5.1. Because the kernel function $K_\tau(r,\lambda,m)$ increases monotonically with radius, but the size distribution generally decreases rapidly with radius, it is convenient to rewrite Eq. (5.4) in terms of a volume size distribution $v_c(\ln r) = dV_c(r)/d\ln r$ representing the volume of particles per unit area per unit log radius interval in a vertical column through the atmosphere. Thus we obtain

$$\tau_{aer}(\lambda) = \int_{r_{min}}^{r_{max}} \frac{3}{4\pi r^3} K_\tau(r,\lambda,m) v_c(\ln r) d\ln r,$$

$$= \sum_{j=1}^q \int_{r_j}^{r_{j+1}} \frac{3}{4\pi r^3} K_\tau(r,\lambda,m) v_c(\ln r) d\ln r, \tag{5.5}$$

$$\approx \mathbf{A}(r,\lambda,m)\mathbf{f}_c(\ln r).$$

In this formulation, the modified kernel function of optical thickness $3K_\tau(r, \lambda, m)/4\pi r^3$ varies as r^3 for small particles (Rayleigh scattering limit) and as r^1 for large particles (geometric optics limit), and is illustrated in Figure 5.1b. The modified kernel function of optical thickness $3K_\tau(r, \lambda, m)/4\pi r^3$ and indicial function of size distribution $v_c(\ln r)$ are approximated by matrices $\mathbf{A}_\tau$ and vector $\mathbf{f}$ as follows:

$$A_{ij} = \int_{\ln r_j - \Delta \ln r / 2}^{\ln r_j + \Delta \ln r / 2} \frac{3}{4\pi r^3} K_\tau(r, \lambda_i, m) d \ln r, \qquad j = 1, 2, \ldots, q, \tag{5.6}$$

$$f_j = v_c(\ln \bar{r}_j).$$

If one has measurements of $\tau_{aer}(\lambda)$ at p discrete wavelengths, and seeks to determine the volume aerosol size distribution $v_c(\ln r)$ at q discrete sizes, one is left with an equation of the form

$$\mathbf{g} = \mathbf{A}\mathbf{f}, \tag{5.7}$$

where

$$g_i = \tau_{aer}(\lambda_i), \qquad i = 1, 2, \ldots, p. \tag{5.8}$$

Equation (5.7) is the mathematical formulation of spectral aerosol optical thickness as a function of aerosol size distribution for a "perfect" measurement with no measurement errors or quadrature errors of integration. In reality, Eq. (5.7) must be rewritten as

$$\mathbf{g} = \mathbf{A}\mathbf{f} + \boldsymbol{\varepsilon}, \tag{5.9}$$

where

$$\varepsilon_i = \varepsilon(\lambda_i), \tag{5.10}$$

represents the vector of errors in $\tau_{aer}(\lambda)$.

As an illustration of the sensitivity of spectral optical thickness to aerosol size distribution, we have computed the spectral aerosol optical thickness to be expected from six model lognormal volume size distributions of the form

$$v_c(\ln r) = \frac{dV_c(\ln r)}{d \ln r} = \frac{C}{\sqrt{2\pi}\sigma} \exp\left(-\frac{1}{2}\left(\frac{\ln r - \ln r_m}{\sigma}\right)^2\right), \tag{5.11}$$

where C is a constant, r_m the mode radius, and σ the width of the distribution, as in Chapter 1. Given such a size distribution (or any other viable aerosol size distribution) plus a reasonable estimate of the complex refractive index, it is straightforward to compute the spectral aerosol optical thickness corresponding to that size distribution.

Figure 5.2 illustrates this *forward problem* whereby a size distribution (Figure 5.2a) can be used to infer the expected spectral optical thickness (Figure 5.2b). The *retrieval problem* is one in reverse, whereby measurements such as those in Figure 5.2b are used to estimate the size distribution that is responsible for producing those measurements. As we will see below, solution of this equation is unstable and nonunique and requires the introduction of "virtual measurements" in the form of constraints, due primarily to the fact that the kernel function of optical thickness, illustrated in Figure 5.1b, is not "orthogonal" and kernels for various wavelengths have considerable overlap and are hence nonunique.

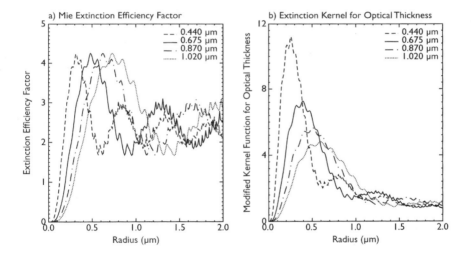

Figure 5.1 (a) Extinction efficiency factor and (b) extinction kernel function $3K_t(r, \lambda, m)/4\pi r^3$ as a function of radius for the inference of aerosol size distribution from optical transmission measurements. A refractive index $m = 1.45 + 0.0i$ was assumed in the calculations, and the calculations were performed for four wavelengths used in the AERONET data processing algorithm.

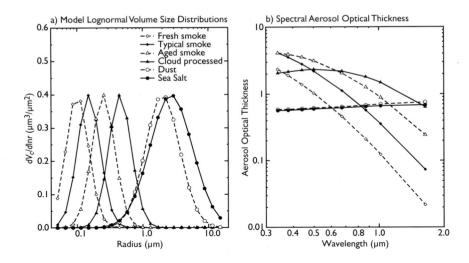

Figure 5.2 (a) Model lognormal volume size distributions for various modes of realistic aerosol particle size distributions characteristic of fresh smoke and pollution ($r_m = 0.1\ \mu$m) to desert dust ($r_m = 2\ \mu$m) and sea salt coarse mode particles ($r_m = 2.7\ \mu$m). (b) Spectral aerosol optical thickness derived from the model size distributions showing varying degrees of curvature and shape variations as a consequence of varying the mode radius of the volume size distributions.

5.2.2 Sky radiance as a function of scattering angle

The atmospheric sky radiance can be modeled by solving the radiative transfer equation for a plane-parallel atmosphere. The angular distribution of diffuse downward radiation can be described by

$$L(\Theta,\lambda) = \frac{E_0\mu_0}{4\pi(\mu-\mu_0)}\left(e^{-\tau/\mu} - e^{-\tau/\mu_0}\right)\left(\varpi p(\Theta,\lambda) + G(\dots)\right), \qquad \mu \neq \mu_0$$

$$L(\Theta,\lambda) = \frac{E_0}{4\pi\mu_0}e^{-\tau/\mu_0}\left(\varpi\tau p(\Theta,\lambda) + G(\dots)\right), \qquad \mu = \mu_0$$

(5.12)

where $L(\Theta,\lambda)$ is the spectral sky radiance measured at different wavelengths and at different scattering angles Θ, μ the cosine of the view zenith angle θ, μ_0 the cosine of the solar zenith angle θ_0, $\tau = \tau^*(\lambda)$ the spectral extinction optical thickness of the atmosphere, and

$p(\Theta,\lambda)$ the phase function at different wavelengths. Note that the first part on the right hand side of Eq. (5.12) is the single scattering approximation, which follows from Eqs (3.68) and (3.69) presented in Section 3.4.5. With these definitions, the phase function satisfies the normalization condition

$$\frac{1}{2}\int_{-1}^{1} p(\Theta,\lambda)d(cos\,\Theta) = 1.$$ (5.13)

The term $G(\ldots) = G(\varpi(\lambda),\tau^*(\lambda),p(\Theta,\lambda),A_g(\lambda),\theta_0,\theta,\varphi)$ describes the multiple scattering effects, where φ is the azimuth angle of the observations and $A_g(\lambda)$ the spectral surface reflectance (albedo).

The above equation is written for an atmosphere in the absence of polarization effects, and there are a large number of publically available software codes that compute the radiation field within and exiting from the atmosphere. A code that we have utilized for this purpose is a discrete ordinates code described by Stamnes et al. (1988) (cf. Section 3.4.2), which subdivides the atmosphere into a number of homogeneous layers with different optical thicknesses, phase functions, and single scattering albedos in the most general case.

The size distribution and complex refractive index of aerosol particles affect the phase function in Eq. (5.12). From the measurement of sky radiance, one can approximate Eq. (5.12) in the case of single scattering ($\tau \ll 1$) as

$$L(\Theta,\lambda) \approx \frac{E_0}{4\pi\mu}\varpi\tau p(\Theta,\lambda),$$

$$\approx \frac{E_0}{4\pi\mu}\left(\tau_{scat}^{aer} p_{aer}(\Theta,\lambda) + \tau_{scat}^{mol} p_{mol}(\Theta,\lambda)\right),$$ (5.14)

where

τ_{ext}^{aer} is the aerosol optical thickness,

τ_{scat}^{aer} the aerosol scattering optical thickness,

τ_{scat}^{mol} the molecular scattering optical thickness,

ϖ the single scattering albedo ($\tau_{scat}^{aer}/\tau_{ext}^{aer}$),

and $p_{mol}(\Theta,\lambda)$ the molecular (Rayleigh) phase function defined by

$$p_{mol}(\Theta,\lambda) = \frac{3}{4}(1 + cos^2\,\Theta).$$ (5.15)

Rewriting Eq. (5.14) we obtain

$$\frac{4\pi\mu}{E_0}L(\Theta,\lambda)-\tau_{scat}^{mol}P_{mol}(\Theta,\lambda)\approx\tau_{scat}^{aer}P_{aer}(\Theta,\lambda),\tag{5.16}$$

where $p_{aer}(\Theta,\lambda)$ is the aerosol phase function and is given by

$$P_{aer}(\Theta,\lambda)=\frac{1}{\tau_{ext}^{aer}}\int_0^\infty \pi r^2 Q_e(r,\lambda,m)p_{aer}(\Theta,r,\lambda,m)n_c(r)dr.\tag{5.17}$$

Substituting Eq. (5.17) into Eq. (5.16) and making the limits of integration finite, we obtain

$$\frac{4\pi\mu}{E_0}L(\Theta,\lambda)-\tau_{scat}^{mol}P_{mol}(\Theta,\lambda)\approx \varpi\int_{r_{min}}^{r_{max}}\pi r^2 Q_e(r,\lambda,m)p_{aer}(\Theta,r,\lambda,m)n_c(r)dr,$$

$$\approx\int_{r_{min}}^{r_{max}}K_{scat}(\Theta,r,\lambda,m)n_c(r)dr,$$

$$\approx\int_{r_{min}}^{r_{max}}\frac{3}{4\pi r^3}K_{scat}(\Theta,r,\lambda,m)v_c(\ln r)d\ln r,\tag{5.18}$$

where the kernel function for scattering $K_{scat}(\Theta, r, \lambda, m) = \varpi\pi r^2 Q_{ext}(r, \lambda, m)p_{aer}(\Theta, r, \lambda, m)$ can be computed from Mie theory for homogeneous spherical particles of known refractive index. In this formulation the modified kernel function for scattering $3K_{scat}(\Theta, r, \lambda, m)/4\pi r^3$ is a product of extinction cross-section and phase function, and depends on particle radius. Rewriting Eq. (5.18) in matrix formulation, we obtain

$$A_{ij}=\int_{\ln r_j-\Delta\ln r/2}^{\ln r_j+\Delta\ln r/2}\frac{3}{4\pi r^3}K_{scat}(\Theta,r,\lambda_i,m)d\ln r,\qquad j=1,2,\ldots,q,\tag{5.19}$$

$$f_j=v_c(\ln\bar{r}_j),\qquad\qquad\qquad\qquad\qquad i=1,2,\ldots,p,$$

$$g_i=\frac{4\pi\mu}{F_0}L(\Theta_i,\lambda_i)-\tau_{scat}^{mol}(\lambda_i)p_{mol}(\Theta_i,\lambda_i).$$

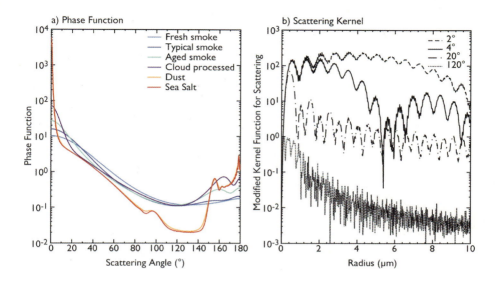

Figure 5.3 (a) Scattering phase function as a function of scattering angle and (b) scattering kernel function $3K_{scat}(\Theta, r, \lambda, m)/4\pi r^3$ as a function of radius for selected scattering angles used for the inference of aerosol size distribution from diffuse sky radiance measurements. A wavelength of 0.675 μm and a refractive index of $m = 1.45 + 0.0i$ was assumed in the calculations, and the sensitivity to various scattering angles is illustrated, where scattering angles near the sun are sensitive to large particles and scattering angles further from the sun are sensitive to progressively smaller particles.

This is once again a Fredholm integral equation of the first kind, and is simplified if the measurements are made in the almucantar for which μ is a constant (and equal to μ_0). Note that in Eq. (5.19) the measurements are obtained for a variety of scattering angles (and wavelengths), and the columnar volume size distribution is once again discretized in ln$r$.

Figure 5.3a illustrates the scattering phase function as a function of scattering angle for the six model lognormal size distributions illustrated in Figure 5.2a, and Figure 5.3b shows the modified kernel function for scattering as a function of radius for various scattering angles and a single wavelength ($\lambda = 0.675 \mu$m). Note that the kernels are sensitive to larger particles the smaller the scattering angle due to Fraunhofer diffraction close to the solar elevation that is dominated by larger particles. The further from the sun the measurements are obtained, the greater the sensitivity to smaller particles and the lower the sensitivity to larger particles.

In the real atmosphere, multiple scattering and surface albedo affect the light scattering of diffuse skylight, and are taken into consideration when this technique is applied. The

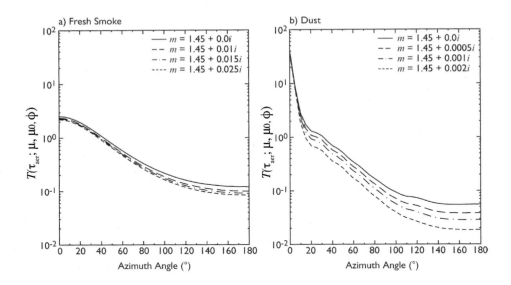

Figure 5.4 Transmission function ($\pi L/\mu_0 E_0$) as a function of azimuth angle in the almucantar where $\theta = \theta_0 = 60°$ for four values of imaginary part of the complex refractive index, an aerosol optical thickness $\tau_{aer} = 0.5$, and a surface albedo $A_g = 0.2$. Panel (a) is representative of an aerosol size distribution having accumulation mode particles similar to smoke and pollution ($r_m = 0.1\,\mu m$), whereas panel (b) is representative of coarse particles similar to desert dust ($r_m = 2\,\mu m$). All computations were performed at a wavelength $\lambda = 0.675\,\mu m$.

single scattering formulation highlighted here serves to illustrate the principles of a linear system of equations and its matrix formulation. Due to multiple scattering effects, which become more significant the greater the aerosol optical thickness and the further from the sun the observations are made, one needs to implement a nonlinear (and iterative) solution of Eq. (5.19).

Figure 5.4 illustrates the sensitivity of the transmission function $T(\tau_{aer}; \mu, \mu_0, \varphi) = \pi L((\tau_{aer}; \mu, \mu_0, \varphi)/\mu_0 E_0$ in the almucantar where $\theta = \theta_0 = 60°$ to the imaginary term m_i of the complex refractive index of the aerosol particles. The refractive index $m = m_r + m_i i$ is a complex number; the real part m_r is the ordinary index of refraction of the material (1.45 in this case), while the imaginary part m_i determines the absorption of electromagnetic radiation.

Although the particular model atmospheres illustrated in Figure 5.4 have $\tau_{aer} = 0.5$, a lognormal size distribution with (a) $r_m = 0.1\,\mu m$ (characteristic of fresh smoke and pollution) and (b) $r_m = 2.0\,\mu m$ (desert dust), a surface albedo $A_g = 0.2$, and a real refractive index $m_r = 1.45$, the obvious sensitivities in the figure are similar for all other model atmospheres

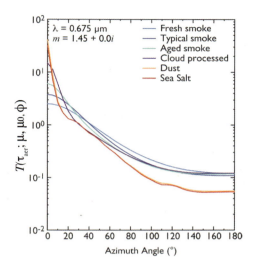

Figure 5.5 Transmission function as a function of azimuth angle in the almucantar where $\theta = \theta_0 = 60°$ for six size distributions shown in Figure 5.2 (a), an aerosol optical thickness $\tau_{aer} = 0.5$, an aerosol refractive index $m = 1.45 + 0.0i$, and a surface albedo $A_g = 0.2$. These computations were performed at a wavelength $\lambda = 0.675 \ \mu m$.

investigated. In general, the solar aureole (the strong intensity region in the vicinity of the sun) is relatively insensitive to m_i, being primarily produced by Fraunhofer diffraction around the particles, whereas the intensity (radiance) at larger scattering angles shows a marked sensitivity to absorption. In this example, where $\theta = \theta_0 = 60°$, the almucantar "measurements" would be obtained for scattering angles from $0°$ to $120°$, representing a large range of scattering angles in the single scattering phase function. It is for this reason that almucantar observations are always obtained when the sun is relatively low in the sky so that a large range of scattering angles can be observed. In the tropics or at high sun, almucantar measurements provide very little useful information on which to characterize aerosol optical and microphysical properties.

Figure 5.5 illustrates the *forward problem* for diffuse skylight whereby a size distribution (Figure 5.2a) can be used to infer the expected sky radiance (Figure 5.5). The *retrieval problem* is once again an inverse problem, whereby measurements such as those in Figure 5.5 are used to estimate the size distribution that is responsible for producing those measurements. As in the case of spectral optical thickness, the solution of this problem is also unstable and nonunique and requires the introduction of "virtual measurements" in the form of constraints, due primarily to the fact that the modified scattering kernel function, illustrated in Figure 5.3b, is not "orthogonal" and kernels for various scattering angles have considerable overlap and are hence nonunique.

5.3 Basic inversion of linear systems

5.3.1 Direct inversion

In Eq. (5.9), the vector $\mathbf{g}$ and matrix $\mathbf{A}$ are known. If we ignore the unknown error vector ε we can solve Eq. (5.9) by multiplying it by the inverse matrix $\mathbf{A}$, namely $\mathbf{A}^{-1}$, to obtain the vector solution

$$\mathbf{f} = \mathbf{A}^{-1}\mathbf{g}, \tag{5.20}$$

which is the direct inversion of Eq. (5.9) in the special case where the number of measurements m is equal to the number of unknowns n. For the matrix $\mathbf{A}$ with linearly independent and non-zero rows, Eq. (5.20) gives a unique solution. It follows from Eq. (5.20) that any uncertainties in the measurement vector ($\Delta\mathbf{g} = \varepsilon$), either due to actual uncertainties in the measurement or quadrature errors in the approximation of an integral with discretized values, gives rise to uncertainties in the solution vector ($\Delta\mathbf{f}$) according to

$$\Delta\mathbf{f} = \mathbf{A}^{-1}\varepsilon. \tag{5.21}$$

In our idealized aerosol optical thickness retrieval, in which the aerosol particles of one size dominate the optical thickness at one wavelength, the rows of the $\mathbf{A}$ matrix are linearly independent and very different, and the inverse of $\mathbf{A}$, and hence the size distribution solution, is stable. On the other hand, in the real world, where the kernel function for an individual measurement is broad and overlapping, the rows of the $\mathbf{A}$ matrix differ little from one another and are nearly linearly dependent. This is the case for aerosol optical thickness, as seen in Figure 5.1b, where an integral over radius for discrete radii intervals (that produces rows of $\mathbf{A}$), leads to tremendous overlap, especially for $r > {\sim}1.5\ \mu$m. This leads to an $\mathbf{A}$ matrix that is nearly singular (small determinant and small eigenvalues), with a corresponding $\mathbf{A}^{-1}$ matrix that has large eigenvalues. Under these conditions, uncertainties are amplified greatly, $\Delta\mathbf{f}$ can be quite large, and the "solution" is unstable.

In spite of the instability of the solution obtained by numerical inversion of Eq. (5.20) in this situation, when substituting back into Eq. (5.9), one would be able to reproduce the measurements within the estimated error. This is characteristic of inversion problems and is symptomatic of inverse problems in which the eigenvalues of the $\mathbf{A}$ matrix are small. This in turn is a result of a high degree of interdependence among some of the kernels $K_i(x)$.

5.3.2 Least-squares solution

One's next reaction to instability is often to seek more data, to argue that if 20 measurements of g_i cannot satisfactorily give 20 tabular values of $f(x_j)$, then perhaps 25, 30, 40 or more can.

The situation when the system of equations is overdetermined, wherein there are more equations, or measurements (p), than unknowns (q), can be used to minimize the solution errors. For such an overdetermined system no one solution can provide exact equality of the right part ($\mathbf{Af}$) to the left part ($\mathbf{g}$), because of the presence of the noise. Therefore, small differences ("fitting errors") between the left and right parts of the linear equation are always present. (If the matrix $\mathbf{A}$ is square, the strict equality is always achieved). If the properties of the noise in the observations $\mathbf{f}$ are known, then the solution of the overdetermined system of equations can be optimized by choosing the solution that generates the "fitting errors" that mimics best the expected distribution of errors in $\mathbf{f}$.

The agreement of "fitting errors" $\Delta\mathbf{g}$ with known error distribution can be evaluated using the known probability density function (PDF) as a function of modeled errors $P(\Delta\mathbf{g})$: the higher the $P(\Delta\mathbf{g})$ the closer the modeled $\Delta\mathbf{g}$ is to the known statistical properties. Thus, the best solution $\hat{\mathbf{a}}^{best}$ should result in modeled errors corresponding to the most probable error realization, i.e. to the PDF maximum

$$P(\Delta\mathbf{g}) = P(\mathbf{g}^*(\hat{\mathbf{a}}) - \mathbf{g}) = P(\mathbf{g}^*(\hat{\mathbf{a}})\,|\,\mathbf{g}) = max, \tag{5.22}$$

where $\mathbf{g}^*(\hat{\mathbf{a}})$ is the vector of "retrieved measurements" using the derived aerosol properties $\hat{\mathbf{a}}$ (aerosol size distribution, spectral real and imaginary refractive indices).

This principle is the essence of the well-known *Maximum Likelihood Method* (MLM). The PDF written as a function of measurements $P(\mathbf{g}^*(\hat{\mathbf{a}})\,|\,\mathbf{g})$ is called the *Likelihood Function*. The MLM is one of the strategic principles of statistical estimation and provides statistically the best solution in many senses. For example, the asymptotical error distribution (for an infinite number of $\Delta\mathbf{g}$ realizations) of MLM estimates $\hat{\mathbf{a}}$ has the smallest possible variance of $\Delta\hat{a}_i$. Most statistical properties of the MLM solution remain optimum for a limited number of observations (James, 2006).

If the error PDF is described by a normal distribution, then the MLM is reduced to a particular case widely known as the "least-squares solution". The basic principle of this method hinges on the fact that the normal (or Gaussian) distribution is the expected and most appropriate function for describing random noise. Hence the normal PDF for each vector $\mathbf{g}$ of measurements can be written in the form

$$P(\mathbf{g}^*(\hat{\mathbf{a}})\,|\,\mathbf{g}) = ((2\pi)^m \det(\mathbf{S}_\varepsilon))^{-1/2} \exp\left(-\frac{1}{2}(\mathbf{g}^*(\hat{\mathbf{a}}) - \mathbf{g})^T \mathbf{S}_\varepsilon^{-1}(\mathbf{g}^*(\hat{\mathbf{a}}) - \mathbf{g})\right), \tag{5.23}$$

where T denotes matrix transposition, $\mathbf{S}_\varepsilon$ is the covariance matrix of the vector $\mathbf{g}$, $\det(\mathbf{S}_\varepsilon)$ denotes the determinant of $\mathbf{S}_\varepsilon$, and p is the dimension of vectors $\mathbf{g}$ and $\mathbf{g}^*(\hat{\mathbf{a}})$.

In the simplest case of only one source of data (e.g., spectral aerosol optical thickness), the principle of maximum likelihood dictates that the best estimate for the solution for the aerosol properties $\hat{a}$ corresponds to the maximum value of Eq. (5.23), which in turn is equivalent to minimizing the term in the exponential. Thus we seek to minimize the square norm Q_1 defined by

$$Q_1 = \varepsilon^T S_\varepsilon^{-1} \varepsilon = \sum_{i=1}^{p} \sum_{j=1}^{p} (g_i*(\hat{a}) - g_i)^T S_{\varepsilon ij}^{-1} (g_j*(\hat{a}) - g_j), \tag{5.24}$$

where ε is the error vector denoting deviations from the measurements and our forward model, and is thus defined by

$$\varepsilon_i = g_i*(\hat{a}) - g_i = \sum_{j=1}^{q} A_{ij} f_j - g_i. \tag{5.25}$$

The minimum value of Q_1 can be determined by setting the partial derivative of Q_1 with respect to each of the unknown coefficients $(f_j, j = 1, \ldots, q)$ equal to zero. Thus we obtain q simultaneous equations of the form

$$\frac{\partial Q_1}{\partial f_j} = \varepsilon^T S_\varepsilon^{-1} \frac{\partial \varepsilon}{\partial f_j} + \frac{\partial \varepsilon^T}{\partial f_j} S_\varepsilon^{-1} \varepsilon,$$

$$= \varepsilon^T S_\varepsilon^{-1} A e_j + e_j^T A^T S_\varepsilon^{-1} \varepsilon,$$

$$= 2 e_j^T A^T S_\varepsilon^{-1} \varepsilon. \tag{5.26}$$

In this expression e_j is a $q \times 1$ vector that has zero in all but the jth element, where the element has a value of unity. Since the second term on the right hand side of Eq. (5.26) is simply the transpose of the first term, and since the transpose of a constant is simply that constant, both terms are equal. Thus we can write

$$A^T S_\varepsilon^{-1} \varepsilon = A^T S_\varepsilon^{-1} (Af - g) = 0,$$

$$A^T S_\varepsilon^{-1} Af = A^T S_\varepsilon^{-1} g,$$

$$f = (A^T S_\varepsilon^{-1} A)^{-1} A^T S_\varepsilon^{-1} g. \tag{5.27}$$

This is the weighted least-squares solution for unknown vector $\mathbf{f}$, given p measurements $> q$ unknowns with unequal measurement uncertainties and correlations among the individual measurements. If we assume there are no correlations and equal measurement error, such that S_ε is a diagonal matrix, then Eq. (5.27) reduces to the unweighted least-squares solution appropriate when the number of measurements and unknowns are different.

The solution given by the least-squares method (Eq. (5.27)) is superior to any other solution of linear system $\mathbf{Af} = \mathbf{g}$ provided that the noise assumption is correct. The covariance matrix for the least-squares method can be derived as follows:

$$S = \langle \Delta\mathbf{f}\,\Delta\mathbf{f}^\mathsf{T} \rangle = \langle (\mathbf{A}^\mathsf{T}\mathbf{S}_\varepsilon^{-1}\mathbf{A})^{-1}\mathbf{A}^\mathsf{T}\mathbf{S}_\varepsilon^{-1}\Delta\mathbf{g}\,((\mathbf{A}^\mathsf{T}\mathbf{S}_\varepsilon^{-1}\mathbf{A})^{-1}\mathbf{A}^\mathsf{T}\mathbf{S}_\varepsilon^{-1}\Delta\mathbf{g})^\mathsf{T} \rangle,$$

$$= (\mathbf{A}^\mathsf{T}\mathbf{S}_\varepsilon^{-1}\mathbf{A})^{-1}\mathbf{A}^\mathsf{T}\mathbf{S}_\varepsilon^{-1}\langle \Delta\mathbf{g}\,\Delta\mathbf{g}^\mathsf{T} \rangle\mathbf{S}_\varepsilon^{-1}\mathbf{A}(\mathbf{A}^\mathsf{T}\mathbf{S}_\varepsilon^{-1}\mathbf{A})^{-1},$$

$$= (\mathbf{A}^\mathsf{T}\mathbf{S}_\varepsilon^{-1}\mathbf{A})^{-1}. \tag{5.28}$$

The LSM estimates $\hat{\mathbf{a}}$ have the smallest variances of random errors and, moreover, the estimate $\mathbf{g}(\hat{\mathbf{a}})$ obtained using $\hat{\mathbf{a}}$ also has the smallest variance determined by Eq. (5.28), i.e. any product $\mathbf{g}(\hat{\mathbf{a}})$ of the LSM estimates $\hat{\mathbf{a}}$ is also optimum. These accuracy limits are related to the definition of Fisher information (James, 2006).

5.3.3 Uniqueness and the nature of *a priori* constraints

It is both mathematically and physically incorrect to solve an underconstrained set of equations, i.e. one in which there are less linearly independent equations/rows than unknowns. There are an infinite number of solutions that satisfy the equations exactly. In addition, the family of possible solutions becomes even broader in the presence of experimental error. Thus it is important to ask the question of whether the problem that is posed is realistic, and in what sense a solution exists.

This problem of nonuniqueness in the solution of the Fredholm integral equation requires that we rephrase the problem to be solved by the measurements as follows.

Given a measurement vector $\mathbf{g}$, the statistics of the experimental error ε, as measured by S_ε, the kernel matrix $\mathbf{A}$, and any other relevant information, how do we determine the most physical solution among the family of solutions that satisfy Eq. (5.9)?

This is the basis of the *estimation problem*, in which a set of ancillary criteria are introduced in order to reduce the infinite number of possible solutions of Eq. (5.9) to the best and most physical solution. The purpose of these criteria is to add additional information (not deriving from the measurements) that enables one of a set of possible $\mathbf{f}$ vectors to be selected. Among the many approaches to this problem, we will consider the following:

(1) Find a smoothed version of the true solution, or one that deviates the least from our expected solution based on prior physical knowledge. These criteria are known as *a priori* constraints.

(2) Characterize the class of all possible solutions by means of its probability density function, and choose some property of the distribution such as its mean or mode as the solution.

Dubovik and King (2000) refer to the introduction of constraints as adding *virtual measurements*, in that the constraints are treated mathematically in an identical way to real measurements. This is an obvious way of reducing the ambiguity associated with an ill-posed problem, but the introduction of erroneous constraints is itself equivalent to adding, along with more information, additional error. That error would have a non-random nature and would result in solution systematic errors, or biases. Thus it is important to add only valid and physically plausible constraints to the possible solutions. It is, after all, possible for a constraint to be too loose, too tight, or simply incorrect.

5.3.4 Measures of smoothness

In most physical problems involving Fredholm integral equations of the first kind, such as those presented in Section 5.2, the indicial function $f(x)$ is generally smooth. That is to say, instabilities and wild oscillations are unphysical. Thus, in order to restrict the solution to only those cases that are stable, it is reasonable to require that the solution be "smooth".

A useful measure of smoothness is that the second-derivative of the solution be small at each quadrature point. Since the second-derivative can be both positive and negative, Phillips (1962) suggested that a useful smoothness constraint would be to minimize the sum of the squares of the second derivatives of the solution points. Alternative measures of smoothness might be to minimize the sum of the squares of the first or third derivatives of the solution points.

If $\mathbf{Kf}$ contains in its elements values that are to be squared and summed to give Q_2, then the square norm to be minimized must in general be of the form

$$Q_2 = (\mathbf{Kf})^T \mathbf{Kf},$$

$$= \mathbf{f}^T \mathbf{K}^T \mathbf{Kf},$$

$$= \mathbf{f}^T \mathbf{Hf} = \sum_{i=1}^{q} \sum_{j=1}^{q} f_i H_{ij} f_j, \tag{5.29}$$

where $\mathbf{H}$ is a symmetric matrix of the form $\mathbf{K}^T\mathbf{K}$.

For a quadrature of equal division, such as the aerosol size distribution that is equally spaced in lnr, the $q \times 1$ vector containing the second derivatives of the solution points is given by

$$\mathbf{Kf} = \begin{pmatrix} 0 \\ 0 \\ f_1 - 2f_2 + f_3 \\ f_2 - 2f_3 + f_4 \\ \cdot \\ \cdot \\ \cdot \\ f_{q-2} - 2f_{q-1} + f_q \end{pmatrix},$$

which requires that the $q \times q$ matrices $\mathbf{K}$ and $\mathbf{H}$ are given by

$$\mathbf{K} = \begin{pmatrix} 0 & & & & \\ 0 & 0 & & & \\ 1 & -2 & 1 & & \\ 0 & 1 & -2 & 1 & 0 \\ & & \cdot & \cdot & \cdot \end{pmatrix}; \quad \mathbf{H} = \begin{pmatrix} 1 & -2 & 1 & 0 & 0 & \cdot & \cdot \\ -2 & 5 & -4 & 1 & 0 & 0 & \cdot \\ 1 & -4 & 6 & -4 & 1 & 0 & \cdot \\ 0 & 1 & -4 & 6 & -4 & 1 & \cdot \\ & & \cdot & \cdot & \cdot & \cdot & \cdot \end{pmatrix}$$

$$(5.30)$$

This is the form of the second-derivative smoothing matrix first derived by Twomey (1963).

Alternative measures of smoothness can be found in Twomey's book (1977b, pp. 124–126). These include:

(1) Minimize the sum of the first differences of the solution points.
(2) Minimize the sum of the third differences of the solution points.
(3) Minimize solutions in which the sum of the squares of the solution points is minimized. This special case results in $\mathbf{H} = \mathbf{I}$, where $\mathbf{I}$ is the identity matrix, and is referred to by mathematicians as *ridge regression*.
(4) Minimize the differences from the average value of all of the f_j solution points.

5.3.5 Constrained linear inversion

As we have seen, the solution of Eq. (5.9) requires that the solution vector $\mathbf{f}$ be obtained by minimizing a performance function Q defined as

$$Q = Q_1 + \gamma Q_2, \tag{5.31}$$

where Q_1 is the square norm of the errors, defined by Eq. (5.24), and Q_2 is the square norm of the smoothness constraint, defined by Eq. (5.29). In this expression γ is a non-negative *Lagrange multiplier*, which serves to weigh the contribution of the smoothness constraint, relative to the contribution of the measurements.

The minimum value of Q can be determined by setting the partial derivative of Q with respect to each of the unknown coefficients $(f_j, j = 1, \ldots, q)$ equal to zero. Thus we obtain q simultaneous equations:

$$\frac{\partial Q}{\partial f_j} = \frac{\partial Q_1}{\partial f_j} + \gamma \frac{\partial Q_2}{\partial f_j} = 0, \tag{5.32}$$

where

$$\frac{\partial Q_2}{\partial f_j} = \mathbf{f}^\mathsf{T} \mathbf{H} \frac{\partial \mathbf{f}}{\partial f_j} + \frac{\partial \mathbf{f}^\mathsf{T}}{\partial f_j} \mathbf{H} \mathbf{f},$$

$$= \mathbf{f}^\mathsf{T} \mathbf{H} e_j + e_j^\mathsf{T} \mathbf{H} \mathbf{f},$$

$$= 2 e_j^\mathsf{T} \mathbf{H} \mathbf{f}, \tag{5.33}$$

for all $j = 1, \ldots, q$. If we substitute Eq. (5.26) and Eq. (5.33) back into Eq. (5.32), we find that all elements of the column vector

$$\mathbf{A}^\mathsf{T} \mathbf{S}_\varepsilon^{-1} (\mathbf{A}\mathbf{f} - \mathbf{g}) + \gamma \mathbf{H} \mathbf{f} = 0.$$

Rearranging the order of the terms we finally obtain the expression for the indicial function $f(x)$:

$$(\mathbf{A}^\mathsf{T} \mathbf{S}_\varepsilon^{-1} \mathbf{A} + \gamma \mathbf{H}) \mathbf{f} = \mathbf{A}^\mathsf{T} \mathbf{S}_\varepsilon^{-1} \mathbf{g},$$

$$\mathbf{f} = (\mathbf{A}^\mathsf{T} \mathbf{S}_\varepsilon^{-1} \mathbf{A} + \gamma \mathbf{H})^{-1} \mathbf{A}^\mathsf{T} \mathbf{S}_\varepsilon^{-1} \mathbf{g}. \tag{5.34}$$

This is the formula for *constrained linear inversion*, as first derived in this form by Twomey (1963) for the special case where the measurement errors are equal and uncorrelated (such that S_ε is diagonal and equal to a constant). It requires a single matrix inverse, whereas Phillips' (1962) solution required the inversion of 2 matrices. The two methods are equivalent and thus this method is also known in the United States as the Phillips-Twomey inversion method. Tikhonov (1963) arrived at the same solution independently, and thus his name is also sometimes associated with this method, especially in Russia, where it is known as regularization.

This form of the constrained linear inversion formula represents a weighted least-squares solution subject to a constraint, and is a natural extension of Eq. (5.27).

If we now consider the situation in which the statistics of the unknown distribution are known, that is to say, we have prior information that the expected (mean) value of the unknown distribution $\mathbf{f}$ is $\mathbf{f}_0$, we may wish to bias the solution vector toward this solution.

In this situation Q_2 can be written as

$$Q_2 = (\mathbf{f} - \mathbf{f}_0)_T(\mathbf{f} - \mathbf{f}_0), \tag{5.35}$$

Taking the partial derivative of Q_2 with respect to each of the unknown coefficients (f_j, $j = 1, \ldots, q$), we obtain q simultaneous equations:

$$\frac{\partial Q_2}{\partial f_j} = (\mathbf{f} - \mathbf{f}_0)^T \frac{\partial \mathbf{f}}{\partial f_j} + \frac{\partial \mathbf{f}^T}{\partial f_j}(\mathbf{f} - \mathbf{f}_0),$$

$$= (\mathbf{f} - \mathbf{f}_0)^T \mathbf{e}_j + \mathbf{e}_j^T(\mathbf{f} - \mathbf{f}_0),$$

$$= 2\mathbf{e}_j^T(\mathbf{f} - \mathbf{f}_0), \tag{5.36}$$

for all $j = 1, \ldots, q$. If we assume that the measurement errors are unequal and correlated, and substitute Eq. (5.26) and Eq. (5.36) back into Eq. (5.32), we find that all elements of the column vector

$$\mathbf{A}^T\mathbf{S}_\varepsilon^{-1}(\mathbf{A}\mathbf{f} - \mathbf{g}) + \gamma(\mathbf{f} - \mathbf{f}_0) = 0.$$

Rearranging the order of the terms we obtain

$$(\mathbf{A}^T\mathbf{S}_\varepsilon^{-1}\mathbf{A} + \gamma\mathbf{I})\mathbf{f} = \mathbf{A}^T\mathbf{S}_\varepsilon^{-1}\mathbf{g} + \gamma\mathbf{f}_0,$$

$$\mathbf{f} = (\mathbf{A}^T\mathbf{S}_\varepsilon^{-1}\mathbf{A} + \gamma\mathbf{I})^{-1}(\mathbf{A}^T\mathbf{S}_\varepsilon^{-1}\mathbf{g} + \gamma\mathbf{f}_0). \tag{5.37}$$

which biases the solution vector $\mathbf{f}$ towards our *a priori* estimate of the solution vector $\mathbf{f}_0$.

5.3.6. Selection of the Lagrange multiplier

The most important aspect to consider when solving the constrained linear inversion formulas is how to select the most appropriate value for the Lagrange multiplier. When $\gamma = 0$, Eqs (5.34) and (5.37) yield the least-squared solution that is highly oscillatory and unstable. This is due to the fact that the kernel functions are generally broad and overlapping, with the corresponding $\mathbf{A}$ matrix (cf. Figures 5.1b and 5.3b), or alternatively $\mathbf{A}^T\mathbf{A}$ matrix, nearly singular. In order to ameliorate this problem, we introduced the constraint matrix $\mathbf{H}$ that, for the second derivative smoothing constraint, is itself singular. However, the sum of two singular matrices (or nearly singular matrices) is not itself singular.

As γ increases towards infinity, the solution tends more and more towards the constraint, placing greater emphasis on the constraint and less emphasis on the measurements. This is of course not reasonable as a solution, since no information on the solution originates from the measurements. Therefore, an optimum value or values of γ must be chosen somewhere between these two extremes, but how does one make an appropriate choice for the Lagrange multiplier? After all, the true solution is not known when analyzing real measurements. It is for this reason that many papers have been published using simulated data with known solutions and known amounts of "error" introduced into the "measurements". Many authors have addressed this very difficult problem (e.g., Rogers, 1976; Twomey, 1977b; King, 1982), and this problem becomes even more challenging when such different parameters as size distribution and complex refractive index are retrieved simultaneously.

Consider a q dimensional space in which each axis represents one element of the solution vector $\mathbf{f}$, viz., f_j for $j = 1, 2, \ldots, q$. For each possible solution vector $\mathbf{f}$, one can readily compute the values of Q_1 and Q_2, from which hypersurfaces for each performance function can be constructed. Figure 5.6 illustrates in two dimensions the situation in q-dimensional vector space. This figure shows the lines of constant Q_1 (solid) and Q_2 (dashed), where Q_1 attains its minimum at solution point "1", and Q_2 attains its minimum at solution point "2".

As γ is increased from zero, the solution represented by the constrained linear inversion moves along the dotted line between the minimum of Q_1 and the minimum of Q_2. Thus Q_1 monotonically increases as the square norm of the errors increases, while the quadratic Q_2 decreases towards the "smooth" solution represented by the minimum of quadratic Q_2. As shown in Figure 5.6, it is in principle possible to increase γ too large and thus to pass the minimum of Q_2, whereby Q_2 would commence to increase. This is an obvious signal of too large a constraint, and seldom happens in reality.

A reasonable criterion for the selection of the Lagrange multiplier is that

$$Q_1 \leq E(Q_1), \tag{5.38}$$

where E denotes the expectation operator. Thus in the schematic illustration presented in Figure 5.6, one would accept any solution that lies within the heavy solid line on the Q_1 hypersurface, and preferably the one at the intersection of the dotted line and the solid line, since this solution is the smoothest one that simultaneously satisfies Eq. (5.38).

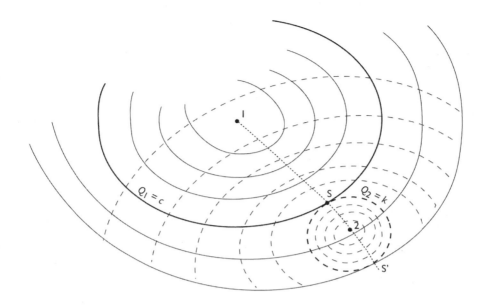

Figure 5.6 Schematic illustration of solution S and the trajectory in function space (dotted) between the minimum of the Q_1 hypersurface (solid curves) and the minimum of the Q_2 hypersurface (broken curves) (adapted from Twomey, 1977b).

Another constraint on the selection of an appropriate value for the Lagrange multiplier appropriate for the majority of physical problems one is likely to encounter in the atmospheric sciences is that all elements of the solution vector are positive. This further restricts the allowable values of the Lagrange multiplier to values of

$$\gamma > \gamma_{min}. \tag{5.39}$$

Criteria (5.38) and (5.39) generally provide sufficient restrictions on the selection of the Lagrange multiplier to allow one to empirically select γ by incrementally increasing γ over a range of values appropriate to a given problem.

5.3.7 Solution covariance matrix

Once an appropriate value of γ has been determined, the solution covariance matrix can readily be obtained. Bevington (1969) shows that uncertainties in multiple regression problems are related to the symmetric matrix $\boldsymbol{\alpha}$, whose elements are given by

$$\alpha_{ij} = \frac{1}{2} \frac{\partial^2 Q}{\partial f_i \partial f_j}. \tag{5.40}$$

The matrix $\boldsymbol{\alpha}$ is called the *curvature matrix* because of its relationship to the curvature of the Q hypersurface in coefficient space. The solution covariance matrix $\mathbf{S}$ is then obtained from the $\boldsymbol{\alpha}^{-1}$ matrix, i.e.

$$\mathbf{S} = \boldsymbol{\alpha}^{-1}, \tag{5.41}$$

which, for the constrained linear inversion solution represented by Eq. (5.34), leads to

$$\alpha_{ij} = \frac{1}{2} \frac{\partial}{\partial f_j} \left[2 \mathbf{e}_i^{\mathrm{T}} \mathbf{A}^{\mathrm{T}} \mathbf{S}_\varepsilon^{-1} (\mathbf{Af} - \mathbf{g}) + 2\gamma \mathbf{e}_i^{\mathrm{T}} \mathbf{Hf} \right],$$

$$= \mathbf{e}_i^{\mathrm{T}} \mathbf{A}^{\mathrm{T}} \mathbf{S}_\varepsilon^{-1} \mathbf{A} \frac{\partial \mathbf{f}}{\partial f_j} + \gamma \mathbf{e}_i^{\mathrm{T}} \mathbf{H} \frac{\partial \mathbf{f}}{\partial f_j},$$

$$= \mathbf{e}_i^{\mathrm{T}} \left[\mathbf{A}^{\mathrm{T}} \mathbf{S}_\varepsilon^{-1} \mathbf{A} + \gamma \mathbf{H} \right] \mathbf{e}_j, \tag{5.42}$$

where we have made use of Eqs (5.26) and (5.34). It readily follows from this expression that

$$\boldsymbol{\alpha} = \mathbf{A}^{\mathrm{T}} \mathbf{S}_{\mathcal{E}}^{-1} \mathbf{A} + \gamma \mathbf{H},$$

$$\mathbf{S} = (\mathbf{A}^{\mathrm{T}} \mathbf{S}_{\mathcal{E}}^{-1} \mathbf{A} + \gamma \mathbf{H})^{-1}. \tag{5.43}$$

This is the equation for estimating the uncertainties in the inversion solution for constrained linear inversion problems, and is the natural extension of Eq. (5.28) for the solution covariance matrix for least-squares problems in the absence of any smoothness constraints.

5.4 Inversion of aerosol size distribution from spectral optical thickness measurements

The method for determining the columnar volume size distribution from spectral optical thickness measurements has a long history of development (King et al., 1978), and has been applied widely throughout the world, including from airborne sunphotometer measurements (Spinhirne and King, 1985; Schmid et al., 2003). Aerosol optical thickness measurements can be derived from well-calibrated sunphotometer measurements, such as those obtained from the ground-based AERONET sun/sky radiometers described by Holben et al. (1998). As an example of this application, AERONET observations obtained in Mongu, Zambia during the dry (biomass burning) season were analyzed using the constrained linear inversion method described above. Since spectral optical thickness is an extinction measurement, it has little sensitivity to refractive index. We assumed the complex refractive index of the aerosol particles was wavelength and size independent and given by $m = 1.45 + 0.0i$.

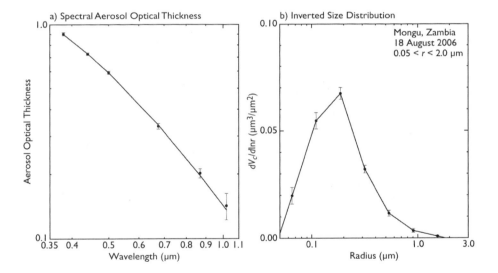

Figure 5.7 (a) Spectral aerosol optical thickness on 18 August 2006 in Mongu, Zambia as measured by the AERONET sun/sky radiometer when the solar zenith angle $\theta_0 = 59.6°$. (b) Columnar volume size distribution derived by inverting the $\tau_{aer}(\lambda)$ spectrum in (a). Also shown is the regression fit to the data using the inverted size distribution. The solid curve in (a) is the $\tau_{aer}(\lambda)$ spectrum computed from the size distribution in (b).

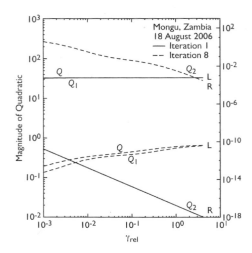

Figure 5.8 Magnitude of quadratics Q_1, Q_2, and Q as a function of the relative Lagrange multiplier $\gamma/(\mathbf{A}^T\mathbf{S}_\varepsilon^{-1}\mathbf{A})_{11}$ for 18 August 2006 over Mongu. The curve labeled R applies to the right-hand scale and all other curves apply to the left-hand scale.

Optical thickness measurements are often made at between six and eight different wave-lengths ranging between 0.38 and 1.64 μm (depending on instrument). Due to both the extinction cross-section (which increases significantly with radius) and the number density of natural aerosol particles (which normally decreases with radius), this spectral region of the attenuation measurements limits the radius range of maximum sensitivity to fine and coarse particles only ($0.05 \leq r \leq 4\,\mu$m), as shown in Figure 5.1. The effective radius range of inversion sensitivity is dependent both on the wavelength range available for the meas-urements and also the size distribution itself, which is not known *a priori*.

Figure 5.7 shows an example of the application of constrained linear inversion to spec-tral optical thickness measurements obtained in Mongu, Zambia on 18 August 2006 at 6:59:11 UTC when the solar zenith angle $\theta_0 = 59.6°$. The observed aerosol optical thick-ness measurements and corresponding standard deviations are shown in the left-hand por-tion of the figure, while the corresponding volume size distribution obtained by inverting these data is shown in the right-hand portion of the figure. Uncertainties in the derived size distribution are shown in the right-hand portion of the figure, and the solid curve in the left-hand portion indicates how the inverted size distribution is able to reproduce the $\tau_a(\lambda)$ measurements (i.e. the direct problem $\mathbf{g} = \mathbf{Af}$).

As discussed above, the critical and rather complex aspect of implementing a viable inversion solution is the selection of the Lagrange multiplier that is necessary to stabilize the solution. We employed the iterative technique described by King (1982), and Figure 5.8 illustrates the magnitude of quadratics Q_1, Q_2, and Q as a function of the relative Lagrange multiplier $(\gamma/(\mathbf{A}^\mathrm{T}\mathbf{S}_\varepsilon^{-1}\mathbf{A})_{11})$, which is the numerical implementation of the schematic shown in Figure 5.6.

The retrieved size distribution shown in Figure 5.7b is characteristic of fresh smoke with small particles, but the uncertainty in the size distribution of small particles is increasingly large due to the reduced sensitivity to small particles using spectral optical thickness measurements alone. This is what has led to the development of inversion of multisource data that includes not only spectral optical thickness but also sky radiance, which contains more information content on both small and large particles as well as some sensitivity to aerosol refractive index, not contained in the spectral optical thickness measurements alone.

5.5 Combination of observations and multi-term LSM solution

As an alternative to solving Eq. (5.9) by constrained linear inversion, we can seek to find the optimum solution vector $\mathbf{f}$, based on a combination of information deriving from our experimental measurements and information deriving from our "virtual measurements".

Formally, both measured and *a priori* data can be written as

$$\mathbf{g}_k = \mathbf{g}_k*(\hat{\mathbf{a}}) + \varepsilon_k, \qquad\qquad k = 1,2,\ldots,K. \quad (5.44)$$

where the vectors $\mathbf{g}_k$ relate to independent sets of measurements. As before, the vector $\hat{\mathbf{a}}$ denotes the aerosol parameters that should be retrieved. The vectors $\mathbf{g}_{k>1}$ may include the values of *a priori* constraints on aerosol parameters or possible accessory data. The independence of the data sets $\mathbf{g}_k$ can be formally expressed by the fact the covariance matrix of the joint data set has the following diagonal array structure:

$$\mathbf{S}_g = \begin{pmatrix} \mathbf{S}_1 & 0 & 0 & 0 \\ 0 & \mathbf{S}_2 & 0 & 0 \\ 0 & 0 & \ldots & 0 \\ 0 & 0 & 0 & \mathbf{S}_k \end{pmatrix}. \quad (5.45)$$

Following the method outlined above for finding the least-squares solution from single source data, the joint probability density function of all inverted data can be obtained by simple multiplication of the probability density function of all vectors $\mathbf{g}_k$ as follows:

$$P(\mathbf{g}_1*(\hat{\mathbf{a}}),\ldots,\mathbf{g}_k*(\hat{\mathbf{a}}) \,|\, \mathbf{g}_1,\ldots,\mathbf{g}_k) = \prod_{k=1}^{K} P(\mathbf{g}_k*(\hat{\mathbf{a}}) \,|\, \mathbf{g}_k)$$

$$\sim \exp\left(-\frac{1}{2}\sum_{k=1}^{K}(\mathbf{g}_k * (\hat{\mathbf{a}}) - \mathbf{g}_k)^{\mathrm{T}}\mathbf{S}_{\varepsilon k}^{-1}(\mathbf{g}_k * (\hat{\mathbf{a}}) - \mathbf{g}_k)\right). \tag{5.46}$$

According to the method of maximum likelihood, the best estimate for the solution of the aerosol properties $\hat{\mathbf{a}}$ corresponds to the maximum value of Eq. (5.46), which in turn is equivalent to minimizing the term in the exponential. Thus we seek to minimize the square norm Q defined by

$$Q = \frac{1}{2}\sum_{k=1}^{K}(\mathbf{g}_k * (\hat{\mathbf{a}}) - \mathbf{g}_k)^{\mathrm{T}}\mathbf{S}_{\varepsilon k}^{-1}(\mathbf{g}_k * (\hat{\mathbf{a}}) - \mathbf{g}_k), \tag{5.47}$$

The minimum value of Q can be determined by setting the partial derivative of Q with respect to each of the unknown coefficients $(f_j, j = 1, \ldots, q)$ equal to zero. Thus we obtain q simultaneous equations of the form

$$\frac{\partial Q}{\partial f_j} = \frac{1}{2}\sum_{k=1}^{K}\left[\varepsilon_k^{\mathrm{T}}\mathbf{S}_{\varepsilon k}^{-1}\frac{\partial \varepsilon_k}{\partial f_j} + \frac{\partial \varepsilon_k}{\partial f_j}^{\mathrm{T}}\mathbf{S}_{\varepsilon k}^{-1}\varepsilon_k\right], \tag{5.48}$$

where

$$\varepsilon_k = g_k * (\hat{\mathbf{a}}) - g_k = \sum_{j=1}^{p}A_{kj}f_j - g_k. \tag{5.49}$$

Substituting Eq. (5.49) into Eq. (5.48) we find

$$\frac{\partial Q}{\partial f_j} = \frac{1}{2}\sum_{k=1}^{K}\left[\varepsilon_k^{\mathrm{T}}\mathbf{S}_{\varepsilon k}^{-1}\mathbf{A}_k\mathbf{e}_j + \mathbf{e}_j^{\mathrm{T}}\mathbf{A}_k^{\mathrm{T}}\mathbf{S}_{\varepsilon k}^{-1}\varepsilon_k\right],$$

$$= \sum_{k=1}^{K}\mathbf{e}_j^{\mathrm{T}}\mathbf{A}_k^{\mathrm{T}}\mathbf{S}_{\varepsilon k}^{-1}\varepsilon_k,$$

$$= \sum_{k=1}^{K}\mathbf{e}_j^{\mathrm{T}}\mathbf{A}_k^{\mathrm{T}}\mathbf{S}_{\varepsilon k}^{-1}(\mathbf{A}_k\mathbf{f} - \mathbf{g}_k), \tag{5.50}$$

In this expression $\mathbf{e}_j$ is a $q \times 1$ vector that has zero in all but the jth element, where the element has a value of unity. Setting the partial derivative of Q with respect to f_j equal to zero and rearranging the terms we obtain the corresponding LSM solution providing a minimum square norm as follows:

$$\sum_{k=1}^{K} \mathbf{A}_k{}^{\mathrm{T}} \mathbf{S}_{\varepsilon_k}{}^{-1} \left(\mathbf{A}_k \mathbf{f} - \mathbf{g}_k \right) = 0,$$

$$\sum_{k=1}^{K} \mathbf{A}_k{}^{\mathrm{T}} \mathbf{S}_{\varepsilon_k}{}^{-1} \mathbf{A}_k \mathbf{f} = \sum_{k=1}^{K} \mathbf{A}_k{}^{\mathrm{T}} \mathbf{S}_{\varepsilon_k}{}^{-1} \mathbf{g}_k \,,$$

$$\mathbf{f} = \left[\sum_{k=1}^{K} \mathbf{A}_k{}^{\mathrm{T}} \mathbf{S}_{\varepsilon_k}{}^{-1} \mathbf{A}_k \right]^{-1} \sum_{k=1}^{K} \mathbf{A}_k{}^{\mathrm{T}} \mathbf{S}_{\varepsilon_k}{}^{-1} \mathbf{g}_k \,, \tag{5.51}$$

and the corresponding solution covariance matrix is given by

$$\mathbf{S} = \left[\sum_{k=1}^{K} \mathbf{A}_k{}^{\mathrm{T}} \mathbf{S}_{\varepsilon_k}{}^{-1} \mathbf{A}_k \right]^{-1}. \tag{5.52}$$

These rather simple formulas generalize the "classical" linear inversion problem, and are an extension of the least-squares solutions given by Eqs (5.27) and (5.28), or the solutions in the presence of a smoothness constraint as outlined in Eqs (5.34) and (5.43). For example, if the data set of the measurements (Eq. 5.44) contains only two equations

$$\begin{cases} \mathbf{g} = \mathbf{A}\mathbf{f} + \varepsilon_g \\ \mathbf{f}^* = \mathbf{f} + \varepsilon_f \end{cases}, \tag{5.53}$$

where the second line denotes *a priori* estimates of the unknown "virtual observations", then this system yields the "multi-term" LSM solution given by (5.51) that is reduced to

$$\mathbf{f} = (\mathbf{A}^{\mathrm{T}}\mathbf{S}_{\varepsilon}^{-1}\mathbf{A} + \mathbf{S}_{\mathbf{f}}^{-1})^{-1} \, (\mathbf{A}^{\mathrm{T}}\mathbf{S}_{\varepsilon}^{-1}\mathbf{g} + \mathbf{S}_{\mathbf{f}}^{-1}\mathbf{f}_0), \tag{5.54}$$

with a corresponding covariance matrix

$$\mathbf{S} = (\mathbf{A}^{\mathrm{T}}\mathbf{S}_{\varepsilon}^{-1}\mathbf{A} + \mathbf{S}_{\mathbf{f}}^{-1})^{-1}. \tag{5.55}$$

These solutions are variously known as the *maximum likelihood solution, optimal estimation, Wiener-Kolmogorov smoothing*, or *Rodgers-Strand-Westwater solution*.

In the special case in which the set of the measurements given by Eq. (5.44) contains two equations of the form

$$\begin{cases} \mathbf{g} = \mathbf{Af} + \varepsilon_g \\ \mathbf{0}^* = \mathbf{Kf} + \varepsilon_k \end{cases}, \tag{5.56}$$

where $\mathbf{Kf}$ denotes the second differences of the solution vector $\mathbf{f}$, assumed equal to zero $\mathbf{0}^*$ with errors ε_k, and further assuming that the covariance matrix of ε_k is diagonal with $\mathbf{S}_k = \mathbf{I}\varepsilon_k^2$, then the "multi-term" LSM solution given by (5.51) reduces to

$$\mathbf{f} = (\mathbf{A}^T\mathbf{S}_\varepsilon^{-1}\mathbf{A} + \gamma\mathbf{H})^{-1}(\mathbf{A}^T\mathbf{S}_\varepsilon^{-1}\mathbf{g}), \tag{5.57}$$

with the corresponding covariance matrix given by

$$\mathbf{S} = (\mathbf{A}^T\mathbf{S}_\varepsilon^{-1}\mathbf{A} + \gamma\mathbf{H})^{-1}, \tag{5.58}$$

where $\gamma = \varepsilon_k^{-2}$.

These solutions were discussed above in Section 5.3.5 and are variously known as the *Phillips-Tikhonov-Twomey* constrained linear inversion formulae.

Thus, "multi-term" LSM solutions naturally generalize the well-known constrained linear inversion equations. This approach allows the use of both *a priori* estimates (second line in Eq. (5.53)) and smoothness constraints (second line in Eq. (5.56)) in a single formulation (it is explicitly shown in Dubovik, 2004). Nonetheless, the main advantage of the "multi-term" LSM approach is not in generalizing the well-known constrained linear inversion formulation, but rather to the fact that this principle can be used to construct new inversion formulations uniting various observations and *a priori* constraints. For example, the AERONET retrieval algorithm developed by Dubovik and King (2000) simultaneously retrieves the aerosol size distribution and the spectrally-dependent real and imaginary parts of the complex refractive index. This algorithm uses different constraints for smoothing the size distribution and spectral dependencies of real and imaginary parts of the complex refractive index. Combining such constraints within the basic constrained linear inversion formulations would not be possible without this generalized formulation. Even more elaborate inversion formulations derived from "multi-term" LSM solutions are used in studies by Dubovik et al. (2008) and Gatebe et al. (2010).

In addition, the "multi-term" LSM solution allows a rather convenient interpretation of the Lagrange multiplier. Specifically, Eq. (5.51) can be rewritten as

$$\mathbf{f} = \left[\sum_{k=1}^{K}\mathbf{A}_k^T\mathbf{W}_k^{-1}\mathbf{A}_k\right]^{-1}\left[\sum_{k=1}^{K}\mathbf{A}_k^T\mathbf{W}_k^{-1}\mathbf{g}_k\right]. \tag{5.59}$$

In this expression we have introduced Lagrange multipliers γ_k and weight matrices $\mathbf{W}_k$ defined as

$$\mathbf{W}_k = \frac{1}{\varepsilon_k^2}\mathbf{S}_{\varepsilon k}, \tag{5.60}$$

where ε_k^2 denotes the variance of errors in the data vector $\mathbf{g}_k$. With this formulation, it is readily apparent that Lagrange multipliers have a clear statistical interpretation as the ratios of variances

$$\gamma_k = \frac{\varepsilon_1^2}{\varepsilon_k^2}. \tag{5.61}$$

Thus, once the variances of *a priori* information are quantified, then the Lagrange multipliers have a clear quantitative interpretation. As shown by Dubovik and King (2000), quantifying the errors in *a priori* constraints is generally a straightforward effort. For example, for applying smoothness constraints on aerosol size distribution, the variance ε_k^2 can be related to the norm of the second derivatives of known mostly unsmooth size distributions. If one would like to directly use *a priori* estimates of the unknowns, the variances ε_k^2 can be derived from known ranges of variability of the sought parameters considering them as confidence intervals.

5.6 Non-negativity constraints

The difficulty of enforcing non-negative solutions is an essential limitation of linear inversion methods. Indeed, the constrained linear inversions defined by Eqs (5.34), (5.37), (5.54), and (5.57) do not have a mathematical structure that allows filtering negative solutions even if the retrieved characteristic is physically positively defined. For example, remote sensing is known to suffer from the appearance of unrealistic negative values for retrieved atmospheric aerosol concentrations that are positive by their nature.

In the framework of a statistical estimation approach, this issue can be addressed by using a lognormal noise assumption in the retrieval optimization. Such an assumption of lognormal noise (instead of conventional normal noise) leads to implementing inversions in logarithmic space, i.e. employing logarithmic transformation of the forward model.

Retrieval of logarithms of a physical characteristic, instead of absolute values, is an obvious way to avoid negative values for positively-defined values. However, the literature devoted to inversion techniques tends to consider this apparently useful tactic as an artificial trick rather than a scientific approach to optimize solutions. Such misconception is probably caused by the fact that the pioneering efforts on inversion optimization by Phillips (1962), Tikhonov (1963), and Twomey (1963) were devoted to solving the Fredholm integral equation of the first kind, i.e. a system of linear equations produced by quadrature.

Considering optical thickness as a function of the logarithm of the aerosol concentrations or temperature profile requires replacing initially linear equations $\mathbf{g} = \mathbf{A}\mathbf{f}$ by nonlinear ones $g_i = g_i(\ln f)$. On the face of it, such a transformation of linear problems to nonlinear ones is not enthusiastically accepted by the scientific community as an optimization. On the other hand, in cases where a forward model is a nonlinear function of parameters to be retrieved (e.g. atmospheric remote sensing in cases when multiple scattering effects are significant), the retrieval of logarithms is readily accepted as a logical approach. Besides, nonlinear Chahine-like iterations (Chahine, 1968; Twomey, 1975) are often considered as efficient alternatives to linear methods for inverting linear system. Rigorous statistical considerations also reveal some limitations of applying Gaussian function for modeling errors in measurement of positively defined characteristics. It is well known that the curve of the normal distribution is symmetrical. In other words, one may affirm that the assumption of a normal PDF is equivalent to the assumption of the principal possibility of obtaining negative results even in the case of physically nonnegative values. For such nonnegative characteristics as intensities, fluxes, etc., the choice of the lognormal distribution for describing the measurement noise seems to be more correct due to the following considerations: (i) log-normally distributed values are positively-defined; (ii) there are a number of theoretical and experimental reasons showing that for positively defined characteristics, the lognormal curve (multiplicative errors, see Edie et al., 1971) is closer to reality than normal noise (additive errors). Also, as follows from the discussion of statistical experiments (Tarantola, 1987), the lognormal distribution is the best for modeling random deviations in non-negatives values.

Similarly to the above considerations of non-negative measurements, there is a clear rationale for retrieving logarithms of unknowns instead of their absolute values (e.g. $\ln(y(x_j))$ instead of $y(x_j)$), provided the retrieved characteristic are positively defined. Although, the MLM does not directly assume a distribution of errors in the final solution, the statistical properties of the MLM solution are well studied (see Edie et al., 1971) and, therefore, can be projected in algorithm developments. In fact, according to statistical estimation theory, if the PDF is normal, the MLM estimates are also normally distributed. It is obvious, then, that the LSM algorithm retrieving $y(x_j)$ would provide normally distributed estimates $y(x_j)$ and, therefore, it can not provide zero probability for $y(x_j) < 0$ even if $y(x_j)$ are positively defined by nature. On the other hand, the retrieval of logarithms instead of absolute values illuminates the above contradiction, since the LSM estimates of $\ln(y(x_j))$ would have the normal distribution of $\ln(y(x_j))$, i.e. lognormal distribution of $y(x_j)$ that assures positivity of non-negative $y(x_j)$. Moreover, the studies by Dubovik and King (2000) suggest considering logarithmic transformation as one of the cornerstones of the practical efficiency of Chahine's iterative procedures. The appendices of the paper by Dubovik and King (2000) demonstrate that these originally empirical equations can be rigorously derived from the MLM method using the assumption of lognormal noise.

Thus, accounting for non-negativity of solutions and/or non-negativity of measurements can be implemented in the retrieval by using logarithms of unknowns ($f_j \rightarrow \ln f_j$) and/or measurements ($g_i \rightarrow \ln g_i$). In many situations, retrieval of absolute values or their logarithms is practically similar. This is because narrow lognormal or normal noise distributions are almost equivalent. For example, for small variations of a non-negative value, the following relationship between Δf and $\Delta f/f$ is valid:

$$\Delta \ln f = \ln(f + \Delta f) - \ln(f) \approx \Delta f / f \; (\text{if } \Delta \ln f \ll 1). \tag{5.62}$$

Thus, if only small relative variations of value a are allowed, the normal distribution of $\ln f$ is almost equivalent to the normal distribution of absolute values f. The covariance matrices of these distributions are connected as:

$$\mathbf{S}_{\ln f} \approx (\mathbf{I}_f)^{-1} \mathbf{S}_f (\mathbf{I}_f)^{-1}, \tag{5.63}$$

where $\mathbf{I}_f$ is a diagonal matrix with the elements $\{\mathbf{I}_f\}_{jj} = f_j$. Hence, for measurements with small relative errors, use of lognormal or normal PDFs with covariance matrices related as given by Eq. (5.63) should give close results. Also, since logarithmic errors can be approximately considered as relative errors, the variances $(\varepsilon_{\ln g})^2$ are unitless and, therefore, Eq. (5.61) defining Lagrange multipliers as variance ratios becomes particularly useful. Practical illustrations of using logarithmic transformations in inversion can be found in the papers by Dubovik and King (2000), and Dubovik et al. (2000, 2011).

5.7 Smoothness constraints in the case of retrieval of characteristic with nonuniformly-spaced ordinates

The utilization of smoothness constraints for a single retrieved function $y(x_i)$ was originated in the papers by Phillips (1962), Tikhonov (1963), and Twomey (1963). Although application of smoothness constraints is usually considered to be an implicit constraint on derivatives, in these original papers and most of the follow-on studies, the solution vector $\mathbf{f}$ was constrained by minimizing the m-th differences Δ^m of the vector $\mathbf{f}$ components:

$$
\begin{aligned}
\Delta^1 &= f_{j+1} - f_j, & (m = 1), \\
\Delta^2 &= f_{j+2} - 2f_{j+1} + f_j, & (m = 2), \\
\Delta^3 &= f_{j+3} - 3\,f_{j+2} + 3f_{j+1} - f_j, & (m = 3).
\end{aligned}
\tag{5.64}
$$

As shown by Eq. (5.30), the corresponding smoothing matrix $\mathbf{H} = \mathbf{K}_m^{\mathrm{T}} \mathbf{K}_m$ was defined using matrices of the m-th differences $\mathbf{K}_m$ (i.e. $\Delta^m = \mathbf{K}_m \mathbf{f}$).

The developments by Dubovik and King (2000) and Dubovik (2004) suggest formulating the smoothness constraints in Eq. (5.56) explicitly as *a priori* estimates of the derivatives of the retrieved characteristic $f_j = y(x_j)$. The values of m-th derivatives z_m of the function $y(x)$ characterize the degree of its non-linearity and, therefore, can be used as a measure of $y(x)$ smoothness. For example, smooth functions $y(x)$, such as a constant, straight line, parabola, etc., can be identified by the m-th derivatives as follows:

$$z_1(x) = dy(x)/dx = 0 \qquad \Rightarrow y_1(x) = C;$$
$$z_2(x) = d^2y(x)/dx^2 = 0 \qquad \Rightarrow y_2(x) = Bx + C; \qquad \qquad (5.65)$$
$$z_3(x) = d^3y(x)/dx^3 = 0 \qquad \Rightarrow y_3(x) = Ax^2 + Bx + C$$

These derivatives z_m can be approximated by differences between values of the function $a_i = y(x_i)$ in N_a discrete points x_i as:

$$\frac{dy(x_{i'})}{dx} \sim \frac{\Delta^1 y(x_i)}{\Delta_1(x_i)} = \frac{y(x_i + \Delta x_i) - y(x_i)}{\Delta_1(x_i)} = \frac{y(x_{i+1}) - y(x_i)}{\Delta_1(x_i)},$$

$$\frac{d^2 y(x_{i''})}{dx^2} \sim \frac{\Delta^2 y(x_i)}{\Delta_2(x_i)} = \frac{\Delta^1 y(x_{i+1})/\Delta_1(x_{i+1}) - \Delta^1 y(x_i)/\Delta_1(x_i)}{(\Delta_1(x_i) + \Delta_1(x_{i+1}))/2}, \qquad (5.66)$$

$$\frac{d^3 y(x_{i'''})}{dx^3} \sim \frac{\Delta^3 y(x_i)}{\Delta_3(x_i)} = \frac{\Delta^2 y(x_{i+1})/\Delta_2(x_{i+1}) - \Delta^2 y(x_i)/\Delta_2(x_i)}{(\Delta_2(x_i) + \Delta_2(x_{i+1}))/2},$$

where

$$\Delta_1(x_i) = x_{i+1} - x_i,$$

$$\Delta_2(x_i) = (\Delta_1(x_i) + \Delta_1(x_{i+1}))/2,$$

$$\Delta_3(x_i) = (\Delta_2(x_i) + \Delta_2(x_{i+1}))/2,$$

$$x_{i'} = x_i + \Delta_1(x_i)/2,$$

$$x_{i''} = x_i + (\Delta_1(x_i) + \Delta_2(x_i))/2,$$

$$x_{i'''} = x_i + (\Delta_1(x_i) + \Delta_2(x_i) + \Delta_3(x_i))/2.$$

In contrast to Eq. (5.64), Eq. (5.66) allows for applying smoothness constraints in more general situations when $\Delta_i(x_i) \neq const$. For example, in the algorithm by Dubovik and King (2000), the retrieved real $m_r(\lambda)$ and imaginary $m_i(\lambda)$ parts of the complex refractive index are functions of λ and the algorithm deals with their values defined for each spectral channel λ_i. Obviously the $(\lambda_i + 1 - \lambda_i) \neq const$ and using the standard definition of differences from Eq. (5.64) for smoothing spectral parameters (e.g. $m_r(\lambda)$ and $m_i(\lambda)$) is not completely correct. Applying the limitations on the derivatives defined by Eq. (5.66) is more rigorous if no significant changes of derivatives of $y(x)$ are expected for different ordinates x_i.

If in the second line of Eq. (5.56), $\mathbf{Kf}$ denotes the second derivatives instead of second differences of the solution vector $\mathbf{f}$, then the corresponding matrix $\mathbf{K}$ and smoothing matrix $\mathbf{H}_z = \mathbf{K}^T\mathbf{K}$ can be defined for second derivatives $(z_2(x) = d^2 y(x)/dx^2)$ as follow

$$
\mathbf{K}_2 =
\begin{pmatrix}
0 & 0 & 0 & 0 & 0 \\
0 & 0 & 0 & 0 & 0 \\
\dfrac{2}{\Delta_1(\Delta_1+\Delta_2)} & \dfrac{-2}{\Delta_1\Delta_2} & \dfrac{2}{\Delta_2(\Delta_1+\Delta_2)} & 0 & 0 \\
0 & \dfrac{2}{\Delta_2(\Delta_2+\Delta_3)} & \dfrac{-2}{\Delta_2\Delta_3} & \dfrac{2}{\Delta_3(\Delta_2+\Delta_3)} & 0 \\
0 & 0 & \dfrac{2}{\Delta_3(\Delta_3+\Delta_4)} & \dfrac{-2}{\Delta_3\Delta_4} & \dfrac{2}{\Delta_4(\Delta_3+\Delta_4)} \\
\cdot & \cdot & \cdot & &
\end{pmatrix}.
$$

$$(5.67)$$

where $\Delta_i = \Delta_i(x_i) = x_{i+1} - x_i$; and

$$
\mathbf{H}_z =
\begin{pmatrix}
\dfrac{4}{(\Delta_1)^2(\Delta_1+\Delta_2)^2} & \dfrac{-4}{(\Delta_1)^2(\Delta_1+\Delta_2)\,\Delta_2} & \dfrac{4}{\Delta_1\Delta_2(\Delta_1+\Delta_2)^2} & \cdot \\[2ex]
\dfrac{-4}{(\Delta_1)^2(\Delta_1+\Delta_2)\Delta_2} & \dfrac{4}{(\Delta_1\Delta_2)^2}+\dfrac{4}{(\Delta_2)^2(\Delta_2+\Delta_3)^2} & \dfrac{-4}{\Delta_1(\Delta_2)^2(\Delta_1+\Delta_2)}+\dfrac{-4}{(\Delta_2)^2\Delta_3(\Delta_2+\Delta_3)} & \cdot \\[2ex]
\dfrac{4}{\Delta_1\Delta_2(\Delta_1+\Delta_2)^2} & \dfrac{-4}{\Delta_1(\Delta_2)^2(\Delta_1+\Delta_2)}+\dfrac{-4}{(\Delta_2)^2\Delta_3(\Delta_2+\Delta_3)} & \dfrac{4}{(\Delta_2)^2(\Delta_1+\Delta_2)^2}+\dfrac{4}{(\Delta_2\Delta_3)^2}+\dfrac{4}{(\Delta_3)^2(\Delta_2+\Delta_3)^2} & \cdot \\[2ex]
0 & \dfrac{4}{\Delta_2\Delta_3(\Delta_2+\Delta_3)^2} & \dfrac{-4}{\Delta_2(\Delta_3)^2(\Delta_2+\Delta_3)}+\dfrac{-4}{(\Delta_3)^2\Delta_4(\Delta_3+\Delta_4)} & \cdot \\[2ex]
0 & 0 & \dfrac{4}{\Delta_3\Delta_4(\Delta_3+\Delta_4)^2} &
\end{pmatrix}.
$$

$$(5.68)$$

One can see that $\mathbf{H}_z$ has a significantly more complex structure than the conventional definition of smoothing matrix given by Eq. (5.30). At the same time, for the special case of uniformly spaced ordinates $\Delta_i = \Delta = const$ the more general matrix $\mathbf{H}_z$ has a straightforward relation to the conventional $\mathbf{H}$ defined by Eq. (5.30):

$$\mathbf{H}_z = \Delta^{-4}\mathbf{H}. \qquad (5.69)$$

Although using Eq. (5.66) leads to a loss of transparency in the definitions of matrices $\mathbf{K}_m$, generating those matrices $\mathbf{K}_m$ and $\mathbf{H}_z$ on an algorithmic level is rather straightforward, and defining the Lagrange parameters γ_k can be logical, based on existing knowledge regarding derivatives of retrieved function $y(x)$ (see Dubovik et al., 2011).

5.8 Inversion of aerosol size distribution and single scattering albedo from spectral optical thickness and sky radiance measurements

The method for simultaneously determining the columnar volume size distribution and spectral refractive index from spectral optical thickness and almucantar sky radiance measurements has been applied to the ground-based AERONET sun/sky radiometers obtained in Mongu, Zambia during the dry (biomass burning) season. The spectral optical thickness data used in this multisource inversion were restricted to only four wavelengths (0.441, 0.674, 0.870, and 1.02 μm). This is because we have angular sky radiance measurements in the almucantar at these same four wavelengths. It is not necessary to restrict the inversion to this limited set of optical thickness observations, but in this case, where spectral almucantar measurements are available over a wide range of scattering angles, sufficient sensitivity to the size distribution of small particles exists making it unnecessary to add spectral optical thickness measurements at shorter wavelengths.

Figure 5.9 shows an example of the application of our multisource constrained linear inversion method to spectral optical thickness and spectral and angular sky radiance measurements obtained in Mongu, Zambia on 18 August 2006 at 6:59:11 UTC when the solar zenith angle $\theta_0 = 59.6°$. The observed aerosol optical thickness measurements and corresponding standard deviations are shown in Figure 5.9a, and the nearly simultaneous spectral and angular transmission measurements and corresponding standard deviations obtained in the almucantar are shown in Figure 5.9b. The almucantar measurements shown here were obtained at 26 discrete azimuth angles (and translated to scattering angles), as shown here with symbols and error bars at each of the four wavelengths for which sky radiance measurements were obtained.

The volume size distribution obtained by inverting these data is shown in Figure 5.9c, together with the spectral values of real and imaginary refractive index shown in Figure 5.9d. Uncertainties in the derived size distribution and refractive index are also shown in Figures 5.9c and 5.9d, and the solid curve in Figure 5.9a indicates how the inverted size distribution is able to reproduce the $\tau_{aer}(\lambda)$ measurements (i.e. the direct problem $\mathbf{g} = \mathbf{Af}$). The smooth curves in Figure 5.9b similarly show how the retrieved size distribution and refractive index permit a very good reproduction of the sky radiance measurements. Comparing Figure 5.9c with Figure 5.7b shows that the retrieved columnar aerosol size distribution for the fine mode aerosols ($r < 1$ μm) is similar in both cases, but that the spectral optical thickness measurements alone were incapable of deriving the second (coarse particle) mode centered at 5 μm. This is not surprising, since the spectral extinction kernels illustrated in Figure 5.1b show virtually no sensitivity to these larger particles. This clear coarse particle peak arises, in this case, from the information contained in the angular sky radiance measurements near the sun (scattering angles less than ~10°) (cf. Figure 5.3b).

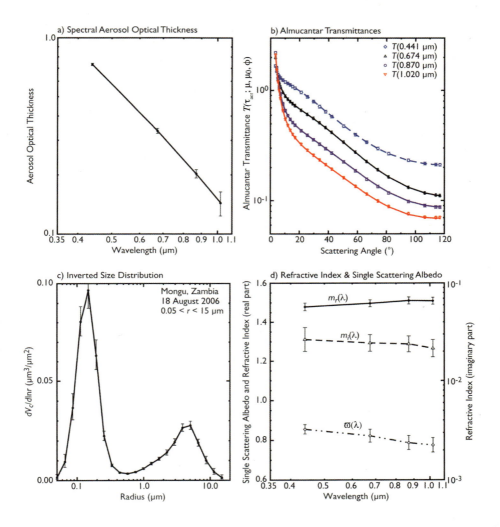

Figure 5.9 (a) Spectral aerosol optical thickness on 18 August 2006 in Mongu, Zambia as measured by the AERONET sun/sky radiometer when the solar zenith angle $\theta_0 = 59.6°$. (b) Transmission function $(\pi L/\mu_0 E_0)$ as a function of scattering angle in the almucantar as measured by AERONET at the same four wavelengths as the spectral optical thickness was measured at in (a). (c) Columnar volume size distribution derived by inverting the $\tau_{aer}(\lambda)$ spectrum together with the spectral almucantar transmission functions in (a) and (b). (d) Real and imaginary parts of the complex refractive index of aerosols, together with the derived single scattering albedo, at four wavelengths derived by inversion of the optical thickness and almucantar measurements shown in (a) and (b). Also shown is the regression fit to the data using the inverted size distribution. The solid curve in (a) is the $\tau_{aer}(\lambda)$ spectrum computed from the size distribution in (c), and the solid curves in (b) are the angular transmission functions computed from the size distribution in (c) and refractive index in (d).

The retrieved size distribution shown in Figure 5.9c is characteristic of fresh smoke with small particles, but the uncertainty in the size distribution of small particles is increasingly large due to the reduced sensitivity to small particles using both spectral optical thickness measurements (as shown in Figure 5.7b) and angular sky radiance measurements (where the information for small particles comes from larger scattering angles). The small uncertainties in the derived size distribution in the coarse particle mode is due exclusively to the fact that the sky radiance measurements were obtained at small scattering angles (3.42°, 4.28°, 5.13°, 5.98°, and 6.84°) and were decidedly wavelength dependent.

Figure 5.9d shows the spectral dependence of both the real and imaginary parts of the complex refractive index of aerosol particles. In this case, the real refractive index of the largely biomass burning aerosols ranged between 1.48 and 1.51 (depending on wavelength), whereas the imaginary refractive index ranged from 0.0265 to 0.0215. The uncertainties in these retrievals, as reflected in the magnitude of the retrieved error bars, is quite small for the real refractive index and somewhat larger for the imaginary refractive index. These uncertainties are very much a function of the magnitude of the aerosol optical thickness, and are smaller (better determined) when the aerosol optical thickness is large, since there is much greater sensitivity to aerosol absorption.

Finally, Figure 5.9d shows the derived spectral single scattering albedo, which is computed (not retrieved directly) based on the real and imaginary parts of the complex refractive index of the aerosol, together with the aerosol size distribution.

In deriving the aerosol size distribution, we used a second-derivative smoothing constraint on the volume size distribution, but for the refractive index we used a first-derivative smoothing constraint based on wavelength dependence. The radii retrieved in the size distribution were equally spaced in $\log r$, and hence the smoothing $\mathbf{H}$ matrix described earlier applies, whereas for the smoothing constraint on spectral refractive index, the wavelengths are not evenly spaced, as discussed above (for the example of second derivative smoothing constraint). This is easy to implement numerically, as done here.

6 Passive shortwave remote sensing from the Ground

Glenn E. Shaw, Brent N. Holben, Lorraine A. Remer

6.1 Introduction

Deducing and attempting to quantify properties of aerosols and trace atmospheric gases from ground-based sensors on the Earth's surface has a long and interesting history that provided the foundation for the general field now known as "Remote Sensing". The method utilizes measurements of the attenuation of light passing through the atmosphere, often taken at multiple wavelengths and sometimes broken down into spectral high resolution. Today, the method based on measuring attenuation of the direct beam is supplemented with measures of the intensity (radiance) or polarization of scattered light from the sun taken at different angles. Using the direct beam measurements to infer aerosol optical thickness is described in Chapter 4. The inversion methods applied to both direct beam and to angular measurements of the diffuse light are described in Chapter 5. In this chapter, we provide a history of ground-based aerosol remote sensing networks, leaving the formalism of the retrievals and inversions to the previous chapters. Not described in detail in this book are methods to retrieve trace gas measurements using absorption features measured by ground-based sensors, even though the history of trace gas measurements and aerosol measurements go hand-in-hand. Remote sensing by ground-based sensors can provide information about the spectral aerosol optical thickness, particle size distribution, aerosol absorption and degree of nonsphericity of the particles. The techniques often impose great demands on precision and accuracy. The history of ground-based remote sensing networks is plagued with good intentions that were stymied by the lack of consistent and reliable calibration that could assure the necessary precision and accuracy.

6.2 Early history

The basis of remote sensing in the atmosphere begins with experiments on the nature of sunlight. In 1666, Isaac Newton found by experimenting with prisms that sunlight itself is by no means pure and basic, but consists of a population of entities that the eye perceives as colors. These "colors" are evidently unalterable. Though a beam of white light can be broken up into a spectrum, Newton found that by selecting one small section of the spectrum and allowing it to pass through a second slit that the then purified component could no longer be broken up with a prism. Thus began the science of spectroscopy, and by 1752 Thomas Melvill discovered in an example of pure curiosity-driven research that chemical salts, when brought to incandescence emit light at specific colors, or wavelengths. In 1802, William Wollaston found accidentally by using a higher resolution spectrometer than Newton's simple prism that the sunlight spectrum had gaps. Joseph Fraunhofer further investigated these gaps in 1814 and identified more than a thousand of them. Joseph Foucault discovered to his amazement in 1849 that two of these gaps were identical to the colors of the split up sodium D emission lines. This important discovery demonstrated that the sun contains compounds identical to those found here on earth. This discovery was made, not by sampling substances on the sun, but by analyzing the radiation field coming from the sun. This discovery heralded what today we call the field of "remote sensing".

In 1855 Anders Ångström (1814–1874), son of a country minister, followed up on these discoveries in Sweden. He investigated the spectrum of hydrogen, which a half century later would go on to be "explained" by a model of the atom introduced by Neils Bohr in 1913. Ångström continued on with his spectroscopic studies, eventually as Professor and Chair of the Physics Department in one of Europe's oldest universities at Uppsala, investigating the solar spectrum of the sun and, a year or so before he died, the spectrum of the aurora. In both of these luminous bodies he found "spectral lines", now identified as light at specific oscillatory frequencies of electromagnetic radiation through the theories of Maxwell and Hertz. He expressed these as wavelengths in terms of the tiny 10^{-10} m, a unit later named in his honor. Again, lines from known substances, especially hydrogen, were found, and also lines of a substance, helium, then not known on Earth, and therefore named after the sun.

Ångström's son, Knut Johan (1857–1910), followed his father's interests and studied at Stockholm University, then went to Uppsala and by 1896 was also Professor of Physics. Like his father, he was interested in the light of the sun, but he also became interested in the "heat" from the sun. By this he meant not only carrying out investigations into solar radiation going into the near infrared, but in trying to evaluate the spectral intensity in terms of the light's "energy" falling on the Earth. For this, in 1893, he invented and described an electrical substitution pyroheliometer. In this device, collimated sunlight passing through an aperture fell on a blackened strip, heating it. Next to this strip was an identical strip shadowed from the sun, but heated by an electrical current. The power necessary to bring the strip to the same temperature as the strip heated by the solar radiation provided a quantitative measure of solar radiation flux in Wm^{-2}. The instrument went through several modifications and various small sources of systematic error, such as edge effects, were investigated and quantified. The instrument was so good that it is still being used today in some observatories. Anders K. Ångström (1888-1981) continued the work of his grandfather and of his father

Despite the advances in instrumentation in the 19th century, the primary remote sensing instrument was and is the human eye. Applying our eyes to viewing the atmosphere we retrieve basic qualitative information: (1) the atmosphere diminishes the intensity of solar radiation; (2) the dimming is larger in the blue or violet than at longer wavelengths ; and (3) the dimming is worse when viewed in polluted places than from mountaintops. Thus at sunset or viewed through a highly polluted atmosphere the sun appears dim and red. The earliest investigations using remote sensing instruments confirmed and quantified these basic facts achieved by simple eye sight. For example, in the 19th century, 200 years after Newton's early work, it was established that the diminishing effect of the atmosphere is actually more complicated than a smooth blue-to-red gradient and that certain wavelengths of sunlight are strongly absorbed by individual trace gases in specific wavelength bands. To complicate things even further, the diminution or "turbidity" varies from one day to another. Assessing this "diminution" nuisance became especially important in quantifying incoming solar radiation reaching the Earth. Quantifying solar radiation had become a subject of great interest because it was suspected that the sun may undergo changes in its luminosity that would modulate climate and weather and the growing of crops.

The science of sun photometry actually emerged by turning this atmospheric "nuisance" correction factor on its head and into information that had interest in and of itself. We call this nuisance correction factor "turbidity". It was early on identified as a useful measure of clarity of the air or of the state of pollution of a particular location.

6.3 The Smithsonian program

In the United States, the Director of the Smithsonian Institute in the late 19th century, Samuel P. Langley, had also become interested in the radiation from the sun and the solar "constant" (which was then widely believed to be anything BUT constant). He invented a detector called the bolometer, which was sensitive enough to detect radiation in a narrow band of frequencies passing through a monochrometer (Langley, 1884). In addition, he and his protégé, Charles Greeley Abbott, came up with a pyranometer that measured the temperature rise of a blackened silver disk upon which collimated sunlight heated. This, in turn, was calibrated against a standard in which flowing water was heated by solar radiation.

Both Langley and Ångström recognized that not only does the atmosphere diminish solar radiation, but that it does so unequally at different wavelengths, what Newton had referred to as the irrefrangibility of the atmosphere. This introduced an enormous complication into the determination of the solar "constant", because to derive accurate measures of the extraterrestrial solar radiation (in Wm^{-2}) it would be necessary to derive, wavelength by wavelength, the atmospheric attenuation. For this, Langley used his bolometer to detect light radiation in narrow spectral regions and invented a graphical method based on Pierre Bouguer's "Essay on the Gradation of Light" (1729), in which he put forward the exponential law of absorption. The Smithsonian "Long Method" (Abbot, 1935; Abbot and Aldrich, 1913) involved taking numerous measurements with the bolometer on top of high mountains at different times during the day, at different solar elevations and at 20 wavelengths, and plotting the logarithm of solar spectral intensity, for each wavelength, against the relative quantity of atmosphere traversed in terms of that traversed by a downward directed beam (the solar air mass). Examples of Langley Plots, the governing equation, definiton of solar air mass, etc., are given in Chapter 4. The slope of the Langley Plots represented the atmospheric extinction and the intercept of the extrapolated linear line with the ordinate is the reading of the bolometer for zero atmospheric air mass. By combining the now known atmospheric transmission with a total solar radiation pyranometer measurement, Abbott and Langley could correct the ground-based, high altitude mountaintop readings and derive accurate estimate of the solar constant (Abbot, 1935; Abbot and Aldrich, 1913). Roosen et al. (1973) re-analyzed the extinction measurements of the Smithsonian data and concluded that, during the first half of the 20th century, the aerosol showed no detectable change over the high altitude observatories.

The Smithsonian work, which extended over a half century, provided thousands of measured values of the atmospheric extinction and represents the first widespread highly quantitative sun photometric measurements in a program whose accuracy and quality have rarely (if ever) been matched since (Roosen and Angione, 1984). The data is available in the Annals of the APO of the Smithsonian Institution.

6.4 Early turbidity investigations

Following the pioneering work by Abbot, Langley and Ångström, improvements were made in portable sun photometer instruments. One improvement was the use of thermopile detectors in conjunction with optical filters, such as the glass colored filters manufactured by Schott Glass in Germany (Ångström, 1970 a,b). These cut-off glass filters, when used with the new electrical thermodynamic detectors, made it relatively easy to sort out, at least roughly, the atmospheric attenuation of light at specific narrow wavelength intervals (Ångström, 1929, 1932). The new now quite portable sun photometers made it possible for people urged by curiosity to measure turbidity from mountaintops and other out-of-the-way places. By the mid 20th century, a large body of data existed on the turbidity of the atmospheric column. In the United States, for instance, Flowers and Viebrock (1965, Flowers et al., 1969) analyzed information from dozens of stations, finding the highest values of optical depth in the East Coast and the lowest in the intermountain West. Volz (1969) published data on the turbidity from 30 stations in Europe, its wavelength variations for a large number of regions and commented on possible trends, which were found to be increasing from increasing industrial activity. Old data from Uppsala and surrounding regions were analyzed and discussed by Volz (1968) who demonstrated turbidity increases also in Europe. Though the number of measurements was increasing, these early instruments were limited in accuracy because of the relatively broad wavelength bands obtained by subtracting readings through glass cut-off filters and because of nonlinearities and drifts in detectors.

Using the thermopile detector/glass filter technology, turbidity measurements were carried out at a wide variety of locations. For example, at Mt Lemmon in Arizona (Dunkleman and Scolnick,1959), in Antarctica (Herber et al.,1993), and from ships at sea. For a review of maritime aerosol extinction measurements observed from cruises spanning 30 years, see Smirnov et al. (2002).

6.5 Technological advances

In the early 1970s, Glenn Shaw, John Reagan and Ben Herman at the Institute of Atmospheric Physics at the University of Arizona introduced the then relatively new PIN-doped optical detectors with high quality low light leakage interference filters manufactured by Barr Associates (Shaw, 1971). This combination led to portable filter-detector units, which have proven to be very stable, drifting in calibration at most usually a few percent per year. However, at very low solar elevation angles, for filters in short wavelength regions, leakage of longer wavelength light can become a problem. Other calibration issues are nonlinearity of the detector and sensitivity to temperature changes by detectors or filter optical characteristics. The new interference filter-solid state detector combination sun photometers could be easily and robustly calibrated using the Langley method at high elevation stations, or by using lamps traceable to a known standard. The standard lamp method is a useful way to transfer calibration from a unit that has been well calibrated by the Langley method to additional field photometers (Schmid et al., 1998). This design was incorporated into a network across India (Krishna Moorthy et al., 1989).

Some mention here might be made of the number and location of wavelength bands suitable for sunphotometry. Bands are chosen to measure solar radiation within "window regions" of the solar spectrum, where water vapor and trace gas absorption is minimum unless measurement of the gas is the goal. Typical sun photometer channels include bands at or near 340, 380, 440, 470, 500, 550, 670, 870, 1020, 1630 nm. The near infrared bands (particularly 1020 silicon detector bands) are susceptible to temperature flucuations and should be thermally isolated if possible. Particular attention to out-of-band blocking is necessary for UV interference filters due to the low energy level in the UV region, especially at 340 nm. The 940 nm channel is also a common sun photometer channel choice. It is not in a spectral window region, but in a water vapor absorption band. Measurements at 940 nm when compared with another channel such as 1020 nm provide information on the light absorption by water vapor and can be related to total column water vapor amounts. To characterize the wavelength dependence of aerosol optical thickness, there is debate concerning the number of wavelength bands to include in measurements. Originally it was demonstrated that 4 wavelength bands distributed over the solar spectrum is sufficient to provide information on aerosol particle size if the aerosol size distribution can be represented by a power law distribution or wide lognormal distribution. Fröhlich (1977) provided guidelines and recommended WMO standard wavelength bands for use in portable sun photometers. More recent studies and modern inversions show that the aerosol is almost always multi-modal and that broad power law or single lognormal distributions are inadequate (Dubovik et al., 2002). There is important information in the curvature of the spectral extinction aerosol optical thickness about the size of the particles (Eck et al., 1999;

O'Neill et al., 2003) so that increasing the number of measured wavelengths into the UV (380 nm) is advised.

6.6 BAPMoN, the satellite era and prototypes for the modern networks

Technological advances resulted in a proliferation of regional networks and individual observations from the 1980s and 1990s that had goals of aerosol characterization from a variety of instruments in many locations around the world. Most were never published and the hand-scribed data will likely never be digitized for analysis on modern computers. Regional networks such as Prodi et al. (1984) (Italy); Ingold et al. (2001) (Switzerland); d'Almeida (1991) (North Africa) ; Krishna Moorthy et al. (1989) (India) were among others that were affiliated with individual national meteorological services participating in the Background Air Pollution Monitoring Network (BAPMoN) program. BAPMoN organized the archiving of data from various sites and networks to present them as a semi-unified global network of sun photometer observations of aerosol optical depth. Despite attempts to minimize measurement variability (Meszaros and Whelpdale, 1985), the global concept was not realized because

> "…serious flaws in both the internal consistency of data from any single station and the comparability among any specific set of station records. Data records prior to 1982, when measurements were made using the Volz sunphotometer, appear to be comparatively consistent although still of questionable value. Data from later periods are fraught with even more substantial problems of internal consistency." (WMO, 1990).

BAPMoN never delivered a global database of consistent and high quality aerosol optical thickness, although the national sun photometry efforts continued through the 1980s and into the 1990s. The problem was a lack of science-driven measurement initiatives to understand basic aerosol properties and how aerosols interacted with the earth–atmosphere system. Without compelling science driving the measurements, the field was in disarray and poorly funded. Although some programs generated data of high quality, overall the anemic aerosol field was fragmented and represented a low point in institutional support for aerosol research. This would change with the advent of the satellite era, when once again aerosols became a "nuisance" that had to be removed to achieve a pure signal of the parameters of interest.

In the 1970s the advent of the ERTS (Landsat) and later AVHRR, satellites gave a new perspective of the Earth's surface and created new disciplines such as satellite oceanography and satellite vegetation monitoring. The goal of this new tool was to obtain quantitative information of the surface reflectance of the Earth (R from Eqs 3.32 and 3.33), and then use this information to derive information about aquatic and terrestrial ecosystems, agriculture, geology, etc. The problem is that the satellites measured reflected radiance at the top of the atmosphere (L_R from Eq. 3.33) that contained information from both the Earth's surface and from the atmosphere that separated the target from the satellite. Clouds created the primary interference, but aerosols were important also. The broad spectral radiance variations in the imagery due to aerosols and their rapid change in temporal and spatial concentrations were a significant cause for confounding Earth surface observations. As more spectral imagery was acquired, scientists with surface applications in mind began to have the luxury to select cloud free and "clear" images of the Earth's surface to better interpret satellite imagery. Strategies were devised, assessed and implemented to minimize the effects of aerosols such as compositing images (Holben, 1986) or removing the aerosol signal from the imagery (Kaufman and Sendra, 1988; Kaufman and Tanré, 1996). However, these strategies could not compete with a quantitative atmospheric correction based on Eq. 3.33 which returned the surface reflectance R, given the radiance measured at the top of atmosphere L_R and knowledge of the aerosol optical depth τ^* and optical properties. The trick was to measure τ^* with sufficient characterization of the aerosol using field measurements.

The first sunphotometer network designed to measure τ^* specifically for atmospheric correction was established in 1983 and ran through 1986 in West Africa as part of NASA's Global Inventory Modeling and Mapping Studies (GIMMS) program to assess biomass productivity in the Sahel zone (Holben et al., 1991). This multi-year period was marked by extreme drought in the Sahel. Observers reported the solar disc was frequently obscured by dust, and annual dust AOD loadings exceeded 1 for most sites during drought years. It was notable that the extreme conditions of the Sahel were a motivation for searching for an automated method of acquiring the ground-based data.

From 1987 to 1995, handheld sunphotometers making use of interference filters and silicon detectors continued to be deployed during field campaigns, mostly in support of atmospheric correction. For example, sunphotometers were deployed during the First ISLSCP (International Satellite Land Surface Climatology Project) Field Experiment (FIFE) (Hall et al., 1991) and the Boreal Ecosystem Atmosphere Study (BOREAS) (Sellers et al., 1995). The primary interest of these experiments was vegetation dynamics and carbon sequestration in different ecosystems. However, researchers made opportunity of the sunphotometer measurements to stress the inherent value of spectral aerosol optical depth measurements to characterize the atmosphere (Bruegge et al., 1992; Markham et al., 1997). Calibration issues were solved with frequent trips to high altitude calibration sites to perform Langley Plots (Chapter 4) and even more frequent inter-comparisons between instruments to transfer calibration. Drift in filter response was monitored and corrected for. With care and proper observer training, these instruments could achieve an accuracy of 0.02 in aerosol optical depth (τ^*).

While the handheld sunphotometers were being used to measure the direct solar beam at the ground and infer the spectral aerosol column extinction, another type of measurement strategy was in development by Maurice Herman and Claude Deveaux (Santer and Herman, 1983; Deuzé et al., 1988; Deveaux et al., 1998) in France and Masayuki Tanaka (Tanaka et al., 1983; Nakajima et al., 1983) in Japan in the early 1980s. This other instrument could measure sky radiance at angles and directions away from the direct beam. Sky radiance contains information of the aerosol scattering properties as the light is scattered out of the direct beam and then down to the observer by particles and gas molecules. The instrument, solidly anchored to the ground, would be accurately pointed at the sun, then under motor control rotated through a series of azimuth angles spanning 360 degrees, keeping the zenith angle constant. This is the definition of an almucantar scan. Various inversion schemes were developed to use this angular information; however, Nakajima et al, (1996 ; Chapter 5) provide the first "optimized" inversion code that allowed reasonably fast processing. The motorized "laboratory" instruments of Tanaka and Herman were adapted to a manual pointing sunphotometer originally developed by James Spinhirne at NASA's Goddard Space Flight Center. Under Yoram Kaufman's measurement strategy, the newly-named sun sky spectral radiometer was deployed in a wide variety of situations and returned the first remote sensing inversion characterization of particle size distribution in selected sites in North America, Western Europe and the eastern Mediterranean. Kaufman et al. (1994) resolved the unusually-sized aerosol mode of stratospheric particles following the Pinatubo eruption in 1992 (Kaufman et al., 1994 ; Holben et al., 1996, and Chapter 7). The angular sky radiance instrument was not reproduced or deployed as a network, but it did serve as a prototype for networks that followed.

The dedicated operators of the handheld sunphotometers and the ground-based sun sky radiometer provided important and necessary aerosol characterization through the 1980s and early 1990s. However, in order to obtain a measurement, people had to go out into the field, often in inhospitable and fatiguing conditions. There was need for the development of a robotically controlled autonomous sun and sky scanning photometer.

6.7 AERONET and other modern networks

A global network of autonomous sun/sky radiometers (developed by Cimel Electronique), supported by NASA's Terrestrial Ecology program, began as two and then twelve instruments in 1993 to provide better atmospherically corrected data and quantify particle size distributions, primarily from biomass burning in the Amazon Basin. These instruments were also deployed during the Sulfate Clouds And Radiation – Atlantic (SCAR-A) experiment in the mid-Atlantic region of the U.S., to not only provide atmospheric correction, but also to characterize aerosol properties (Remer and Kaufman, 1998). Similar sun/sky radiometers were deployed in West Africa and France in 1992, making up the PHOTONS network, administered by the Laboratoire d'Optique Atmosphérique (LOA) at the University of Lille in France.

These deployments in the early to mid-1990s demonstrated the value of a well-maintained standardized sun/sky radiometer network for atmospheric correction, characterizing aerosol properties, and most importantly to serve as ground-truth for satellite retrievals. The mid-1990s was a time when the Earth Observing System (EOS) satellite program was in development. The science team for the MODerate resolution Imaging Spectroradiometer (MODIS) instrument, scheduled to fly on NASA's EOS Terra and Aqua satellites, made a strong argument for investment in a global autonomous sun/sky radiometer network for the purposes of aerosol algorithm development before launch, and validation of aerosol and atmospheric correction products after launch. Thus, the Aerosol Robotic Network was created from the initial 12 instruments deployed in 1993. The PHOTONS network soon linked with AERONET, as did AEROCAN, a Canadian network.

The most commonly-used autonomous sun/sky radiometer was developed by the Cimel Electronique company in France. Different instruments have different wavelength configurations. However, there is relative standardization so that a 440 nm channel in one instrument, for example, can be compared with a 440 nm channel in another instrument. The most common configuration is $340, 380, 440, 500, 675, 870, 940$ and 1020 nm, and all instruments have the $1020, 870, 675$ and 440 nm channels. The filter bandwidth is 10 nm for all visible and near-infrared channels and from 2 to 4 nm for the UV channels. Some Cimels include a 1640 nm channel and some have fewer wavelengths, but include polarization; however, these single spectral channel (870 nm) polarization instruments have been decommissioned from the network.

Rapid growth in the number of AERONET sites, improved data quality and additional available data products have helped make the network a gold standard of global aerosol research, branching beyond a support mechanism for satellite products to being an essential provider of aerosol characterization for a wide variety of applications.

The AERONET design was meant to overcome the failings of BAPMoN and the difficulties of the handheld and manually operated networks. The foundation of AERONET was based on imposing standardization of:

- **Instrumentation** – All instruments would be of the same design, employ automatic data collection and transmission, be functionally autonomous and solar powered for remote sites, and use an identical measurement protocol

- **Calibration** – Published and otherwise scientifically accepted calibration procedures are standardized and rigorously applied to traceable reference standards at AERONET calibration centers. This results in uniform high accuracy calibration.

- **Data processing** – All data are processed according to scientifically published and accepted algorithms at the AERONET processing center. This results in uniform near real-time processing and analysis of the data.

These three points were sufficient to overcome the failings of previous networks; however, AERONET adopted the philosophy of a federated network and a public domain database available through the internet. Federation allows any entity to participate as long as they meet the standardization requirements and agree to make their data available to the scientific community. No exchange of funds is required to participate in AERONET, thus the program is able to grow with scientific resources from all entities interested in participating, and this has allowed AERONET to build a large data archive of quality-assured products. The AERONET concept of a federation with an open data base has been copied by other networks having different objectives and using different measurement techniques. Such copies include MPLNet lidars (http://aeronet.gsfc.nasa.gov/available_mpl.html), the Marine Aerosol Network (MAN) that is a subsidary of AERONET but based on shipboard measurements (Smirnov et al., 2009), and SOLRAD-NET ground-based solar flux measurements (http://solrad-net.gsfc.nasa.gov/index.html).

Spectral aerosol optical thickness ($\tau^*(\lambda)$) and fine and coarse components of the aerosol optical thickness are computed from the direct beam measurements (Figure 6.1). A suite of aerosol physical and optical products is also derived from the spectral-angular measurements of the sky radiance (Figure 6.2), and these are discussed below and in Chapter

5. Aerosol optical thickness is reported to an accuracy of 0.01 in most channels (Eck et al., 1999), based on nearly yearly inter-calibration versus reference instruments that are calibrated by the Langley method at the high altitude observatories of Mauna Loa (3400 m) and Izana (2400 m). Most importantly, one could compare aerosol characteristics measured at a station in the eastern United States with the same parameters measured in the Amazon basin of South America and know that the standardization made the two sets of measurements comparable. AERONET began in 1993 with small permanent and temporary network deployments in South America and along the North American mid-Atlantic coast. The network has grown steadily through federation with many partners in 86 countries and territories at roughly 450 sites since 2009. Nearly 1000 sites have been operated since 1993. Some permanent stations have been recording aerosol information for close to 18 years (as of 2011). Figure 6.3 shows the distribution of AERONET station locations in 1993, 2000 and the total accumulative location of stations from 1993 to 2010.

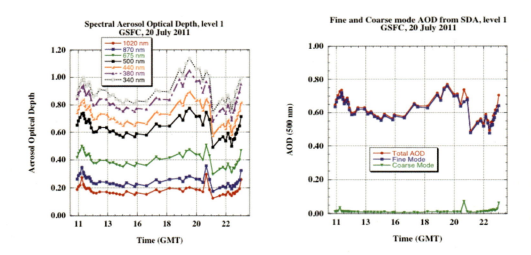

Figure 6.1 Examples of AERONET products derived from direct sun observations. Left, spectral aerosol optical depth. Right, fine, coarse and total aerosol optical depth derived from the Spectral Deconvolution Algorithm (SDA) described by O'Neill et al. (2003). All derived from observations at the Goddard Space Flight Center (GSFC) site on 20 July 2011, and plotted as a function of time of day with time units in GMT. Shown are Level 1 products before cloud screening and quality assurance are applied. Level 1 data shows high frequency variability that may later be attributed to cloudiness and eliminated. For example the isolated peak at 20:45 GMT that is more pronounced at longer wavelengths and in the coarse mode AOD may be cloud contamination. Note that Ångström exponent can be easily derived from the spectral AOD (left), and is not explicitly shown here although it is available as a primary product.

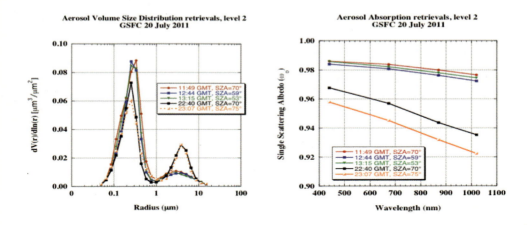

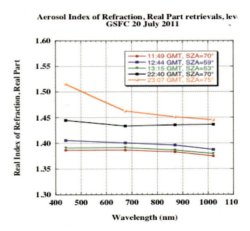

Figure 6.2 Examples of AERONET inversion products from the scanning sky radiance measurements. Top left: volume size distribution. Top right: spectral single scattering albedo. Bottom: spectral real part of the refractive index. All derived from observations at the Goddard Space Flight Center (GSFC) site on 20 July 2011. There were six retrievals made that day and the five independent retrievals that met the quality criteria (AOD440 > 0.4, solar zenith angle > 50° and minimum residual error < 5%) are plotted with a different colored line. The time of each retrieval and solar zenith angle associated with each retrieval are given in the boxed legends in each panel. The two afternoon retrievals suggest a different aerosol type has moved into the area, characterized by a stronger coarse mode component, more light absorption and higher real part of the refractive index.

Developed independently and in parallel with AERONET are several other networks that have achieved a high level of standardization, calibration and transparency. Notable networks that meet these criteria are the Global Atmospheric Watch (GAW) Precision Filter Radiometers (PFR) network, and the Skynet Network (http://atmos.cr.chiba-u.ac.jp/; Kim et al., 2004) of sun and sky radiometers. These networks deploy instruments similar to the AERONET instruments that measure the direct solar irradiance through a collimator, and in the case of Skynet also measure the narrow field of view sky radiance over a range of angles. The GAW PFR network is managed by the World Optical Depth Research Center (WODRC) in Davos, Switzerland with the data archive at the Norwegian Institute for Air Research (NILU). Aerosol optical depth is measured at four wavelengths, 368, 412, 500 and 862 nm, and there are currently 16 stations distributed widely across the globe, mostly in pristine locations to measure background aerosol (http://www.pmodwrc.ch/worcc/index.html).

Skynet is the most similar network to AERONET, using the Prede Skyradiometer to measure both the direct solar beam and angular sky radiance in seven wavelengths from 340 to 1020 nm and some models to 2100 nm (http://www.prede.com/skyradiopom011.pdf). Skynet is managed from Chiba University in Japan, and the software that can be applied to Skynet measurements to retrieve aerosol parameters, called Skyrad.pack, is available from the University of Tokyo (Nakajima et al., 1996). As of the writing of this book, there are 23 sites listed, mostly in Japan and east Asia. The SKYNET network sites are composed of instrument clusters of active and passive sensors to characterize aerosol properties. The Prede sky radiometer is an integral part of the cluster. Although SKYNET imposes standardization of instrumentation, the calibration procedures are unique and linked to processing by the local investigator. Comparisons between AERONET and the Prede SKYNET retrievals show very good comparisons for τ but not as good comparisons for ϖ and other inversion products (Sano et al., 2003; Kim et al., 2008). Rigorous comparisons are planned to validate all sky radiometer products. Due to the intensive supersite suite of measurements of SKYNET, the Prede sky radiometer is considered more of a research network than an operational one.

Until recently, the AERONET and other networks have focused on placing instruments and establishing stations in a sampling strategy that aimed to cover the most important aerosol types across the globe. Rarely would multiple instruments be placed in close proximity to each other, as that was viewed as unnecessary redundancy and misuse of limited resources. Recently, a different sampling strategy is emerging, in which stations are established on a pseudo-grid of approximately 10 km spacing. Called Distributed Regional Aerosol Gridded Observation Networks (DRAGON), such a grid was enacted during the summer of 2011 in the mid-Atlantic region of the United States. Figure 6.4 shows the 2011 DRAGON grid and the resulting AOD and single scattering albedo measurements made at 23:07 GMT on 20 July 2011. The fine resolution spacing of the instruments reveals important small-scale structures in the aerosol characteristics that would have been entirely ignored by the one, two or possibly three instruments that might have been deployed under the previous sampling strategy. DRAGON networks will be an important future direction of AERONET and other networks.

Another important network organized under the AERONET umbrella is the Marine Aerosol Network (MAN) (Smirnov et al., 2009). MAN represents the first attempt to provide

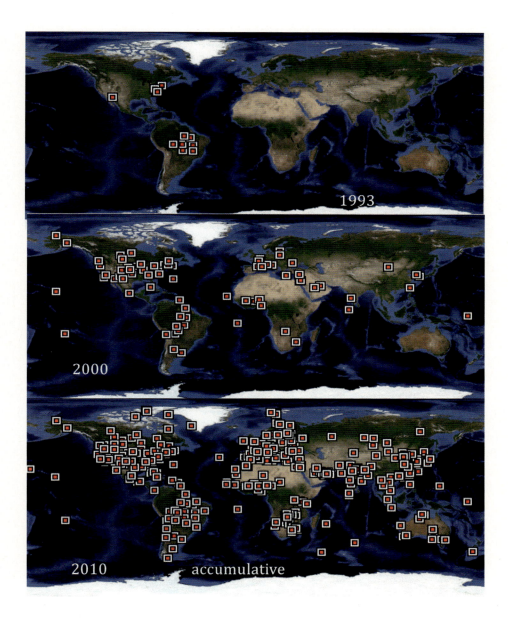

Figure 6.3 Locations of AERONET stations in 1993, 2000 and the accumulative location of stations (1993–2010).

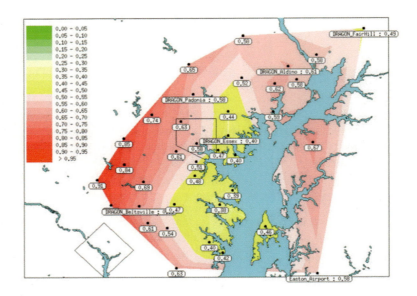

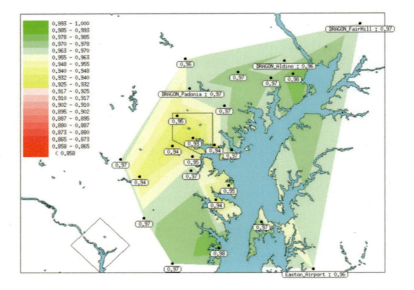

Figure 6.4 Aerosol optical depth (top), and single scattering albedo (bottom), derived from AERONET observations on 20 July 2011, as part of the Distributed Regional Aerosol Gridded Observation Networks (DRAGON) in the midAtlantic region of the U.S. eastern seaboard. The small scale grid allows resolution of large detail aerosol events. For example, on this day we see heavier loading and more absorbing aerosol to the west, which later moves over the entire region.

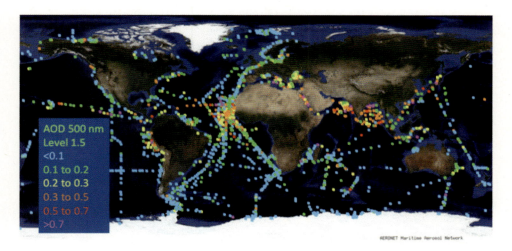

Figure 6.5 Aerosol optical depth at 500 nm from the accumulation of all Marine Aerosol Network (MAN) cruises, 2004–2011.

high quality total column spectral aerosol optical depth observations across the world's oceans. Following the AERONET philosophy of standardized instrumentation, calibration, measurement and data processing protocols, and an accessible database, MAN acquires data from trained volunteers aboard research vessels and ships of opportunity. The standard instrument of MAN is the Microtops II sun photometer, a handheld device measuring aerosol optical depth in one of two configurations: 340, 440, 675, 870, 936 nm or 440, 500, 675, 870, and 936 nm (Ichoku et al., 2002a; http://www.solarlight.com/products/sunphoto.html). All MAN instruments are calibrated at the AERONET facility at the NASA Goddard Space Flight Center using a transfer calibration procedure from the AERONET reference instrument that has been calibrated at Mauna Loa (see Chapter 4). The Microtops II instruments provide the same direct sun products shown in Figure 6.1, with slightly less accuracy than can be expected from the automated land-based AERONET instruments. At the time of writing, almost 150 cruises have been completed, covering all of the world's oceans (see Figure 6.5). MAN data is being used to characterize marine aerosol and develop maritime aerosol models (Kleidman et al., 2011; Lapina et al, 2011; Moorthy et al., 2010; Wilson et al., 2010,) and to validate satellite-based remote sensing of aerosols over ocean (Adames et al., 2011; Bevan et al., 2011; Kahn et al., 2010; Sayer et al., 2010; Smirnov et al., 2011 among others). MAN represents a significant forward step in using ground-based remote sensing to characterize aerosol by beginning to cover the 70% of the globe previously inaccessible from other ground-based networks.

Not described explicitly in this book is the important contribution made from airborne sunphotometers. These instruments are based on the same principles as a ground-based sunphotometer but have overcome the difficulty of finding and pointing to the sun automatically while in flight (e.g. Matsumoto et al., 1987; Livingston et al., 2005). Airborne instruments provide a freedom that ground-based instruments cannot. In particular, they provide vertical profiles of aerosol extinction (e.g. Redemann et al., 2000; Russell et al., 2007), aerosol optical depth above clouds (e.g. Coddington et al., 2010), fine horizontal spatial resolution that is important in cases such as smoke plumes from active fires (e.g., Pueschel et al., 1988; Pueschel and Livingston,1990), and measurements over the ocean (e.g. Redemann et al., 2009; Livingston et al., 2009). An airborne counterpart to the ground-based sky radiometers that are used to invert sky radiance has made several successful test flights in 2010 and 2011, and will collect science data for the first time in July 2012.

6.8 Aerosol inversions

A major innovation of the modern networks of AERONET and SKYNET is the systematic measure of spectral-angular sky radiance and the production of inversion products from these measurements. From the time of the Smithsonian measurements through BAPMoN and the limited handheld deployments, the measurements were exclusively of the direct solar beam and the remote sensing products were limited to extinction aerosol optical thickness, Ångström exponent and total column trace gas amounts such as precipitable water vapor (Chapter 4). Inversion could be made of the spectral extinction optical thickness to derive size distribution (King et al., 1978; Yamamoto and Tanaka, 1969; and Chapter 5), but such inversions require *a priori* physical constraints, were infrequent and not part of any systematic network. Recent work by O'Neill and his collaborators (2001, 2003) demonstrates that the curvature of the spectral aerosol extinction optical depth provides sufficient information to derive the fraction of the aerosol optical depth attributed to the fine mode in an assumed bimodal distribution. The "fine mode fraction of optical depth" becomes an important aerosol metric that can be used to ascertain aerosol type when more detailed inversion products are not available, or to separate aerosol from cloud optical depths. The so-called Spectral Deconvolution Algorithm fine mode fraction, based on the O'Neill inversion and derived from direct sun measurements, has become a standard product in the AERONET database (Figure 6.1).

When scattered sunlight as a function of angle is employed *in addition* to the spectral extinction coefficients, then the information content available to the inversion increases considerably. By using sky radiance and spectral extinction measurements in tandem one can perform high quality spectral inversions (Shaw, 1979; Nakajima et al., 1996; Dubovik and King, 2000; Dubovik et al., 2000; Sinyuk et al., 2007; and Chapter 5). Products resulting from these inversions available in the AERONET data archive, for example, include volume size distributions with no assumption of lognormality or number of modes, complex refractive index in four wavelengths, and the percentage of spherical and nonspherical particles. From these basic inversion products, calculations are made of the effective radius divided into fine and coarse modes, spectral single scattering albedo, spectral phase function, spectral and broad band fluxes. The ground-based remote sensing inversion products are widely used and often provide a "ground truth" for satellite-based inversion products beyond spectral aerosol optical depth. However, caution should be exercised. While the ground-based spectral-angular inversions provide a high quality retrieval, their products are still an inversion requiring assumptions about the aerosol. This makes the inversion products less certain than the direct beam extinction aerosol optical depth. For example, in the AERONET inversion, it is assumed that the same complex refractive index holds for the entire size range of the aerosol particles when often fine and coarse mode particles are generated by different sources and have different chemical compositions.

6.9 Examples of science from ground-based remote sensing

Discovery of Arctic haze. During the Arctic Ice Dynamics Joint Experiment in the spring of 1972 (http://psc.apl.washington.edu/aidjex/toc.html), measurements made by a small sunphotometer at Barrow in Alaska revealed the surprising result that the turbidity in this pristine region was larger than that measured in Tucson, Arizona. Moreover, the strong wavelength dependence of the spectral aerosol optical depth showed that the "haze" consisted of small, mostly submicron, particles. The source of these particles was a complete mystery since there were no pollution sources nearby and, furthermore, the computed back trajectory followed back to the north polar region over the pack ice. Following up on this mystery, a Cessna-180 aircraft was modified to allow for handheld sunphotometer measurements through a cut plastic window and measurements of turbidity were made at different altitudes up to 12,000 feet. The only previous report of a strange Arctic Haze had been made earlier by Murray Mitchell during military weather reconnaissance flights in the Arctic during the 1950s, but Mitchell's observations were visual only. Building on the quantitative spectral measurements using remote sensing, haze particles were collected on filters, and from these in situ measurements it was possible to build a case claiming that the Arctic becomes generally polluted by industrial activity mostly in Eurasia during the late winter and spring (Rahn et al., 1977; Kerr, 1979; Heintzenberg, 1980; Shaw, 1995).

Trans-Pacific transport. In the 1970s, haze layers were noticed above the high altitude Mauna Loa Observatory on the Big Island of Hawaii. This is normally a pristine location used to measure Langley plots and calibrate instruments. By driving up and down the mountain to take sunphotometer measurements at different altitudes a vertical profile of the spectral aerosol optical depth was obtained. The origin of these haze layers was unclear. There was little connection between the aerosol layers and atmospheric temperature or humidity structure based on the thermodynamic soundings from Hilo. The back trajectory pointed to an air mass origin towards the west of the Hawaiian Island chain where there is nothing but ocean. From relatively simple hand-made measures of sky radiance along the line of constant elevation angle that passes through the sun (the almucantar) it could be deduced that the particles causing the haze were concentrated in a diameter size range of a half to one micron. It was probable that these were particles that had transported across the Pacific Ocean from Asian sources (Shaw, 1980). This inference suggested that dust

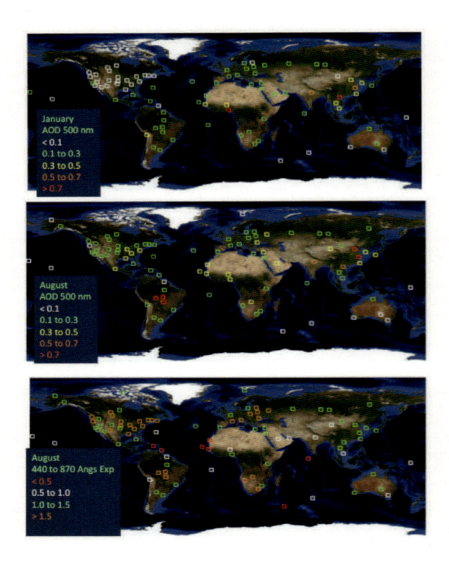

Figure 6.6 Aerosol climatology from the AERONET network. Top: long-term monthly mean aerosol optical depth at 500 nm for January. Center: same but for August. Bottom: long-term monthly mean Ångström exponent for August. Top and center panels show the seasonal shift from very clean to moderately-loaded conditions in North America and the onset of the biomass burning season in South America. Bottom panel shows the low Ångström exponent dust from the Sahara crossing the Atlantic and the high Ångström exponent smoke from biomass burning in South America and southern Africa.

and East Asian pollution routinely crosses the Pacific, a startling result at the time. With satellite images (Husar et al., 2001; Yu et al., 2008) and careful in situ chemical measures (Jaffe et al., 1999) we now have much better understanding of intercontinental transport of aerosol. However, it was sun photometry that first helped to shift the paradigm.

Emerging aerosol climatology. The key contribution of the ground-based remote sensing networks has been the creation of an aerosol climatology that is continuously updated, and which provides insight and understanding of the global and regional aerosol environment (Holben et al., 2001). The network of instruments, automatically working day after day, creates a data base with ample opportunity to ascertain aerosol optical depth and Ångström exponent. Analyses of these data reveal regional and seasonal characteristics, and relationships between the parameters in various aerosol regimes. For example, Figure 6.6 shows the global aerosol climatology for the months of January and August for 92 selected sites. The analysis shows a strong seasonal shift in aerosol optical depth in North America and the onset of biomass burning in South America. The Ångström exponent in August identifies the cross-Atlantic transport of Saharan dust and the high Ångström exponent regions of biomass burning smoke in South America and southern Africa. Updating the climatology is a continuous procedure done automatically and displayed at http://aeronet.gsfc.nasa.gov/new_web/climo_maps/climo_maps.html.

Characterization of total column ambient aerosol types. In addition to the climatology of aerosol optical depth, a second key contribution of the networks such as AERONET is a characterization of aerosol particle properties of various aerosol types. This goes beyond the derivations from the direct sun measurements of aerosol optical thickness, Ångström exponent and precipitable water vapor. The inversions applied to the sky radiances from the scanning radiometer (Chapter 5) yield a wealth of information including volume size distribution, spectral real and imaginary refractive indices and single scattering albedo (Dubovik and King, 2000; Dubovik et al., 2000). Figure 6.7 shows some of the results of the sky inversion (Dubovik et al., 2002). These results are constructed from accumulation of many individual retrievals at specific stations or logically grouped stations. With a network of automatic instruments collecting data every day, sufficient statistics can be acquired to sort and bin the data according to certain parameters, such as aerosol optical thickness or precipitable water vapor. In this way, aerosol properties in certain locations can be modeled dynamically, changing as aerosol loading or humidity varies (Remer and Kaufman, 1998; Remer et al., 1998). The advantage to these aerosol models constructed from sun/sky radiometer networks such as AERONET is that the results represent the total column ambient aerosol properties that are very difficult to measure from in situ instruments, even if airborne. Such total column ambient characterization of the aerosol is necessary to derive direct clear sky aerosol radiative forcing and also to represent aerosol particle properties in retrieval algorithms from satellite sensors.

Single scattering albedo of dust-smoke mixtures. Multi-year data from several stations in the AERONET network that are embedded in complex aerosol situations where both desert dust and combustion-type aerosols are found in mixtures have been used to investigate how those mixtures combine in total column ambient aerosol properties (Eck et al., 2010). The results show that spectral single scattering albedo properties generally combine linearly, as a function of the fine mode fraction of the mixture with the single scattering albedo of each component remaining constant. The linear combination applied to mixtures with pollution and dust in India, biomass burning and dust in the Sahel south of the Sahara and pollution and dust in China. The Indian and African sites also show that the absorption aerosol optical depth $((1-\varpi\tau^*)$ is linear in wavelength in logarithmic components, but the Chinese sites are not. This points to a more absorbing component in Chinese coarse mode than normally is expected from mineral dust and could have an industrial source. Only with the implementation of the modern networks of ground-based remote sensing, with their long data collection record and standardization between stations, could this analysis be possible.

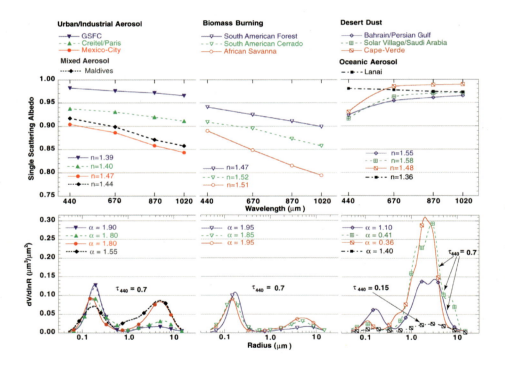

Figure 6.7 Examples of single scattering albedo, real part of the refractive index (n), Ångström exponent (α) and volume size distribution for various AERONET stations, grouped according to aerosol type. Volume size distributions are shown for a common value of aerosol optical depth ($\tau440$), as indicated. From Dubovik et al. (2002), reprinted with permission from the American Meteorological Society.

6.10 Summary

Despite the proliferation of satellites retrieving aerosol parameters, aerosol remote sensing began with ground-based instruments and continues to the present. Using both methods that measure narrow-band light extinction in specified channels across the solar spectrum and inversion methods that retrieve multiple particle properties from measures of the spectral-angular radiation field, networks of ground-based remote sensing instruments continue to make a significant contribution to aerosol science. Historically, while a single researcher with a single instrument could produce aerosol optical depths with high accuracy, the first international efforts to provide this information as a network failed because of inconsistency, improper calibration methods, lack of transparency and accessibility to the data. For a number of years, interest in aerosol optical depth measurements lay only in atmospheric correction of satellite images to retrieve surface properties.

With a better understanding of the role of aerosols in climate change dynamics and the advent of the Earth Observing System satellite constellation, resources became available to develop better ground-based aerosol remote sensing networks. These networks were to be autonomous, robotic installations with standardized configurations and protocols. Calibration would meet scientific standards and be delivered uniformly across the network. All data, algorithms, calibration records, etc. would be publicly available. The current standard instrument is a sun/sky scanning radiometer that measures both the collimated direct beam for aerosol optical depth measurements and the spectral angular radiance for inversions of columnar optical properties that has proliferated to become a global force for aerosol studies in the modern era. Today, much of aerosol research makes use of the data from one of these existing networks, whether it is to validate satellite retrievals or obtain scientific understanding of aerosol from the measurements themselves.

7 History of passive remote sensing of aerosol from space

Omar Torres and Lorraine A. Remer

7.1 First observations from space

The earliest views of aerosol from space came from Russian cosmonauts who took hand-held photographs of Earth and Earth's atmosphere through the windows of orbiting space-craft in the early 1960s (Lazarev et al., 1987). From these photographs we could see for the first time the bluish haze that covered polluted regions and the dust emitted from deserts. The pictures showed that these hazes were inhomogeneous and temporally inconsistent, but quantitative information was missing until the first spectroscopic measurements of the atmosphere were obtained in 1970 by Soyuz-9 cosmonauts using handheld spectrometers. Thus, the era of space-based aerosol remote sensing had begun.

The first stratospheric aerosol measurement on record was obtained during the joint Soviet-American mission in 1976 using a handheld sunphotometer. On this mission the Stratospheric Aerosol Measurement (SAM) was carried out as a proof-of-concept by the Apollo-Soyuz Test project that made use of solar occultation observations to quantify stratospheric aerosols in terms of the vertical distribution of aerosol extinction coefficient (Pepin and McCormick, 1976; see Chapter 4).

Aerosol measurements from unmanned platforms in space were obtained for the first time by the Multi Spectral Scanner (MSS) sensor onboard the Earth Resources Technology (ERTS-1) satellite launched in 1972 (Griggs, 1975; Fraser, 1976; Mekler et al., 1977). Observations from the Advanced Very High Resolution Radiometer (AVHRR) on the TIROS-N satellite deployed in October 1978 were used to generate the first operational aerosol product. Since then, a long list of spaceborne sensors have been used to detect and characterize aerosols. For three decades (1970–2000) a majority of satellite-based aerosol observations were obtained by sensors designed and deployed for other purposes. The creative use of these measurements to infer aerosol properties paved the way for the eventual development and deployment of more accurate sensors specifically designed to measure aerosol properties.

This chapter describes the different methods to retrieve aerosol information applied to the "heritage sensors", those in flight before the onset of the Earth Observation System (EOS) era. The "modern sensors", aboard satellites that began collecting data in 2000 are described in Chapter 8. Many of the heritage sensors are still in operation, still extending time series of aerosol measurements, as of 2012.

7.2 Measuring stratospheric aerosols using solar occultation measurements

The solar occultation technique is a simple method of measuring vertical profiles of atmospheric extinction from Earth orbit using the sun as a light source. This is analogous to ground-based sunphotometry. See Chapter 4 and Figure 4.4 for a description of how the instrument observes the attenuated sunlight through the upper atmosphere for every sunset and sunrise observed. Typically the spacecraft orbits the Earth approximately once every 90 minutes or 16 times per day, depending on orbital parameters. Since each orbit provides two measurement opportunities, the sensor can acquire 32 separate measurements during each 24-hour period at different geographical locations over the Earth depending on the spacecraft orbit. The number of measurement opportunities and the geographical coverage can be increased if measurements are made during both lunar and solar occultation events.

Following the successful SAM test (Pepin and McCormick, 1976), the first satellite-based long-term monitoring program of the atmospheric aerosol load started with the deployment in 1978 of the SAM II sensor aboard the Nimbus 7 spacecraft. SAM II provided vertical profiles within the stratosphere of aerosol extinction at 1.064 μm in both the Arctic and Antarctic polar regions. The SAM II data coverage began on 29 October 1978 and extended through 18 December 1993, at which time SAM II was no longer able to track the sun. The discovery of Polar Stratospheric Clouds (PSCs) (McCormick et al., 1982), and the observation of the formation and evolution of the stratospheric aerosol layer in the aftermath of the El Chichon eruption are among the main scientific contributions of the SAM II program. The first multiyear climatology of PSCs was developed based on SAM II observations (Poole and Pitts, 1994). Nitric acid trihydrate PSCs were later found to play a critical role in the heterogeneous chemistry process leading to the formation of the stratospheric ozone hole (Molina, 1991 and references therein).

The Stratospheric Aerosol and Gas Experiment (SAGE I) on the Applications Explore Mission-2 (AEM-2) satellite was designed to monitor the evolution of the stratospheric ozone layer using solar occultation. To account for aerosol effects on the ozone inversion process, the vertical distribution of the stratospheric aerosol load was also measured in terms of particle extinction coefficient at 0.450 and 1.00 μm. SAGE I operated between 1979 and 1981 and provided a description of the global background stratospheric aerosol load (Kent and McCormick, 1984) that complemented the SAM II measurements polar coverage. SAGE I provided the reference state of the stratospheric aerosol load prior to the series of volcanic eruptions that started in 1982 with the eruption of El Chichon in April 1982.

The short-lived but successful SAGE I stratospheric monitoring program was followed by the launch of the SAGE II sensor on the Earth Radiation Budget Satellite (ERBS) in March 1984. SAGE II provided the scientific community with a global depiction of the distribution of stratospheric aerosol, ozone, water vapor and nitrogen dioxide over a period of 21 years. The SAGE II record of aerosol extinction at four wavelengths from the UV to the near IR is the longest continuous aerosol record from the same sensor. During this lengthy time series the sensor observed the decay of the El Chichon aerosol layer as well as the formation and dissipation of the Pinatubo aerosol layer following the 1991 eruption. The SAGE III/Meteor-3M satellite mission operated as a joint partnership between NASA and the Russian Aviation and Space Agency (RASA). SAGE III was launched onboard a Meteor-3M spacecraft in December 2001. It measured aerosol extinction coefficients at nine wavelengths between 0.384 and 1.545 μm (Thomason et al., 2010 and references therein). The Meteor-3M mission, along with the SAGE III mission, was terminated in March 2006.

7.3 El Chichon and Pinatubo volcanic eruptions

The combined set of solar occultation observations by SAGE (1979–1981), SAM II (1978–1993) and SAGE II (1984–2005) were used to continuously characterize the evolution of the stratospheric aerosol load during a 27-year period during which the two largest volcanic eruptions of the last century took place. SAGE and SAM II measurements characterized the conditions of minimum stratospheric aerosol load prior to the April 1982 eruption of El Chichon. SAM II and SAGE II monitored the decay of the El Chichon stratospheric sulphuric acid aerosol layer. SAGE II mapped the entire lifetime of the stratospheric aerosol layer resulting from the Pinatubo eruption in June 1991, the largest volcanic eruption of the last 30 years, that greatly perturbed the stratospheric aerosol layer (Lambert et al., 1997).

Other instruments with aerosol sensing capability onboard the Upper Atmospheric Research Satellite (UARS) launched in September 1991 about three months after the massive Pinatubo volcanic eruption (Reber et al., 1993) contributed to the characterization of the Pinatubo aerosol layer. UARS, a stratospheric chemistry mission included the Cryogenic Limb Array Etalon Spectrometer (CLAES), the Halogen Occultation Experiment (HALOE), and the Improved Stratospheric and Mesospheric Sounder (ISAMS) sensors equipped to measure vertical distribution of aerosols in the stratosphere to account for aerosol scattering effects in the retrieval of concentrations of chemical species in the upper atmosphere. CLAES measured aerosol extinction profiles at 8 IR channels and operated between 1991 and 1993 (Roche et al., 1993). HALOE operated from 1991 to 1997 and was one of the most successful UARS instruments. Making use of the solar occultation approach to sound the stratosphere, mesosphere, and lower thermosphere, HALOE measured vertical profiles

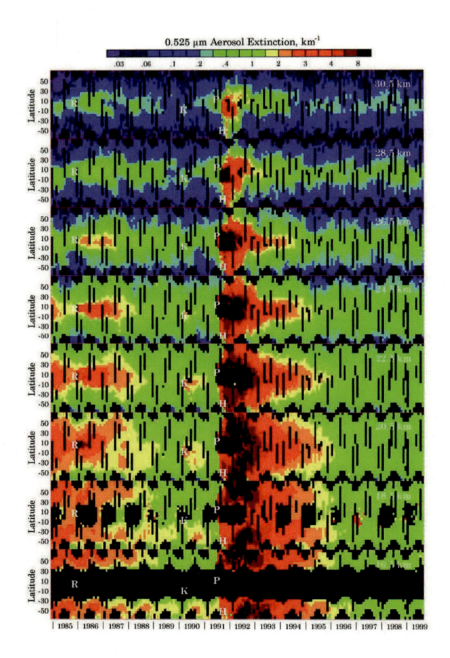

Figure 7.1 Zonally averaged 0.525 μm aerosol extinction at 30.5, 28.5, 26.5, 24.5, 22.5, 20.5, 18.5 and 16.5 km. The letters repeated in each frame mark the latitude and time of the Ruiz (R), Kelut (K), Pinatubo (P) and the Hudson (H) eruptions. From Bauman et al. (2003b), reprinted with permission from the American Geophysical Union.

of atmospheric ozone (O_3), hydrogen chloride (HCl), hydrogen fluoride (HF), methane (CH_4), water vapor (H_2O), nitric oxide (NO), nitrogen dioxide (NO_2), temperature, and IR aerosol extinction profiles (Russell et al., 1993). The ISAMS (Taylor et al., 1993) instrument measured aerosol extinction at 6.21 and 12.1 μm from September 1991 to July 1992.

The 15-year [1985–1999] record of the global stratospheric aerosol load as seen by the SAGE II and CLAES sensors (Bauman et al., 2003) is shown in Figure 7.1. The record starts with the remaining effects of the 1982 El Chichon eruption, enhanced in 1986 by the El Ruiz eruption, a much less significant but still important source of statospheric aerosol. The effect of the massive 1991 Pinatubo eruption is readily apparent at all levels between 16.5 and 30 km. The return to background levels takes place by early 1997 at the lowest altitude shown.

Attempts were also made to monitor the evolution of the Pinatubo aerosol layer using nadir observations by the AVHRR and TOMS sensors (algorithms described below). Stowe et al. (1992) and Long and Stowe (1994), studied the progression of the global Pinatubo aerosol layer using NOAA-11 AVHRR observation at 0.63 μm. The Pinatubo aerosol layer also had a significant effect on the remote sensing of atmospheric parameters such as the total column ozone amount derived by the Total Ozone Mapping Spectrometer (TOMS) on the Nimbus-7 satellite. The large signal of the aerosol effect on the retrieved ozone amounts (Bhartia et al., 1993) was used to indirectly estimate the magnitude of the tropical stratospheric aerosol load in the UV as a function of time (Torres et al., 1995).

7.4 Total column aerosol retrieval from reflected visible and near IR observations

As described in the previous section, aerosol characterization using solar occultation measurements have provided descriptions of the temporal and spatial variability of stratospheric aerosols, most of which are of natural origin. To properly account for the total atmospheric aerosol, however, it is necessary to include the troposphere where most anthropogenic aerosols reside. Thus, downward-viewing satellite measurements are required for total column aerosol retrieval.

Downward-viewing satellite retrievals are fundamentally different from solar occultation measurements in that they measure reflected solar radiation, not the direct beam attenuated through the upper atmosphere. This is represented by the radiance, rewritten here from Eq. (3.33a),

$$L(0,\mu,\varphi) = L_0(0,\mu,\varphi) + \frac{RF_0(\tau^*)T_t(\tau^*,\mu)}{(1-RS)}, \tag{7.1}$$

where $L(0,\mu,\varphi)$ is the measured upwelling radiance at the top of the atmosphere, composed of two parts. $L_0(0,\mu,\varphi)$ is called the path radiance and it represents radiance scattered into the satellite sensor's field of view from atmospheric constituents (aerosols or gases). The path radiance has had no interaction with the surface beneath and would be the upwelling radiance of the atmosphere for a black surface. The second term on the right expresses the interaction of the radiance between the atmosphere and the surface (see Figure 3.4) that is later scattered into the sensor's field of view. Here $F_0(\tau^*)$ is the downward flux reaching a black surface, and T_t the total transmission function representing the total amount of direct plus diffuse radiation transmitted. R is the Lambertian surface reflectance and S the reflected flux at the bottom of the atmosphere called the spherical albedo of the atmosphere for illumination from below (Tanré et al., 1979).

The measured reflected radiance at the top of the atmosphere includes the combined effect of gaseous absorption and molecular scattering, aerosol scattering and absorption, cloud scattering and surface reflection. From this intertwined mixture of information the aerosol signal must be isolated and the competing effects must be accurately accounted for or altogether avoided in order to achieve accurate measurements of total column aerosol load.

The spectral choice and width of each spectral band to be used in the retrieval and the spatial resolution of the measurement are of great importance for aerosol retrieval. By choosing wavelengths in narrow spectral bands that correspond to the window regions of the solar spectrum, gas absorption can be avoided. Furthermore, avoiding other spectral bands that introduce uncertainty from surface features (i.e. ocean color) helps in the isolation of the aerosol signal. Since cloud interference with the observation must also be avoided, the spatial resolution must be sufficiently fine so that aerosols can be observed through clear holes in otherwise cloudy skies. As the technology of spaceborne sensors becomes more sophisticated, expanding to a broader spectral range, multiple angles and polarization, different strategies to avoid misinterpreting reflectance from clouds as reflectance from aerosol can be used instead of simply trying to avoid clouds. However, historically, cloud masking has been a critical component of all downward-viewing retrievals.

A major hurdle in aerosol retrieval is the treatment of bright surfaces whose contribution to the satellite-measured signal is significantly larger than the aerosol contribution. The separation of the small aerosol component from a significantly larger surface term is a large source of error if the surface reflective properties are not accurately characterized. For that reason the most accurate retrievals have been done over water surfaces at visible and near IR wavelengths, taking advantage of the very low reflectance of water in this spectral range. Retrieval algorithms that use low-reflecting backgrounds to infer aerosol properties from space observations are generically referred to as "dark-target" approaches and are physically based on interpreting *path radiance* in terms of aerosol optical depth (Kaufman and Sendra,1988). As $R \to 0$ in Eq. (7.1), $L(0,\mu,\varphi) \to L_0(0,\mu,\varphi)$. A parallel method developed to retrieve aerosol over bright land surfaces is based on adjacency effect or contrast reduction and is physically based on using *transmission* to infer aerosol properties (Tanré et al.,1988), described below in Section 7.4.3.

Aerosol retrieval algorithms that use observations of reflected radiance have been based on pre-calculated top-of-atmosphere reflectances assuming aerosol models characterized in terms of aerosol particle size distribution and composition (i.e. refractive index). A set of forward radiative transfer calculations is made for a variety of pre-defined aerosol models for the range of solar and sensor view angles expected to be encountered by the satellite observations. These calculations are made for a range of aerosol loadings. The result of the calculations is a set of reflected radiances at top-of-atmosphere, indexed by wavelength, aerosol loading, solar and view geometry and by aerosol model, $L^{LUT}(\lambda,\tau,\mu_o,\varphi_o,\mu,\varphi,\text{model})$. The results of the forward calculations are generically referred to as look-up tables (LUT) and are permanently stored. During a retrieval, after a series of checks and modifications of the measured reflected radiance to obtain the aerosol reflectance clear of influence of clouds and gases, the procedure is to match the modified measured radiance, $L(0,\mu,\varphi)$ at wavelength, λ, to the output of the stored LUT for the solar and view angles of that particular measurement. When a match is found, the retrieval identifies the aerosol loading and, in some cases, certain parameters of the aerosol model that are not prescribed *a priori*. The main retrieved parameter is aerosol optical depth, which is the same as the vertically integrated extinction coefficient. Depending on how many wavelengths, angles and polarization states are used in the retrieval, the aerosol optical depth may be a function of wavelength. Aerosol retrievals using these LUTs are very sensitive to the choice of aerosol microphysical properties (i.e. particle size distribution and complex refractive index) as they ultimately determine the amount of scattered radiance reaching the satellite-borne sensor. The future of aerosol remote sensing will include more sophisticated sensors returning much more information of the viewed scene. There should be sufficient information in these multiwavelength, multiangle polarimeters to mathematically invert the aerosol characteristics without the burden of *a priori* assumptions about aerosol particle size distribution and complex refractive index. Historically, simply returning an accurate measure of aerosol loading, the aerosol optical depth, even in one wavelength, was a major step forward.

7.4.1 Retrieval of aerosols over the oceans using the dark target approach

The astronaut photographs of the Earth clearly demonstrated the severe impacts that the atmosphere would have on interpreting images of the Earth's surface from space and retrieving quantitative information on surface reflectance. Important satellite-based measures of Normalized Difference Vegetation Index (NDVI) and ocean color that had significant important applications for estimating terrestrial and marine bio-productivity, crop yields and early famine warnings, would be confused by changing aerosol in the atmosphere above the scene. These scenes would require atmospheric correction to obtain the surface reflectance, and fundamental to atmospheric correction would be the quantification of the aerosol loading. As atmospheric correction improved, so would remote sensing of aerosol. Aerosol remote sensing over ocean thus began with the need for atmospheric correction to retrieve true water-leaving radiances from the apparent reflectance. The basic work here was pioneered by Gordon and Wang (1994), and Fraser et al. (1997), and made use of earlier work by Griggs (1975). The by-product of oceanic atmospheric correction produced an aerosol product. Atmospheric corrections were applied to the Coastal Zone Color Sensor (CZCS) initially and then to the Sea-viewing Wide field of view Sensor (SeaWiFS). The same methods were adapted to specifically focus on aerosol retrieval and applied retroactively over the global oceans using satellite sensors never intended for aerosol retrievals.

The Earth Resources Technology Satellite – 1 (ERTS-1) and later Landsat-2 carrying the multispectral scanner (MSS) provided the first opportunity to derive aerosol optical depth over ocean from space (Griggs, 1975). ERTS-1 was launched in 1972 and the Landsat series continued the tradition with the last Landsat-7 launched in 1999 and the Landsat continuation mission scheduled for launch in late 2012. Griggs (1975) was able to retrieve aerosol in three wavelengths: 0.55 μm, 0.65 μm and 0.75 μm, each matched independently to pre-calculated radiances in a LUT with *a priori* assumptions of aerosol size distribution, complex refractive indices and rough ocean surfaces. Other techniques were explored with MSS data including contrast methods that will be described in Section 7.4.3. Continuing efforts were made to derive aerosol from MSS and subsequent Landsat-borne instruments based on the Griggs (1975) dark target approach. None of these efforts ever produced an operational product. ERTS-1 and the later Landsat satellites were designed to observe vegetation and geological targets. Spatial resolution of the MSS sensor was 80 m, with later sensors such as the Thematic Mapper (TM) having a broader spectral range and even finer resolution. The result of this fine resolution imagery was a narrow swath of less than 200 km, insufficient to deliver a daily picture of the global aerosol system. Yet the fine resolution and multiple channels overwhelmed the electronic archiving capabilities of the time. Images were collected, but not stored unless a paying customer was identified. Almost all customers had land-based applications in mind for the imagery. Most views of the rela-

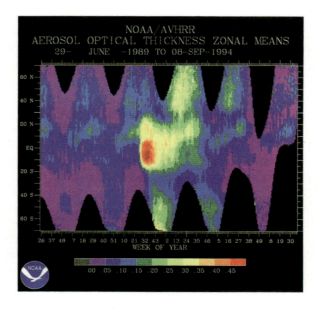

Figure 7.2 Latitudinally resolved time series of total aerosol optical depth at 0.50 μm retrieved from the AVHRR satellite using the first generation single channel retrieval. From Stowe et al. (1997). Reprinted with permission from the American Geophysical Society.

tively boring ocean were never saved, making a retroactive processing of Landsat data for ocean aerosol retrievals impossible. In the later years of the Landsat program, images were saved on a routine basis, but the motivation to retrieve aerosol information from these data on a routine basis never gained momentum, being replaced by the success of retrievals of sensors that could provide wide swath data and nearly global views of the aerosol system on a daily basis.

The Advanced Very High Resolution Radiometer (AVHRR) was the first sensor to support operational aerosol retrievals over oceans and provide daily views of the global aerosol system. AVHRR is a five-channel sensor (nominally 0.63 μm, 0.83 μm, either 1.6 μm or 3.7 μm, 11 μm and 12 μm) designed as a surface and cloud imager by NOAA/NESDIS for weather applications. Since its inception in 1978, three generations of AVHRR sensors have been flown onboard TIROS-N and the NOAA polar operational environmental satellites (POES). The solar reflectance bands centered at 0.63 and 0.83 μm have remained unchanged, providing the desirable temporal continuity for long-term environmental monitoring. AVHRR sensors have flown onboard TIROS-N, NOAA-6, through NOAA-12 and NOAA-14 to 17 satellites. The global coverage and moderate spatial resolution (1 km) made AVHRR an attractive instrument for operational aerosol retrievals. Even so, the instrument was not designed for aerosol retrievals and hurdles had to be overcome. The most significant challenge for AVHRR aerosol retrievals was to account for drift in the instruments' radiometric calibration. Various vicarious calibration methods were used to standardize instrument calibration (Holben et al., 1990; Vermote and Kaufman, 1995; Rao and Chen, 1996). The incentive to derive quantitative aerosol information from AVHRR increased once a reliable calibration was available.

Almost all of the aerosol information available from AVHRR is contained in the first two channels: 0.63 μm and 0.83 μm. There have been several strategies to make use of this aerosol information that we categorize as either single channel or two channel retrievals.

Single channel AVHRR retrieval approach

The single channel retrieval makes use of AVHRR Channel 1 (0.63 μm). The original algorithm assumes a Lambertian ocean surface with surface albedo of 0.015, and one aerosol model represented by a modified Junge particle size distribution equivalent to an Ångström exponent of 1.5 and complex refractive index of 1.5–0.0i (Rao et al., 1989; Stowe et al., 1990; Stowe et al., 1997). Note that the aerosol is assumed to be completely non-aborbing with ϖ_o=1 because the imaginary part of the refractive index is 0.0. The Ångström exponent of 1.5 is higher than the values of pure maritime aerosol reported by Smirnov et al. (2002). The retrieval procedure followed the typical dark target oceanic methods: masking clouds and sun glint, matching measured radiances in the 0.63 μm channel to pre-calculated values in the LUT for the particular solar and view angles of the scene in question, and retrieving the aerosol optical depth that was used to calculate the LUT top-of-atmosphere radiance. This algorithm was used to characterize the aerosol optical depth over the global oceans, identifying regions of heavy aerosol loading. An example of this product for the monthly mean aerosol optical depth in 1995 is shown in Figure 7.2. The algorithm applied to the long AVHRR time series that began in 1978 produced a data set that fully characterized the total aerosol optical depth of the Pinatubo eruption, similarly as to how the solar occultation instruments had characterized the Pinatubo aerosol in the stratosphere (Figure 7.1).

The next generation single channel AVHRR retrieval kept the same basic procedure as at the beginning, but made several adjustments to the *a priori* assumptions used by the retrieval. First the Lambertian surface albedo was decreased considerably to 0.002, and a diffuse glint correction (Viollier et al.,1980; Gordon and Morel, 1983) was applied in lieu of the more modern usage of a non-Lambertian rough ocean surface model. Second, the Junge size distribution was replaced by a single mode lognormal size distribution that with a slightly modified complex refractive index produced an Ångström exponent of 0.6, more in line with measurements of maritime aerosol (Smirnov et al., 2002). The aerosol model remained entirely non-absorbing (Stowe et al., 1997; Ignatov et al.,1995b). This algorithm is used for generating the operational AVHRR aerosol products on a 1-degree global grid, known as AVHRR Pathfinder-Atmosphere (PATMOS) (Stowe et al., 2002; Jacobowitz et al., 2003) and is publicly available.

A continuing concern for all aerosol retrievals is the prospect of systematic cloud contamination in the retrieval product. The single channel algorithms use a cloud-clearing scheme that includes a series of radiance thresholds and a spatial variability test that examines the range of radiances in each 2 by 2 set of pixels. If the spread of radiances is two high in this set of 4 pixels, the 4 pixels are not used in the aerosol retrieval.

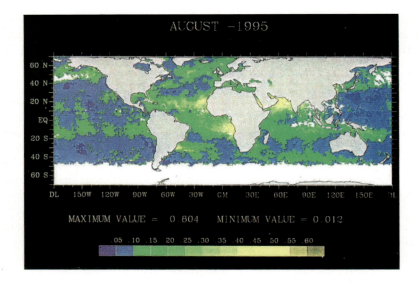

Figure 7.3 Total aerosol optical depth at 0.50 μm retrieved from the AVHRR satellite using the first generation single channel retrieval. From Stowe et al. (1997). Reprinted with permission from the American Geophysical Union.

Two channel AVHRR retrieval approach

A different approach for retrieving aerosol characteristics from AVHRR makes use of both solar channels on the AVHRR sensor (0.63 and 0.83 μm). Theoretically, the use of two channels allows for the calculation of the Ångström exponent or other parameter, which gives a qualitative measure of particle size. Dual-channel AVHRR algorithms have been developed by several groups (i.e. Higurashi and Nakajima, 1999; Mishchenko et al., 1999a). The methods include a LUT indexed by solar and view angles, aerosol optical depth and either Ångström exponent or a ratio between two particle modes that is directly proportional to Ångström exponent. These methods employ a rough ocean surface model with a globally fixed wind speed and a fixed complex refractive index (1.50 + 0.005i). Note the assumption here are that the particles are slightly absorbing with a nonzero imaginary part of the refractive index. All particles are assumed spherical in all cases. In one method the size distribution is composed of two lognormals, each fixed in mode radius and width, but allowed to increase or decrease in volume. The ratio between the two mode volumes is a free parameter in the retrieval (Higurashi and Nakajima, 1999). In another method the size distribution is a single mode modified power law with the exponent of the power law the free parameter (Mishchenko et al., 1999a). Sensitivity studies show little difference between the single mode and two-mode assumptions for particle size distributions (Mishchenko et al., 1999a). The Mishchenko et al. (1999a) algorithm was applied to the AVHRR record beginning in 1983 to produce the Global Aerosol Climatology Product (GACP) (Geogdzhayev et al., 2002), a publicly available time series of monthly mean aerosol optical depth and Ångström exponent on a 1 degree grid over the global oceans. Figure 7.4 shows the time-series of global mean aerosol optical depth derived by application of the dual-wavelength algorithm to AVHRR observations for the 1983-2001 period (Geogdzhayev et al., 2004).

The GACP product makes use of a modified ISCCP cloud detection in which the infrared tests are made more conservative (more likely to categorize the pixels as "cloudy") and a test is added based on the ratio of the 0.63 μm and 0.83 μm radiances. The ratio test is based in the physics that clouds with their larger hydrometeors should be more spectrally neutral than the finer particle aerosols. In addition to the continuing concern about cloud contamination in the aerosol product, an important source of uncertainty in the two-wavelength AVHRR algorithm is the relative calibration error since there is no onboard calibration of the instrument.

7.4.2 Retrieval of aerosols over land using the dark target approach

The retrieval of aerosols over land is a much more difficult problem than its counterpart over ocean. The land surface tends to be both brighter and more variable than the ocean surface. Thus, while aerosol retrievals over oceans can model the radiance contribution from the surface with a high degree of certainty, the same is not true over land. One method used to derive aerosol over land is to find land surfaces that mimic the advantages of the ocean: relatively dark and homogeneous. Dark dense vegetation provides such an opportunity.

Kaufman and Sendra (1988) demonstrated an atmospheric correction technique for Landsat MSS images over land that would retrieve aerosol amounts using a dark target

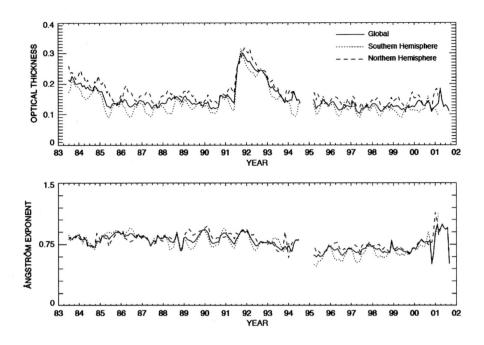

Figure 7.4 Multi-year AOD climatology over the global oceans derived by the two-channel AVHRR algorithm (top), and resulting Ångström exponent (bottom). From Geogdzhayev et al., 2004. Reprinted with permission from the Journal of Quantitative Spectroscopy and Radiative Transfer.

approach. The basis of a dark target approach is that aerosol will normally brighten a dark scene, causing the satellite-measured reflectance, $L(0,\mu,\varphi)$, to be larger than the surface reflected radiance (with reflectance R), mostly because of the large path radiance, $L_o(0,\mu,\varphi)$. This dark target approach identified the darkest pixels in the image, assumed the surface reflectance of these pixels (R), and then derived the aerosol amount based on that assumption. The approach then assumed that aerosol varied less over the image than the surface reflectance so that the aerosol amount determined from the darkest pixels could be applied to the whole image. The dark target approach worked because the error introduced by the surface reflectance assumption might be *relatively large*, but because of the small reflectances would be small in an absolute sense. There were two tricks to employing this method: finding the dark dense vegetation in the image and assuming the surface reflectance at the wavelengths of the retrieval.

The Normalized Difference Vegetation Index (NDVI) (Jordan, 1969; Tucker, 1979) uses the simple combination of the two reflective AVHRR channels to identify vegetation in the AVHRR image.

$$\text{NDVI} = \frac{\rho_{0.83} - \rho_{0.63}}{\rho_{0.83} + \rho_{0.63}}, \tag{7.2}$$

where ρ_λ is the reflectance given for a specific geometry as

$$\rho_\lambda = \frac{\pi L(\lambda)}{\mu_o E_o}, \tag{7.3}$$

for wavelength, λ, and where μ_o is $\cos(\theta_o)$ and E_o is the irradiance at top of the atmosphere. This is the reflectance defined in Eq.(3.62), denoted by S in Chapter 3. NDVI can be used to identify the dark vegetated pixels from which to derive the aerosol. The problem with NDVI is that as aerosol loading increases the measured reflectances increase. Because aerosol creates a spectrally dependent signature in the measured reflectances the value of $\rho_{0.83}$ increases more slowly than $\rho_{0.63}$ as aerosol loading increases. The result is that NDVI decreases as loading increases and it becomes difficult to identify the dark targets through the haze. To overcome this problem, techniques made use of the 3.7 μm band on AVHRR. The measured 3.7 μm reflectance includes components from thermal emission and from solar reflectance. The emissive part has to be removed from the measured signal in order to make use of the channel to find the dark targets. This is done using information from the AVHRR thermal band at 11 μm (Roger and Vermote, 1998). The advantage of using $\rho_{3.7}$ is that this relatively long wavelength is insensitive to fine aerosol types and can penetrate through the haze and 'see' the surface (Kaufman and Remer, 1994). Figure 7.5 shows insensitivity to aerosol by longer wavelength channels. For Landsat-4 Thematic Mapper (TM) and subsequent imagers such as MODIS, the 2.1 μm channel was used instead of the 3.7 μm channel because it had all the advantages of being insensitive to fine particles and yet did not require correction for thermal emission (Kaufman et al., 1997a).

The dark target approach was used with AVHRR radiances for specific studies (Kaufman and Nakajima., 1993; Vermote et al., 1996), but never adapted to produce an operational over land product. However, later it became the basis for one of the MODIS aerosol over land products (Kaufman et al., 1997b).

7.4.3 Retrieval of aerosols over land using the adjacency effect

The dark target approach will not work over scenes with bright surfaces, such as deserts. An alternative idea that was applied to deserts is based on how aerosol scatters light into the field of view from adjacent targets, this process is called the adjacency effect. The initial interest in the adjacency effect grew from the need to account for it as part of atmospheric correction. To obtain the true reflectance of a specific surface target, the light scattered into the scene from surrounding scenes would have to be removed. The amount of light scattered in the scene is dependent on the contrast between the target and surrounding targets, and also the aerosol loading and particle properties (Mekler and Kaufman, 1980; Kaufman and Fraser, 1983a and 1983b).

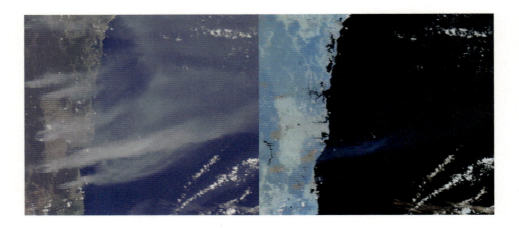

Figure 7.5 MODIS satellite sensor image of Australia, 25 December 2001. Left panel shows smoke from burning wildfires using the wavelength combination of $0.66, 0.55, 0.47$ μm to create a true color image. Right panel shows same image but constructed from the wavelength combination of 1.24, $1.63, 2.13$ μm. The longer wavelengths are insensitive to the smoke and "see" the surface without interference.

The direct relationship between the adjacency effect blurring the image and the amount of aerosol in that image led to the development of the contrast reduction algorithm. The contrast reduction or the decrease in the difference of apparent reflectance, ρ (as defined in Eq. 7.3) between pixels a certain distance apart (d) is a function of the difference in surface reflectance of the two pixels (ΔR), the total transmission (T_t) and the extinction from the total optical depth of the atmosphere $(\exp(-\tau^*/\mu))$:

$$\Delta\rho_{Ri,j}(\mu_0, \mu, \varphi) = \rho_{Ri,j+d} - \rho_{Ri,j} = \Delta R_{i,j} T_t(\mu_0) \exp(-\tau^*/\mu), \qquad (7.4)$$

where i and j denote the indices that define the gridded pixel location, μ_0 and μ, are cosines of the solar zenith and view zenith angles, and φ is the relative azimuth angle. In Eq. 7.4, we assume that the two pixels are on the same row i and there is a distance d between the two columns $j+d$ and j.

Assuming that the land surface does not vary from day to day, then $\Delta R_{i,j}$ will remain constant and the differences in $\Delta\rho_{i,j}$ from image to image will depend on $T_t(\mu_0)\exp(-\tau^*/\mu)$. The larger the optical depth (τ^*), the smaller the contrast in apparent reflectance ($\Delta\rho_{i,j}$). The procedure requires satellite images from multiple days having the same viewing geometry. The method was applied to deserts using Landsat TM images with identical sun-surface-sensor geometry, with at least one clean image. The hazier images were blurrier. The clean image had more distinct features. These could be compared quantitatively with point-spread functions and the aerosol retrieved using a LUT and assuming particle properties including size distribution and complex refractive index (Tanré et al., 1988). Besides the ability to retrieve over deserts where the dark target method could not, the contrast reduction method did not require the same level of calibration accuracy because it relied on radiance differences, not the absolute value of the radiance. However, the method requires a fairly fine pixel size. Tanré et al. (1988) demonstrated the technique using 30 m resolution data. AVHRR with its 1 km resolution suffers a 40% reduction in contrast sensitivity, but is still useable according to Holben et al. (1993).

Contrast reduction techniques based on the adjacency effect require multiple images of the same scene and cannot be used in near real time. The scene has to display contrasts at a scale consistent with the sensor resolution and has to be constant during the multiple acquisitions. Perhaps those were the reasons why the algorithm was never applied in any consistent or operational manner. This method was revisited using Landsat and MODIS imagery by Lyapustin et al. (2004).

7.5 Aerosol remote sensing using near UV observations

7.5.1 The UV Aerosol Index

The near-UV capability of aerosol remote sensing using satellite observations is a relatively recent development (Hsu et al., 1996; Herman, J.R. et al., 1997; Torres et al., 1998). As many important scientific discoveries, the UV method was the fortunate unintended result of refinements to the algorithm that since 1978 has been used for the retrieval of atmospheric total column ozone amount using measurements of backscattered ultraviolet (BUV) radiation by the TOMS sensor[1].

The UV Aerosol Index (UVAI) is basically a residual quantity resulting from the comparison between measured and calculated radiances (L_λ^M and L_λ^{cal} respectively) in the range 330–390 nm where gas absorption effects are small. The calculated radiances are obtained using a simple model of the Earth–atmosphere system consisting of a molecular atmosphere bounded at the bottom by a Lambert Equivalent Reflector (LER) (Dave and Mateer, 1967). In this approximation the upwelling UV radiance, L, at the top of the atmosphere is given by the expression

$$L^M(\Omega,\mu,\mu_0,\varphi,p_0) = L_0(\Omega,\mu,\mu_0,\varphi,p_0) + \frac{RF_0 T_t(\Omega,\mu,\mu_0,p_0)}{1 - RS(\Omega,p_0)}. \tag{7.5}$$

---

[1] The Aerosol Index was the direct result of Dr Pawan K. Bhartia's relentless effort to improve the TOMS total ozone retrieval algorithm. Although Dr Bhartia has never written a first author paper on the subject, he deserves full recognition for the discovery and refinement of this valuable aerosol remote sensing tool.

Equation (7.5) is the same as Eq. (7.1) except for the explicit dependency of the radiative transfer in the UV range on ozone content (Ω) and surface pressure (p_o). A key assumption in this model representation is that the reflectivity R of the column atmosphere's lower boundary is wavelength independent in the near UV. This hypothetical surface approximation is intended to account for the effects of surface, clouds, and aerosols which are not explicitly included in the radiative transfer calculations.

Although the Nimbus 7 TOMS sensor had been in operation since October 1978, the validity of the LER approximation to reproduce the spectral dependence of the measured near-UV radiances was directly tested for the first time in the mid-1990s in the transition from version 6 to version 7 of the TOMS algorithm. The test initially consisted of the calculation of the Lambertian reflectance term by solving for R in Eq. (7.5) at wavelengths in the range 330–380 nm where ozone absorption effects are very small. The Lambertian reflectance of the hypothetical surface accounting for the effects of clouds and aerosol but corrected for Rayleigh scattering is calculated from the expression

$$R_{\lambda_0} = \frac{L^M_{\lambda_0} - L_{o\lambda_0}}{E_{o\_\lambda_0} T_{t\_\lambda_0} + S_{\lambda_0}(L^M_{\lambda_0} - L_{o\lambda_0})}, \tag{7.6}$$

where the dependence variables have been dropped for simplicity. The terms L_o, T_t, E_o and S are obtained from Rayleigh scattering calculations. The term R is calculated at two wavelengths λ_0 and λ, generally 380 and 340 nm, respectively. According to the assumed non-wavelength dependence of the hypothetical Lambertian surface the residual quantity,

$$\Delta R = R_{\lambda_0} - R_\lambda, \tag{7.7}$$

should be zero when surface reflection and cloud scattering processes as well as particle scattering and absorption are adequately represented by the simple LER model. On the other hand, non-zero differences are produced when any of the above radiative transfer effects are not adequately represented by the LER approximation as predicted by the early work of Dave et al. (1978).

Examination of global maps of ΔR indicated that UV-absorbing aerosols were by far the most important source of positive differences (Hsu et al., 1996). Hence the term Absorbing Aerosol Index (AAI) or simply, Aerosol Index (AI), was coined to refer to this residual quantity.

The Aerosol Index, initially defined in terms of a reflectivity difference, was later modified to its current form in terms of radiances (Herman, J.R. et al., 1997) to eliminate large angular dependencies at mid-latitudes that made difficult the interpretation of the observed reflectivity differences.

In radiance terms, the AI is calculated as the difference between the measured and the calculated spectral contrast expressed as

$$AI = -100\left[\log\left(\frac{L^M_\lambda}{L^M_{\lambda_0}}\right) - \log\left(\frac{L^{cal}_\lambda}{L^{cal}_{\lambda_0}}\right)\right] \tag{7.8}$$

The term $L_{\lambda_0}^{cal}$ is obtained from Eq. (7.5) using R_{λ_0} as calculated from Eq. (7.6) so that

$$L_{\lambda_0}^{cal} = L_{\lambda_0}^M.$$

Thus, Eq. (7.8) reduces to

$$AI = -100 \log \left[\frac{L_{\lambda}^M}{L_{\lambda}^{cal}} \right]. \tag{7.9}$$

The log representation is used for historic reasons associated with the ozone retrieval algorithm. A unit AI value represents a 2.3% radiance change. Thus, the AI represents the error in estimating the satellite radiance at λ from radiance measurements at λ_0 assuming a Rayleigh-only atmosphere bounded at the bottom by a spectrally invariant Lambert-Equivalent Reflector.

7.5.2 Aerosol Index properties

The AI is generally positive for absorbing aerosols. The magnitude of the AI associated with absorbing aerosols depends on optical depth, particle size distribution, single scattering albedo, and height of the aerosol layer above surface (Herman, J.R. et al., 1997; Torres et al., 1998). The AI detects absorbing aerosols over water and all terrestrial surfaces including deserts and ice/snow covered surfaces (Hsu et al., 1999). The AI also detects aerosols in cloud–aerosol mixtures, and above cloud decks (Torres et al., 2011).

Non-absorbing aerosols yield negative AI values and their magnitude depends mainly on optical depth and, to a lesser extent, on particle size distribution with small size aerosols (radius less than 0.4 μm) yielding a larger magnitude of AI than do larger particles [Torres et al., 1998].

Since other non-aerosol- related geophysical effects yield a residue (Eq. (7.7)), it is necessary to make use of threshold values to separate the actual aerosol signal from those from other sources. This is particularly true over the oceans where both positive and negative residues of magnitude less than unity are often associated with ocean color effects. For instance, pure water absorption and colored dissolved organic matter in the remote oceans may yield values as high as 0.7, whereas chlorophyll absorption in coastal waters produces negative residues of about the same magnitude. These overlapping effects over the oceans are not a serious limitation for the detection of absorbing aerosols since the magnitude of the AI associated with the long-range transport of smoke and dust is often several times larger than the background ocean signal. However, the use of the negative AI as a proxy of non-absorbing aerosols over the oceans is severely limited as the background oceanic signal is as large (or even larger) than the aerosol related signal, because aerosol optical depths of oceanic non-absorbing aerosols are generally small. The background effect over land areas is not as large as over the oceans although a threshold detection limit must still be used to separate the aerosol signal from that associated with the wavelength dependent surface albedo in the near-UV region especially over arid and semi-arid regions. The physical meaning of the residue at solar zenith angles larger than about 70° remains uncertain.

The absorbing aerosol index has been used in a variety of applications including the mapping of the global distribution of the frequency of occurrence of carbonaceous and desert aerosols (Herman, J.R. et al., 1997), volcanic ash (Seftor et al., 1997) determination of sources of dust aerosols (Prospero et al., 2002; Yoshioka et al., 2005; Washington et al., 2003), validation of Chemical Transport Models (CTM) (Ginoux et al., 2001; Zender et al., 2003), characterization of biomass burning emissions (Duncan et al., 2003), correction of aerosol effects on total ozone retrieval (Torres and Bhartia, 1999), longterm analysis of global aerosol load (Li et al., 2009), and many more.

7.5.3 Benchmark aerosol events

In addition to the seasonal aerosol activity associated with the annual cycles of agriculture-related biomass burning and desert dust mobilizations across the oceans, the UV Aerosol Index captured synoptic scale episodic aerosol events of natural or anthropogenic origin. A few historic aerosol events that took place prior to the development of the UV remote sensing technique are briefly described here. The satellite depiction of these events in terms of the AI was not observed in real time but the impact on the global environment and the nature of their origin made headlines around the world as major environmental perturbations.

The 1987 Great China Fire

The great China Fire, one of the most destructive fires over the last three decades (Cahoon et al., 1994), was first detected on May 2, 1987 in the Heilongjiang Province of northeastern China. It burned for about three weeks, affecting an area of 1.3 million acres of prime forest and resulting in the loss of over 200 lives and 5000 homes (Cahoon et al., 1994). The Nimbus 7 TOMS Aerosol Index record shows that the synoptic scale smoke plume was transported eastward across the Pacific Ocean, then across Canada and the Atlantic Ocean, reaching as far as central Poland on May 15. Figure 7.6 depicts the spatial extent of the smoke plume on May 10th in terms of the Aerosol Index. The observed discontinuity of the aerosol plume over the Northern Pacific is related to the 24 hr difference between the satellite observations on both sides of the International Date Line.

The 1988 Yellowstone Park Fires

Another historic conflagration of continental proportions was the fires in the United States' Yellowstone National Park located in the states of Wyoming, Montana and Idaho. The Yellowstone Park fires occurred in 1988, the year when the driest summer in the park's recorded history took place. About 793,000 acres or 36% of the park's area burned over a two month period from July to September (Romme and Despain, 1989). In spite of the huge efforts in containing the fires, it was only the first snowfall of early September that began dampening the fires although the last of the smoldering flames were not extinguished until November. Over much of this period, a dense smoke layer covered a large fraction of the Continental US, Canada and Northern Mexico. The Nimbus 7 TOMS Aerosol Index image in Figure 7.7 depicts the re-circulated carbonaceous aerosol layer on September 9th. Nim-

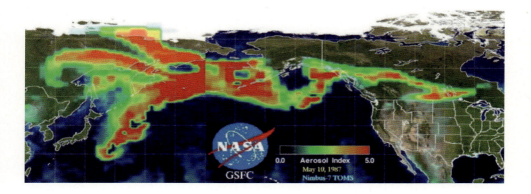

Figure 7.6 May 10, 1987 snapshot of the aerosol layer generated by the Great China Fire in terms of the TOMS Aerosol Index.

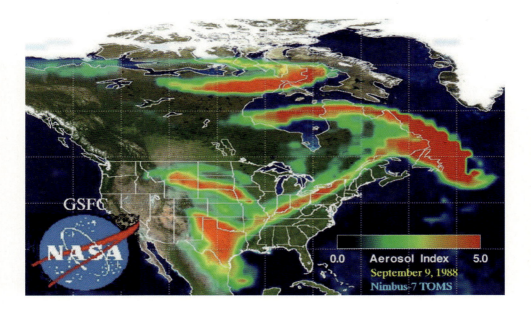

Figure 7.7 Geographical extent of the Yellowstone Park fires plume registered by the Nimbus7 TOMS aerosol Index on September 9, 1988.

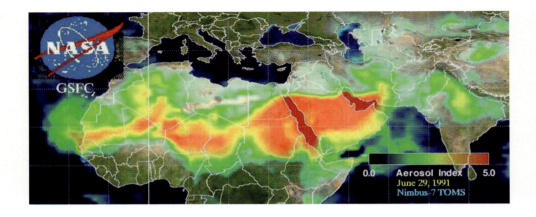

Figure 7.8 The east-to-west extent across Northern Africa of the Kuwaiti Oil fires on June 29, 1991 as seen by TOMS.

bus 7 TOMS data shows that smoke from the Yellowstone Park fires was also transported across the Atlantic Ocean, reaching Western Europe.

The 1991 Gulf War oil fields fires

In the aftermath of the January 1991 armed conflict in Kuwait, over 700 oil wells were set ablaze. The resulting environmental catastrophe produced hundreds of oil spills across the Kuwaiti desert. The oil fires burned for over a 10-month period. The last fire was reportedly extinguished in November 1991. Although it was initially believed that most of the smoke was confined to the Persian Gulf region (Cahalan, 1992), Aerosol Index observations by the Nimbus-7 TOMS sensor would show that the massive smoke plume was actually transported thousands of kilometers away from the source region across Northern Africa following the same path in which Saharan dust is annually mobilized out of the Saharan desert and across the North Atlantic Ocean to the Americas. The plume from the Kuwaiti oil fires may have possibly reached the Pacific Ocean as suggested by Lowenthal et al. (1992) based on observations in Hawaii. The long-range westward transport of aerosol material generated in Kuwait is clearly shown in Figure 7.8 on June 29, 1991.

7.5.4 Application to other sensors

Since its inception, the UV Aerosol Index has been implemented as either an operational or research product in inversion algorithms applied to instruments with UV observing capability such as GOME, GOME-2, SCIAMACHY, and OMI. Some of these products and the applications stemming from their use are described for GOME (Gleason et al., 1998; de Graaf et al., 2005), SCIAMACHY (de Graaf et al., 2007; Penning de Vries et al., 2009), OMI (Torres et al., 2007; Ahn et al., 2008; Dirksen et al., 2009). The near-UV advantages for aerosol detection, i.e. sensitivity to aerosol absorption and low surface albedo, are used in a retrieval algorithm as described in the following section.

7.5.5 Quantitative use of near-UV observations

The information content of the Aerosol Index is turned into quantitative estimates of aerosol extinction optical depth (τ) and single scattering albedo(ϖ) at 380 nm by application of an inversion algorithm to TOMS near-UV observations (Torres et al., 1998). The τ and ϖ quantities are derived using a standard inversion algorithm that uses pre-computed reflectances for a set of assumed aerosol models. Three major aerosol types are considered: desert dust, carbonaceous aerosols associated with biomass burning, and weakly absorbing sulfate-based aerosols. Each aerosol type is represented by seven aerosol models of varying single scattering albedo, for a total of 21 microphysical models. For a chosen aerosol type, represented by a sub-set of seven aerosol models, the extinction optical depth and single scattering albedo are retrieved by examining the variability of the relationship between the 340–380 nm spectral contrast and the 380 nm reflectance. Since the retrieval procedure is sensitive to aerosol vertical distribution, the aerosol layer height must be assumed. The choice of vertical distribution varies with aerosol type and location (Torres et al., 2002). Surface reflectance is prescribed using an existing climatological database derived from the long-term TOMS reflectivity record (Herman and Celarier, 1997). The near-UV retrieval algorithm was applied to the Nimbus 7 (1979–1992) and Earth Probe (1996–2001) TOMS records (Torres et al., 2002) as part of the Global Aerosol Climatology Project. The TOMS aerosol climatology was the first global record of atmospheric aerosol load that provided data over both the oceans and the continents. Figure 7.9 shows the combined Nimbus 7-Earth Probe TOMS aerosol optical depth record from 1979 to 2001. No TOMS measurements were available during the 1993–1996 three-year gap.

The Deep Blue algorithm (Hsu et al., 2006) was inspired by the proven retrieval capability in the UV over arid and semi-arid areas. Although strictly not in the UV, at the MODIS 412 nm channel deserts appear sufficiently darker than at the other available MODIS wavelengths allowing the retrieval of aerosol properties. The Deep Blue application, currently used in the operational MODIS algorithm for the retrieval of aerosol optical depth over deserts, is discussed in detail in Chapter 8.

7.6 Multiangle, polarization and geosynchronous capabilities

7.6.1 Multiangle

The above discussion of the evolution of the occultation, dark target and UV methods outlines the historical progression of the three main methods of satellite aerosol remote sensing that resulted in operational or semi-operational products, producing a long time series. In addition to these three historical lines were other methods making use of additional capabilities inherent in other satellite sensors. Primarily, these include multiangle observations of the same pixel, polarization and multiple observations within the same day. These additional capabilities were explored during the same time period when data from AVHRR and TOMS were being used to derive aerosol information. Here, we give a brief overview of these early explorations that in some cases became predecessors to the EOS era satellite sensors that were designed specifically with aerosol in mind.

Although the concept was proposed earlier (Martonchik and Diner, 1992), multiangle aerosol retrievals began with the Along-Track Scanning Radiometer-2 (ATSR-2) (Stricker et al., 1995). This instrument, launched in 1995 on the European Remote Sensing Satellite (ERS) provided the first opportunity to view each Earth scene from multiple angles. ATSR-2 used a conical scanning mechanism that views a particular scene first at approximately 56° in the forward direction and then 2 minutes later from nadir. The instrument measured in seven wavelengths, three in the infrared and not used for aerosol retrievals, and four at 0.555, 0.659, 0.865, and 1.6 µm that were used for aerosol retrieval. Algorithms to derive aerosol from these multiwavelength multiangle views were developed by J.P. Veefkind and his co-authors (Veefkind et al.,1998, 1999). In part based on work by Flowerdew and Haigh (1996), the method relies on the ATSR-2 natural viewing geometry to avoid the surface hotspot from vegetation and assumes a Lambertian surface. The retrieval is based on the radiance measurement at the top of the atmosphere given by Eq. (7.1). Because there are only two geometries, forward and nadir, the geometrical dependency can be denoted by 'f' and 'n' superscripts. Furthermore, the method assumes that the surface reflectance

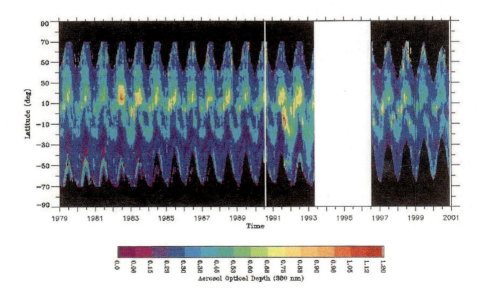

Figure 7.9 TOMS 380 nm AOD 1979–2001 global record from Torres et al. (2002). Reproduced with permission from the American Meteorological Society.

viewed at the forward angle is directly proportional to the reflectance viewed at nadir, such that

$$R^f(\lambda) = k \bullet R^n(\lambda),\qquad(7.10)$$

where R is the Lambertian surface reflectivity of Eq. (7.1), but designated either for the forward view R^f or the nadir view R^n, and k is the ratio between forward and nadir surface reflectivity, which is assumed to be invariant with wavelength. The method assumes that the atmosphere dominated by fine mode aerosol is effectively invisible to radiation in the short wave infrared (Figure 7.4 and Kaufman et al., 1997a), and uses the measured top-of-atmosphere radiances in the 1.6 μm channel to calculate k. The k parameter is then applied to the visible wavelengths. Thus, the surface reflectance model is known even if the surface reflectance is not, and the difference in measured radiance of the two geometries is dependent only on parameters that are a function of the aerosol optical depth, τ. The method requires an assumed aerosol model that provides the intrinsic aerosol optical properties, and a Look-Up Table to match with the measured radiance differences.

The ATSR-2 method was applied and validated on the east coast of the United States (Veefkind et al., 1998). Subsequent applications of the dual-view method included aerosols over Europe, India and Africa (Gonzales et al., 2000; Robles-Gonzales and Leeuw, 2006, 2008). A similar method was adapted to the Advanced Along Track Scanning Radiometer AATSR (Grey et al., 2006). While the ATSR-2 retrievals never created a long time series of global aerosol measurements, as did AVHRR and TOMS, the ATSR-2 retrievals did pioneer the path towards multiangle measurements in general. Later, the Multiangle Imaging Spectro Radiometer (MISR) would continue the tradition of using multiple angles to derive aerosol information and create a long-term operational data base of aerosol products.

7.6.2 Polarization

Polarization adds significant capability to aerosol retrieval, especially when combined with multiangle and multiwavelength measurements. Early studies based on ground-based measurements and planetary observations suggested polarization could play an important role in aerosol remote sensing (Deuzé et al., 1988; Hansen and Hovenier, 1974). Aerosol retrievals using polarization were explored theoretically and were found to provide significant advantage in terms of information content (Mishchenko and Travis, 1997) and in being much less sensitive to surface properties (Leroy et al. 1997) than intensity measurements alone.

The first Earth-viewing space instrument employing polarization was the Polarization and Directionality of the Earth Reflectance (POLDER) that was launched in 1996. POLDER is a wide field-of-view instrument that images ±43° and ±51° along-track and cross-track, respectively, and produces a pixilated 242×274 pixel image using a CCD detector array (Deschamps et al., 1994). The result is a series of images as the instrument moves along its orbit, with several images partially overlapping, creating multiple views of the same pixel. A filter wheel provides measurements at 9 channels from 443 nm to 910 nm, with three of those channels each measured at three polarization states (Deschamps et al., 1994). The aerosol retrieval algorithm makes use of this information to derive aerosol optical depth and Ångström exponent over ocean (Deuzé et al., 1999) and the fine mode aerosol optical depth over land (Herman et al., 1997; Deuzé et al., 2001). There have been three POLDER instruments in space. The first two were of relatively short duration, August 1996 to June 1997 and December 2002 to September 2003, respectively, due to malfunctions on the ADEOS satellites that provided the platform for these instruments. The third POLDER was launched on the PARASOL satellite in December 2004. PARASOL joined the international afternoon constellation of satellites in polar orbit and remained in this constellation until December 2009. As of this writing, PARASOL continues collecting and transmitting data. Because POLDER/PARASOL is a contemporary of the other Terra and A-Train sensors, a more detailed description of the retrieval algorithm is left for Chapter 8.

7.6.3 Retrievals from geosynchronous satellites

Geosynchronous satellites offer fine temporal resolution observations of aerosols on a regional basis so that particle plumes can be monitored over the course of the day. The historical difficulty with using these sensors for aerosol retrievals has been the uncertain calibration. Fraser et al. (1984) used a vicarious calibration method to calibrate the visible channel on GOES-1, and Knapp and Vonder Haar (2000) did the same to calibrate GOES-8. Fraser et al. (1984) and Knapp et al. (2002) describe a technique that uses a composite method to fix the surface reflectance for every time of day of observation. The minimum reflectance for each pixel for each time of day of a two-month period was corrected for gaseous absorption and a minimum background aerosol optical depth. The result was an estimate of the surface reflectance that was assumed invariant over the entire period of interest. Measured reflectance that exceeded this minimum reflectance was assumed to be due to aerosol loading. A Look Up Table with an assumed aerosol model provided the means to retrieve the aerosol optical depth, using a basic dark target approach, but with a known surface reflectance. Other similar attempts include Wang et al. (2003) for GOES and Popp et al. (2007) for the Spinning Enhanced Visible and Infrared Imager (SEVIRI) on the geosynchronous Meteosat Second Generation (MSG) satellite. The procedure was made operational at NOAA under the name GOES Aerosol/Smoke Product (GASP) (Prados et al., 2007).

7.7 Summary

The primary space-based passive methods to derive aerosol are:

(1) Occultation methods that measure the extinction of solar radiation and provide vertical profiles of aerosol extinction through the stratosphere. These methods have a long history and make possible a multi-decadal time series that has illustrated the effect of explosive volcanic eruptions on stratospheric aerosol.

(2) Dark target methods that retrieve the aerosol optical depth through the total column of the atmosphere. These methods have to overcome issues with calibration of the sensor, and they require *a priori* assumptions of aerosol optical properties and surface reflect-

ance. At best, the dark target retrieval can retrieve aerosol optical depth and some information of spectral dependence or particle size. Applied to AVHRR, dark target methods have provided a long time series over the oceans that suggest trends in global aerosol loading, but questions concerning sampling, drift and merging of records from different satellites prevent the long-term AVHRR trend from being universally accepted.

(3) UV methods also retrieve through the total column of the atmosphere, although their standard product is a robust Aerosol Index (AI) that provides a useful qualitative measure of aerosol loading and aerosol light absorption. Retrievals of quantitative aerosol optical depth can also be derived, but with *a priori* assumptions of aerosol optical properties, surface reflectance and aerosol layer height. However, the UV method is less sensitive to surface reflectance than the dark target methods. The UV method applied to the TOMS data record provides our longest existing record of aerosol optical depth information.

Besides these three main methods applied to the SAM/SAGE, AVHRR and TOMS measurements, respectively, to produce long aerosol time series, other methods have been developed to make use of multiangle, polarization and geosynchronous capabilities. These satellite aerosol products applied to data collected in the 1980s and 1990s made way for the era of the modern sensors that began with the launch of Terra in late 1999. Occultation, dark target and UV methods, embellished with better calibration, polarization, more wavelengths and more angles provide the basis of EOS-era satellites of the 2000s.

8 Recent instruments and algorithms for passive shortwave remote sensing

Lorraine A. Remer, Colette Brogniez, Brian Cairns, N. Christina Hsu,
Ralph Kahn, Piet Stammes, Didier Tanré, Omar Torres

8.1 Introduction

Passive remote sensing of aerosol using the shortwave spectrum draws on a long heritage of experience that began with three main techniques described in Chapter 7: occultation methods, dark target approaches, and spectral ultra-violet (UV) algorithms. Beginning in the 1970s, these techniques have been applied to instruments flown on a series of different satellite platforms and have produced important time series of aerosol parameters that span decades. This heritage is especially valuable given the fact that the early downward-viewing sensors used for aerosol retrieval – the Advanced Very High Resolution Radiometer (AVHRR), the Total Ozone Mapping Spectrometer (TOMS) and even the Geostationary Operational Environmental satellites (GOES) – were designed for purposes other than retrieving aerosol. However, the success of using these instruments for aerosol characterization motivated the development of sensors designed with aerosol retrievals in mind. Improved spatial resolution, narrower spectral channels, increased spectral range and density, enhanced capability in terms of multiple angular views of the same scene and polarization are some of the specific improvements designed into the sensors flying during the 2000s that were intended to provide better aerosol retrievals than AVHRR and TOMS.

There is no clear-cut division between heritage sensors and 'modern' sensors in terms of time period, as seen in Figure 8.1. The occultation instruments (SAM2, SAGE, SAGE2, SAGE3, HALOE, POAM2 and POAM3) have been making a continuous data record that overlaps specific sensors from 1979 to 2006, thus neatly bridging from the heritage era of the 1970s and 1980s to the modern era of the 2000s. The AVHRR aerosol record spans 1978 to the present and TOMS aboard first NIMBUS 7 and then EarthProbe (EP) begins in 1978 and extends with one interruption to 2006. This chapter introduces some of the sensors and aerosol algorithms that are creating operational data records in the 2000s. By 'operational' it is implied that the sensors, algorithms, products and validation are supported by various space agencies for use by the entire research community, public agencies, applied scientists, the press and even average citizens. Not described in this chapter are the geostationary sensors, the Advanced Along Track Scanning Radiometer (AATSR), MEdium Resolution Imaging Spectrometer (MERIS) and the Sea-viewing Wide Field of view Sensor (SeaWIFS), even though these sensors have produced aerosol products into the same time period.

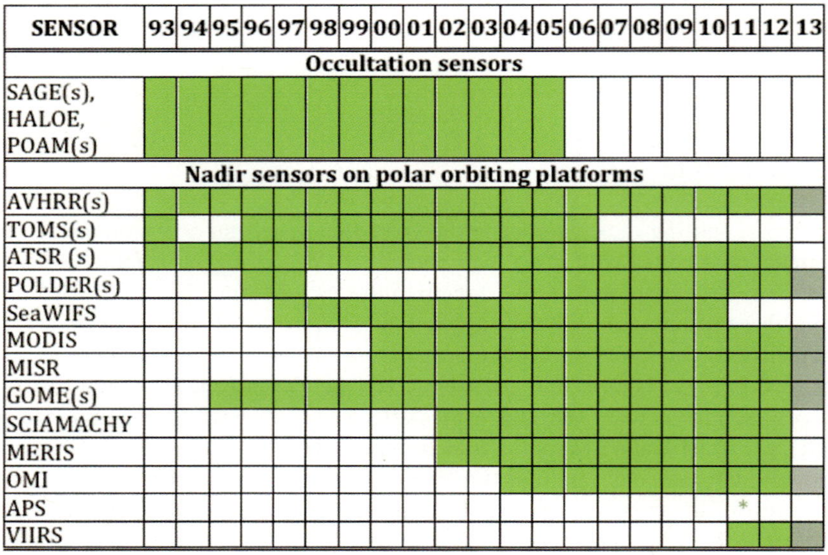

SENSOR	93	94	95	96	97	98	99	00	01	02	03	04	05	06	07	08	09	10	11	12	13
Occultation sensors																					
SAGE(s), HALOE, POAM(s)																					
Nadir sensors on polar orbiting platforms																					
AVHRR(s)																					
TOMS(s)																					
ATSR (s)																					
POLDER(s)																					
SeaWIFS																					
MODIS																					
MISR																					
GOME(s)																					
SCIAMACHY																					
MERIS																					
OMI																					
APS																	*				
VIIRS																					

Figure 8.1 Time line of passive satellite sensors providing aerosol retrievals grouped into occultation sensors that observe stratospheric and upper tropospheric aerosol and nadir-viewing sensors aboard polar orbiting satellites. The time line begins in 1993, although the occultation sensors, AVHRR, TOMS and ATSR began contributing before 1993 and as early as 1978. Sensors still functional as of 2012 and expected to continue operating show extended functionality as gray boxes in 2013. ATSR(s), MERIS and SCIAMACHY ended their operation in April 2012 when their platform, ENVISAT, lost communication and ended its mission. VIIRS was successfully launched in October 2011. APS, the Aerosol Polarimetric Sensor, was launched in March 2011, but unfortunately did not reach orbit.

The geostationary sensor, GOES, produces a publicly available aerosol data record, but was not designed for aerosol retrievals and is considered a heritage instrument that is described in Chapter 7. The Spinning Enhanced Visible InfraRed Imager (SEVIRI) is another geostationary instrument, but with better aerosol capability. It is not included in this chapter because its publicly available data set (the GlobAerosol product at http://www.globaerosol.info/) is not in wide use (Thomas et al., 2009; Bulgin et al., 2011), while its research-level products are not operational (Govaerts et al., 2010; Wagner et al., 2010). AATSR is a direct descendant of ATSR and ATSR-2, which are described in Chapter 7. A detailed look at AATSR aerosol products can be found in Holzer-Popp et al. (2002), Grey et al. (2006) and in the AATSR product handbook at envisat.esa.int/handbooks/aatsr/CNTR2.htm. MERIS and SeaWIFS were primarily designed to support remote sensing of ocean ecology and not aerosol. Aerosol algorithms applied to these sensors were primarily for atmospheric correction of the imagery (see Chapter 7). Scenes with high aerosol loading were simply masked from the aerosol retrieval (Myhre et al. 2005). Recently, new algorithms have been applied to the SeaWIFS data record, which have produced a better representation of aerosol loading over oceans. These new algorithms are patterned on the MODIS aerosol retrieval over oceans, described below, but applied to an instrument with much narrower spectral range (Sayer et al., 2011).

In this chapter we provide an overview of the instruments, aerosol algorithms and products associated with several of the sensors that have produced an operational data record into the 2000s. Included in our discussion will be the occultation instruments, Stratospheric Aerosol and Gas Experiment (SAGE(s)) and Polar Ozone and Aerosol Measurements (POAM(s)). Then we survey the downward-viewing polar-orbiting sensors: the MODerate resolution Imaging Spectroradiometer, (MODIS) with its wide spectral range, the Multiangle Imaging SpectroRadiometer (MISR) applying its multiple angle looks at the same scene, the Ozone Monitoring Instrument (OMI) with its channels in the ultraviolet range, Global Ozone Monitoring Experiments (GOME(s)) and SCIAMACHY (Scanning Imaging Absorption Spectrometer for Atmospheric Chartography) with their hyperspectral capabilities, the Polarization and Anisotropy of Reflectances for Atmospheric Science coupled with Observations from a Lidar (PARASOL) with its multiangle looks and polarization, and finally the Aerosol Polarimetry Sensor (APS) offering highly accurate multiangle polarimetry. Some of these sensors are still flying and taking data at the time of this writing. Some have retired, and APS unfortunately was lost on launch in early 2011. It is included here because it still represents an interesting application of remote sensing theory laid out in Chapters 2, 3 and 5 and a reflight of a twin instrument is being considered.

Many of the 'modern' sensors are producing an operational product that is validated, archived and easily available to the public. The research and applied science communities have responded enthusiastically to the use of these data. Satellite remote sensing aerosol products are used to constrain climate models, to study the importance of heterogeneous processes in the atmospheric chemistry, to find relationships between clouds and aerosols, to quantify transport of particulates across oceans, to provide information in public health studies, and to improve air quality and visibility forecasts. Besides the operational products, exploratory new products and new capabilities are being explored, including combining information from several sensors that fly in formation. These cutting-edge research products are addressed in the last section of the chapter.

8.2 SAGE(s)-POAM(s): Occultation instruments

8.2.1 Missions and instrument description

The Stratospheric Aerosol and Gas Experiment (SAGE) series of instruments performing occultation measurements, including the SAM II experiment, began in 1978, as described in Chapter 7. These were NASA instruments and NASA experiments designed to explore the composition of the stratosphere and later the upper troposphere. SAM II operated at a unique wavelength and was dedicated to aerosol detection (McCormick et al., 1979), but the SAGE instruments, focused also on ozone measurements, were multi-channel instruments (Mauldin, 1985, Chu et al., 1989).

All SAGE experiments followed the same procedure. The pointing system tracked the center of brightness of the solar disk in azimuth, while scanning rapidly across the sun vertically relative to the Earth's limb. The scanning process was performed during the entire sunrise/sunset event enabling all tangent altitudes in the atmosphere to be sampled several times while viewing different parts of the solar disk. Transmissions were computed using the unattenuated irradiance coming from the same position on the sun (see Chapter 4). SAGE II and III fields of view are 300 arcseconds horizontally and 30 arcseconds vertically, which translates to about 0.5 km at the tangent altitude. Transmissions are grouped in boxes 1 or 0.5 km wide (for SAGE II and SAGE III, respectively) in altitude and an average transmission profile is derived along with uncertainty estimates on a 1 or 0.5 km grid. Such transmission profiles are produced in all channels.

Channel, nm	Species contribution
384.5	Aerosol, O_3, NO_2
448.5	Aerosol, O_3, NO_2
520.3	Aerosol, O_3, NO_2
601.2	Aerosol, O_3, NO_2, H_2O
675.6	Aerosol, O_3, NO_2
755.4	Aerosol, O_3
869.3	Aerosol, O_3, H_2O
1021.6	Aerosol, O_3
1545.2	Aerosol, CO_2, H_2O

Table 8.1 Species contribution in the SAGE III aerosol channels for the solar mode. The omnipresent contribution of the Rayleigh scattering is not indicated.

SAGE II was a spectrometer making measurements in seven channels in the 386 to 1020 nm range. It was launched in October 1984 into a 57-degree inclination orbit yielding a geographical coverage from about 80°N to 80°S. The SAGE II mission terminated in August 2005. SAGE III was an improved version of its predecessor with a CCD array detector comprising 809 pixels in the 290–1020 nm range, with a spectral resolution of about 1–2 nm. Out of these pixels only 87 were transmitted to the ground, some of them grouped in "resolved channels": 19 pixels between 433 and 450 nm dedicated to NO_2 retrieval and 10 3-pixel averages between 562 and 621nm for ozone. In addition, there is a photodiode at 1545 nm dedicated to aerosol characterization. A mobile attenuator permitted both solar and lunar occultation measurements. It was launched in December 2001 into a sun-synchronous polar orbit allowing observations in middle and high latitudes in the solar occultation mode. The SAGE III mission ended in March 2006.

Table 8.1 gives the list of the aerosol extinction channels only for the solar mode, since there is no aerosol product for the lunar mode. The main target species and interfering species are specified.

8.2.2 Algorithm description

The official solar retrieval algorithms for SAGE II and SAGE III both start with the species separation followed by the vertical inversion. Other inversion techniques exist, employing a reverse strategy, but they have been applied only to case studies or for validation purposes and have not been applied to the whole archive (Chu et al., 1989; Rusch et al.; 1998, Brogniez et al., 2002). The official algorithms for SAGE II and SAGE III follow similar approaches but are not identical (Chu et al., 1989; SAGE III ATBD, 2002). Only the main steps of the SAGE III algorithm are briefly described below.

At each wavelength the slant path optical depth (SPOD) can be considered as the sum of the SPOD of several species, and can be expressed in the general form

$$\tau^{SP}_{total}(Z_t,\lambda) = \tau^{SP}_{Ray}(Z_t,\lambda) + \sum_{i=1}^{N} \tau^{SP}_{gas-i}(Z_t,\lambda) + \tau^{SP}_{aer}(Z_t,\lambda), \tag{8.1}$$

where $\tau^{SP}_{Ray}(Z_t,\lambda)$, $\tau^{SP}_{gas-i}(Z_t,\lambda)$, and $\tau^{SP}_{aer}(Z_t,\lambda)$ are the Rayleigh (molecular), *gas-i* species and aerosol SPOD respectively, N being the number of absorbing gases at the considered wavelength. Z_t is the tangent altitude as defined in Chapter 4 and Figure 4.4. Deriving aerosol requires removing Rayleigh and gas contributions from the total SPOD.

Each SPOD can be expressed as a function of extinction as

$$\tau^{SP}(Z_t,\lambda) = 2 \int_{Z_t}^{Z_{top}} \sigma_e(\lambda,Z)\, d\ell\,(\lambda,Z_t,Z), \tag{8.2}$$

where Z_{top} is the top of the atmosphere altitude, $\sigma_e(\lambda,Z)$ is the extinction coefficient, $d\ell$ is the geometric slant path between the altitudes Z and $Z+dZ$, accounting for the refraction. In case of an absorbing gas, the extinction coefficient is the product of the absorption cross-section with the number density $\sigma_e(\lambda,Z) = \sigma_a^g(\lambda,Z).n(Z)$.

The effects of refraction and the Rayleigh SPOD, $\tau_{Ray}^{SP}(Z_t, \lambda)$, are computed using temperature and pressure profiles available from the National Center for Environmental Prediction (NCEP) reanalyses. Ozone and nitrogen dioxide slant path number density profiles are first retrieved from the resolved channels using a multilinear regression technique. Aerosol SPOD is then derived in the nine channels after removal of all gas contributions using a set of spectral cross-sections. The availability of channels identified to retrieve NO_2 and O_3 enables the aerosol retrieval to avoid using a functional form for aerosol spectral variation as was necessary in the SAGE II aerosol retrieval.

The last step, the spatial inversion, employs a modified-Chahine method to derive gas number density and spectral aerosol extinction vertical profiles.

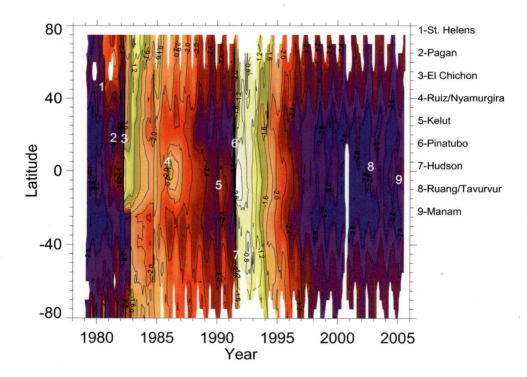

Figure 8.2 Stratospheric aerosol optical thickness ($\log_{10}$) at 1000 nm from SAM II, SAGE I, SAGE II (version 6.2) and SAGE III (version 4). The gaps due to missing observations and to El Chichon and Pinatubo's perturbations have been filled with other measurements such as lidar data following the method described in Hamill and Brogniez (2006). The main volcano eruptions are indicated. (Courtesy L. Thomason, NASA LaRc).

The retrieval algorithm provides uncertainty estimates on all products, which account for: (i) the total SPOD measurement uncertainties; (ii) the Rayleigh optical depth estimate uncertainty; and (iii) the uncertainties resulting from the removal of contributions by other interfering species. It is thus obvious that the accuracy of the products is dependent on knowledge of all gas absorption cross-sections.

Such a retrieval algorithm means that aerosol extinction coefficients are a by-product of ozone and nitrogen dioxide retrievals. Therefore, aerosols are better retrieved when the gaseous absorption contribution of the interfering species is relatively weak. This condition is not fulfilled when the aerosol loading is low, as it was during the SAGE III lifetime, making the proper choice of the gaseous cross-sections critical.

Details of the current versions of the species separation algorithms can be found for SAGE II in Thomason et al. (2001) and for SAGE III in Thomason et al. (2010).

As stated in Section 7.3, the SAGE series of instruments has permitted the establishment of the longest record of stratospheric aerosol extinction coefficient at ~1000 nm, enabling the characterization of the temporal and spatial evolution of the stratospheric aerosol loading. Figure 8.2 shows a color map contours of the derived aerosol optical depth. The gaps due to missing observations and to El Chichon and Pinatubo's perturbations have been filled with other measurements such as lidar data following the method described in Hamill and Brogniez (2006). One can clearly see that very low values (about 0.001) were occurring just before the El Chichon eruption and from 1997 up to the end of the missions (background level corresponding to periods without strong volcanic activity). On the other hand, very large values occurred following Pinatubo's eruption (0.1).

The Polar Ozone and Aerosol Measurement (POAM) II and III were developed by the Naval Research Laboratory to study the polar stratosphere (Glaccum et al., 1996). The POAM measurement procedure is different from that of SAGE as the instrument does not scan across the solar disk but stares at the center of its brightness. Then, as for the SAGE series, transmissions are computed using the unattenuated irradiance coming from the same position on the sun. The FOVs are about 3000 arcseconds horizontally and 50 arcseconds vertically, which translates to about 0.8 km at the tangent altitude. Transmission profiles are produced in all channels along with uncertainty estimates on a 1 km grid.

The POAM II instrument makes measurements in nine channels in the 352 to 1060 nm range. It was launched in September 1993 into a sun-synchronous polar orbit, yielding a geographical coverage in latitude bands about 54–71°N and 63–88°S. The POAM II mission terminated in November 1996. POAM III launched in March 1998 into an identical orbit, thus with the same latitudinal coverage, was similar to its predecessor except that the 1060 nm channel was shifted to 1020 nm, as in the SAGE II&III instruments (Lucke et al., 1997). The POAM III mission ended in December 2005.

The POAM II&III official retrieval algorithms are identical; they start with the species separation followed by the vertical inversion (Lumpe et al., 1997, 2002), as for SAGE but the technique is different. They use a nonlinear optimal estimation technique to retrieve simultaneously from the total SPOD, the gas and aerosol SPOD. For that reason they assume that the spectral variations of the aerosol SPOD follow the empirical law

$$ln\tau_{aer}^{SP}(\lambda) = \mu_0 + \mu_1 \, ln(\lambda) + \mu_2 \, (ln(\lambda))^2, \tag{8.3}$$

where μ_i are the effective aerosol coefficients to be retrieved in the algorithm. Of course, in this processing refraction effects are accounted for. The spatial inversion employs a linear optimal estimation technique to derive gas number density and spectral aerosol extinction vertical profiles. As can be seen, an important difference between SAGE III and POAM aerosol retrievals is that the former does not assume a specific spectral variation of the aerosol extinction coefficients.

HALOE, also developed by NASA, focused on several gas species (ozone, methane, HCl, H_2O...) and aerosols. The HALOE mission operated from September 1991 to November 2005 onboard the UARS platform (Russell et al., 1993). Measurements were performed in eight infrared bands and aerosol extinction coefficients were provided at four infrared wavelengths (2.45 to 5.26 μm) (Hervig et al., 1995).

8.2.3 Size distribution retrieval and derived products

Retrievals of size distribution using occultation techniques are difficult and many techniques have been developed to infer the aerosol size distribution and the higher order moments such as aerosol surface area density (SAD) or volume density and effective radius (Wang et al., 1989; Steele and Turco, 1997; Thomason et al., 1997; Anderson et al., 2000; Yue et al., 2000; Timofeyev et al., 2003; Bingen et al., 2004). The wavelength range (visible and near-infrared) of available aerosol extinction measurements limits the performances of all techniques. Indeed, the study of the sensitivity of aerosol extinction to aerosol size ranges (via kernels, see Chapter 5) shows that visible to near infrared channels provide useful information for aerosols in about the 0.1–0.8 μm range, contrarily to smaller or larger aerosols. Consequently, SAGE operational products do not include size distribution characteristics (number density and mean radius) but only higher order moments, SAD and effective radius.

The reliability of the SAD inferred from aerosol extinction coefficients depends of course on the precision of the extinction coefficients and also on the aerosol amount. Thomason et al. (2008) have shown that SAGE II SAD is poorly reliable in cases of low stratospheric aerosol loading. According to Thomason et al. (2010) SAGE III SAD should be used with caution.

8.3 MODIS/AQUA/TERRA

8.3.1 Mission and instrument description

The MODerate resolution Imaging Spectroradiometers (MODIS) were launched aboard NASA's Terra and Aqua satellites in December 1999 and May 2002, respectively. The MODIS mission is cross-disciplinary addressing the Earth System from land, ocean and atmosphere perspectives. The atmospheric focus encompasses aerosols, clouds, water vapor and atmospheric temperature soundings, the interaction between these parameters and across disciplines, especially in terms of obtaining a better understanding of how these parameters affect the Earth as a system and its climate. MODIS on Terra is collocated with the Multiangle Imaging Spectro Radiometer (MISR) and on Aqua it flies in formation with a constellation of complementary sensors. Initially, MODIS aerosol algorithms were stand-alone procedures, but increasingly the advantages of combining information from several sensors are providing better retrievals and new products.

Terra crosses the equator in a descending orbit at nominal local time of 10:30 am, while Aqua crosses in an ascending orbit three hours later. Each sensor is independent, but working together the twin sensors on two different platforms with equator crossing times 3 hours apart allow for some analysis of diurnal signal, but more importantly provide a cross check on sensor calibration drift and artifacts. MODIS has 36 channels spanning the spectral range from 0.41 to 15 μm, and representing three spatial resolutions: 250 m (2 channels), 500 m (5 channels), and 1 km (29 channels). The aerosol retrieval makes use of eight of these channels (0.41–2.13 μm) to retrieve aerosol characteristics, and uses additional wavelengths in other parts of the spectrum to identify clouds and river sediments. MODIS scans cross track, observing each target at only one angle per orbit. Swath width is 2330 km, which nearly covers the globe each day and provides multiple daily views of high latitude locations.

8.3.2 Algorithm description

The MODIS aerosol algorithm is actually three separate algorithms: Dark Target over ocean, Dark Target over land and Deep Blue over land. The Dark Target algorithms over land and ocean follow from the AVHRR heritage of aerosol retrievals and are the original aerosol algorithms that were applied immediately after Terra launch. The Deep Blue algorithm is built on the SeaWiFS heritage and was implemented in 2006.

The Dark Target Ocean algorithm is aptly described in Tanré et al. (1997) and Levy et al. (2003) with updates in Remer et al. (2005). It is based on translating the elevated apparent reflectance of a dark surface target into aerosol information. Following Tanré et al. (1996), we know that the six reflectances measured from MODIS and used in the ocean retrieval (0.55–2.13 μm) contain three pieces of information about the aerosol. From this information we derive three parameters: the optical thickness at one wavelength $(\tau_{0.55}^{tot})$, the reflectance weighting parameter at one wavelength $(\eta_{0.55})$, which is defined below, and the effective radius, which is the ratio of the 3rd and 2nd moments of the aerosol size distribution. The MODIS inversion attempts to minimize the difference between the observed spectral radiance in the six MODIS channels and radiance precomputed in a look-up table (LUT). The look-up table models the total reflectance observed by satellite, which includes not only aerosol contributions, but also spatially and temporally constant atmospheric (Rayleigh) and ocean surface (chlorophyll, foam, whitecaps, and sun glint) contributions. The inversion uses the observed geometry and assumes that the total aerosol contribution is composed of a single fine and single coarse mode, and allows the inversion to choose from a set of four fine mode aerosol models and five coarse mode models. All size distributions are lognormal. The fine and coarse modes from the LUT are combined using η as the weighting parameter (Wang and Gordon, 1994),

$$\rho_{\lambda}^{LUT}(\tau_{0.55}^{tot}) = \eta\,\rho_{\lambda}^{f}(\tau_{0.55}^{tot}) + (1\text{-}\eta)\,\rho_{\lambda}^{c}(\tau_{0.55}^{tot}). \tag{8.4}$$

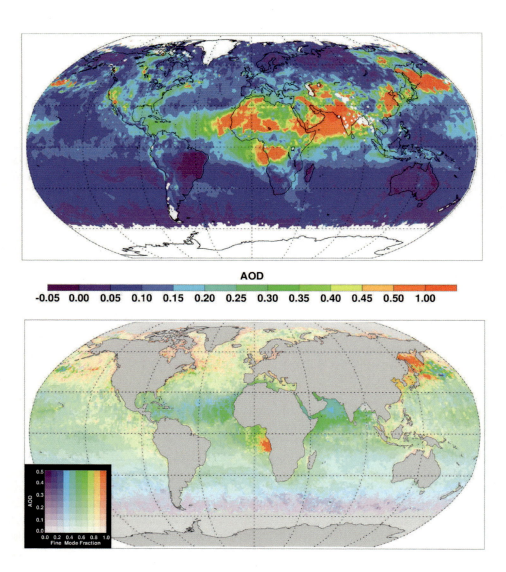

Figure 8.3 (Upper) Monthly mean aerosol optical depth from the three combined operational aerosol algorithms applied to the MODIS observations during July 2008. (Lower) Monthly mean fine fraction size parameter, η, defined in Equation 8.4 and representing the fraction of the total $\tau_{0.55}$ contributed by the fine mode component of the aerosol for July 2008. Graduations in color scale from purple to red represent changes in fine fraction from 0.0 to 1.0, respectively. Graduations in color intensity represent $\tau_{0.55}$ from 0.0 to 0.80. Size parameter products are not recommended for quantitative use when $\tau_{0.55} < 0.15$ over ocean. Size parameter products over land should be regarded as a diagnostic of the algorithm rather than a physical quantity and are therefore not shown in the lower plot.

Here "f" refers to fine mode and "c" to coarse mode, LUT refers to the reflectance from the look-up table. The spectral reflectance measured from the satellite that corresponds to the LUT value, ρ_λ^{LUT} ($\tau_{0.55}^{tot}$) for the determined values of η and $\tau_{0.55}^{tot}$, is a weighted average of the reflectance values for an atmosphere with a pure fine mode "f" and optical thickness $\tau_{0.55}^{tot}$ and the reflectance of an atmosphere with a pure coarse mode "c" also with the same $\tau_{0.55}^{tot}$. For each of the 20 combinations of one fine mode and one coarse mode, the inversion finds the pair of $\tau_{0.55}^{tot}$ and η that minimizes the differences between the measured reflectances and those from the LUT.

The Dark Target Land algorithm also compares measured reflectances against LUT values. The version used since 2006 is markedly different from the version that was first employed at Terra launch (Kaufman et al., 1997b). The current algorithm is described by Levy et al. (2007). Like the Dark Target Ocean algorithm, the land algorithm is based on retrieving aerosol optical thickness, $\tau_{0.55}^{tot}$, from the elevated apparent reflectance over a dark surface target. Over ocean, the algorithm makes simple assumptions of the spectral ocean surface reflectance. Over land, the surface reflectance is too variable for a simple assumption of reflectances. Instead, the algorithm assumes dynamical relationships between surface reflectance bands modified by geometry and the amount of vegetation in the scene. An absolute error of 0.01 of surface reflectance in a visible channel will result in an approximately 0.10 error in retrieved $\tau_{0.55}^{tot}$. The brighter the target scene, the greater the absolute error introduced into the retrieval. Even for scenes dark to our eye, some channels will be highly reflective. Thus, the retrieval restricts itself to three channels (0.47 μm, 0.66 μm and 2.13 μm) expected be relatively dark over natural land surfaces such as moist soils and vegetation. Some surfaces such as deserts, snow and ice are inherently too bright in any channel, and are avoided by the algorithm. The procedure assumes an aerosol model based on location and season, and gives some flexibility as to whether the model is fine mode dominated or coarse mode dominated. This flexibility does not translate into the same quality of information contained in the Dark Target Ocean's η parameter and should be considered a diagnostic of the algorithm and not a physical parameter. The procedure performs a simultaneous inversion of the three channels, returning $\tau_{0.55}^{tot}$ as its primary product.

The Deep Blue algorithm follows from a different heritage than the Dark Target algorithms. It is based on translating the deviation of the apparent reflectance from expected Rayleigh scattering values into aerosol information and, unlike the Dark Target approach, can be applied to bright desert surfaces and is sensitive to aerosol absorption as well as aerosol extinction. The Deep Blue algorithm takes advantage of the fact that surface reflectances over desert and semi-desert regions are much darker over the blue part of the visible light spectrum compared to the longer wavelengths. Since the optical thickness for Rayleigh scattering is much smaller at the blue wavelength compared to the UV part of the spectrum, the corresponding TOA reflectance of blue light is much less sensitive to the altitude of aerosol layers. The algorithm is applied to cloud-free scenes, mixes smoke and dust aerosol models, using a nonspherical model for dust aerosol, and bases its surface reflectance assumptions on a precalculated surface reflectance database. The algorithm makes use of the 0.412 μm, 0.490 μm, and 0.670 μm bands to provide spectral information in finding the correct mix of smoke and dust (Hsu et al., 2004). The resulting output includes both aerosol optical depth, Ångström exponent, and dust single scattering albedo.

Combining the three MODIS algorithms provides a global quantitative view of the Earth's aerosol system over most of the land and ocean regions of the Earth. Figure 8.3 shows an example of monthly mean aerosol optical depth using a combination of the three algorithms described above. Figure 8.3 also shows an example of monthly mean size parameter, η, for the same month. The size parameter is retrieved only quantitatively by the Dark Target ocean retrieval.

8.3.3 Data use considerations

Some, but not all, of the MODIS aerosol products have been compared extensively against similar measurements made by instruments with known higher accuracy. In the MODIS case, this means comparison against data measured by the AERONET instruments (see Chapter 6). By comparing with higher accuracy instruments, the uncertainty of MODIS aerosol optical depth, has been quantitatively characterized and validated (Remer et al., 2005; Levy et al., 2010). Validated does not indicate a perfect retrieval, nor does it mean an unbiased retrieval. Also, validation is performed either globally or over broad geographical regions. A specific region may introduce biases when surface conditions and/or aerosol characteristics deviate from the assumptions used in the global retrieval. Careful analysis and empirical corrections can produce an unbiased product (Zhang et al., 2008).

One significant bias not due to regional inconsistency with retrieval assumptions is the 0.015 offset between Terra and Aqua $\tau_{0.55}^{tot}$ over oceans (Remer et al., 2008). This offset has been traced to calibration issues, with Terra appearing to be too high. Other calibration issues have introduced artificial trends in the MODIS time series (Levy et al., 2010; Zhang and Reid, 2010)

While the MODIS aerosol product provides the community with well-characterized, easily accessible information on aerosols, care must be taken to consider Quality Assurance (QA) flags at all data levels. Global and regional mean statistics are sensitive to choice of aggregation procedures that translate the nominal 10 km product based on orbital geometry to the so-called Level 3 gridded 1 degree product (Levy et al., 2009). It is preferable for users to always acquire daily data and produce their own monthly means, and thereby have control over the aggregation procedures employed.

Because the MODIS aerosol product is operational, it continuously undergoes analysis and changes are implemented. Documentation on algorithm changes and errata are kept current in the MODIS Dark Target Algorithm Theoretical Basis Document (ATBD) available at http://modis-atmos.gsfc.nasa.gov/reference_atbd.html.

8.4 MISR/TERRA

8.4.1 Mission and instrument description

The Multiangle Imaging SpectroRadiometer (MISR) was launched aboard the NASA Earth Observing System (EOS) Terra spacecraft in December 1999, into a sun-synchronous orbit that crosses the equator at about 10:30 local time, descending on the day side of the planet.

It is unique among the EOS-era satellite instruments in having a combination of high spatial resolution (up to 275 m/pixel), a wide range of along-track view angles, and high-accuracy radiometric calibration (~3% absolute, ~1.5% channel-to-channel) and radiometric stability (Diner et al., 1998). Near-global coverage is obtained about once in nine days at equatorial latitudes, and about once every two days in polar latitudes up to 82°.

MISR measures upwelling short-wave radiance from Earth in four spectral bands centered at 446, 558, 672, and 866 nm, at each of nine view angles spread out in the forward and aft directions along the flight path, at 70.5°, 60.0°, 45.6°, 26.1°, and nadir. Over a period of seven minutes, as the spacecraft flies overhead, a 380-km-wide swath of Earth is successively viewed by each of MISR's nine cameras. As a result, the instrument samples a very large range of scattering angles – between about 60° and 160° at mid latitudes – providing information about aerosol microphysical properties. These views also capture air-mass factors ranging from one to three, offering sensitivity to optically thin aerosol layers, and allowing aerosol retrieval algorithms to distinguish top-of-atmosphere reflectance contributions from the surface and atmosphere.

The MISR data are used to study Earth's surface, clouds, and aerosols. Total-column aerosol optical depth (τ) is retrieved over dark water and most land, including bright desert (Martonchik et al., 2004; Kahn et al., 2010). Sensitivity to aerosols varies with surface albedo, scene heterogeneity, and the range of scattering angles observed. As with most passive aerosol remote sensing techniques, MISR aerosol retrievals are not obtained under cloudy conditions, over complete snow or ice cover, mountainous terrain, and some coastal regions (Case 2 waters), where runoff, pollution, biological activity, or shallow water make the ocean surface reflective in the MISR red and near-infrared wavelengths. A description of the MISR aerosol algorithm quality flags, and comparisons with near-coincident MODIS retrievals of τ and Ångström exponent, is given by Kahn et al. (2009a).

At least over dark water, for good but not necessarily ideal viewing conditions (i.e. cloud-free, mid-visible τ larger than about 0.15, adequate range of scattering angles), MISR can also distinguish about 3–5 groupings based on particle size, 2–4 groupings in single-scattering albedo (SSA), spherical vs. nonspherical particles, and thin cirrus from most types of aerosol (Chen et al., 2008; Kahn et al., 2001; Kalashnikova and Kahn, 2006; Pierce et al., 2010). In addition, MISR stereo height retrievals are used to obtain aerosol injection height in the source regions for wildfire smoke, desert dust, and volcanic plumes, where there are sufficient contrast-features in the plumes to perform multiangle image matching (Kahn et al., 2007).

8.4.2 Algorithm description

The MISR Standard Aerosol Retrieval algorithm runs in an operational, fully automatic mode. It reports τ and aerosol type derived over 17.6×17.6 km retrieval regions, by analyzing the MISR top-of-atmosphere radiances measured in the 16×16 1.1 km sub-regions within each region (Martonchik et al., 1998; 2002). Separate algorithms are applied to observations taken over dark water and over heterogeneous land. Both aerosol retrieval algorithms can be divided into three stages (Diner et al., 2006). Stage 1 involves pre-processing the radiometrically and geometrically calibrated radiances, and performing out-of-band spectral, ozone and water vapor absorption corrections. In addition, screening is performed in Stage 1 for missing radiance data, complex terrain over land, low solar-zenith-angle cases, sun-glint-contaminated views over water, and cloudy locations; the remaining radiance data are converted to equivalent reflectances. In Stage 2 of the processing, the type of algorithm used to perform the retrieval (land or water) is determined based on the data that pass the Stage 1 tests. Finally, in Stage 3, acceptance criteria are used to identify simulated top-of-atmosphere equivalent reflectances from look-up tables that match those observed by the instrument.

The land–water differences are most prominent in Stage 2 of the algorithm, after acceptable retrieval regions have been selected. Over water, only the red and near-infrared channels are used, to minimize water reflectance. The dark ocean surface is modeled as a Fresnel reflector, with standard, wind-dependent surface roughness and whitecap models;

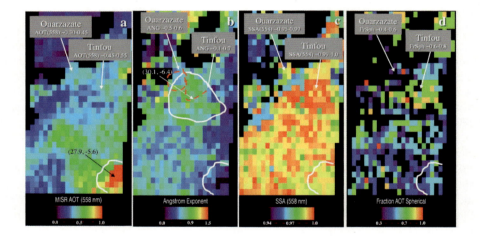

Kahn et al., Tellus 2009

Figure 8.4 MISR Standard aerosol retrieval products acquired on 04 June 2006 at 11:11 UTC, over the SAMUM Field Campaign ground sites at Ouarzazate and Tinfou, Morocco. (a) Mid-visible Aerosol Optical Thickness; (b) Ångström exponent; (c) Mid-visible Single-scattering albedo (SSA); (d) Fraction Optical Thickness of Spherical particles. A dust-laden density flow appears in the southeast (lower right) corner of these images, and an airmass containing higher SSA, higher Ångström exponent (smaller), spherical particles compared to the background desert aerosol is found at the center of the region, outlined in white in panel (b). The dashed red line in this panel traces the flight of the DLR Falcon F20, which verified the presence of two thin layers of small, spherical particles in the highlighted region; back-trajectories indicate that this represents pollution particles transported from the Algerian coast. (From Kahn et al., 2009b. Reprinted with permission from the International Meteorological Institute in Stockholm)

in initial processing, near-surface winds are adopted from a monthly, global climatology whereas about one month after data acquisition, the retrievals are rerun using measured winds.

The over-land algorithm is a two-step process, both of which are intrinsically multiangle (Martonchik et al., 2009). In Step 1, the idea that the angular signature of surface reflectance is approximately spectrally invariant is used to filter out those aerosol mixtures from among those assumed in the algorithm that grossly disagree with the observed radiances (Diner et al., 2005). In Step 2, a principal components analysis is performed on the available multiangle, multi-spectral top-of-atmosphere (TOA) MISR radiances (Martonchik et al., 1998). A scatter matrix is constructed as the sum over all sub-regions within the 17.6 km retrieval region that pass the Stage 1 tests, of camera-by-camera differences between TOA equivalent reflectances:

$$C_\lambda(i,j) = \sum_{x,y}\left[L_{x,y,\lambda}^{MISR}(i) - L_{bias,\lambda}^{MISR}(i)\right]\left[L_{x,y,\lambda}^{MISR}(j) - L_{bias,\lambda}^{MISR}(j)\right], \tag{8.5}$$

where i and j are camera indices, x and y are spatial indices, λ is the wavelength index, L are the TOA reflectances, and the "*bias*" subscript references the camera and band-appropriate reflectances for the sub-region within the retrieval region having the darkest green-band reflectance. As the atmosphere is assumed homogeneous over the retrieval region field-of-view (taking account of angular viewing), the atmospheric path radiance cancels from each term, and the scatter matrix depends only on surface properties. Scene contrast must be sufficient to effect this surface–atmosphere separation. The spatially dependent surface reflectance contribution is then constructed from the EOFs and eigen values of the C matrix, and the atmospheric path radiance is obtained as the difference between the MISR TOA reflectances and the calculated surface contribution. The optical depth is determined to minimize the difference between the observed and reconstructed reflectances. This process is repeated for all mixtures that pass the angular similarity filter of Step 1, and all those that meet the acceptance criteria are reported as matching the observations to within the sensitivity of the measurements.

Critical to the success of this process is providing the algorithm with aerosol component optical models and mixture options that reflect those commonly found in the atmosphere, at a level-of-detail appropriate to the sensitivity of the measurements. A combination of theoretical sensitivity studies, field campaign results, and statistical comparisons with surface validation data, stratified by expected aerosol type, has been used to develop the aerosol component and mixture climatology (Kahn et al., 2001; 2005, 2009b; Chen et al., 2008; Kalashnikova and Kahn, 2006]. At least two-thirds of MISR-retrieved aerosol optical depths fall within 0.05 or 20% of coincident validation values, whichever is larger. Sensitivity to particle size, shape, and SSA varies with conditions. So aerosol air mass type, a classification based on the aggregate of MISR particle microphysical property constraints, is being investigated as a robust way to summarize the retrieved aerosol type information. The limited particle property validation data available indicate that about a dozen aerosol air mass types can be distinguished from the MISR data under good observing conditions. Aerosol type discrimination, in a situation where field data were available for validation, is illustrated in Figure 8.4, and references to a range of aerosol applications papers that use the MISR data are given in Kahn et al. (2011). Improvements to the Version 22 MISR aerosol product, including the addition of medium-mode and more absorbing aerosol components, as well as mixtures of smoke and dust analogs missing from the current algorithm climatology but common at some locations in some seasons, are planned for future product releases (Kahn et al., 2010).

8.5 POLDER/PARASOL

8.5.1 Mission and instrument description

The PARASOL (Polarization and Anisotropy of Reflectances for Atmospheric Science coupled with Observations from a Lidar) mission (Tanré et al., 2011) is the second in CNES's Myriade (Centre National d'Études Spatiales) line of microsatellites. PARASOL was launched on December 18, 2004 and was part of the A-Train. It was in a sun-synchronous orbit at an altitude of 705 km with 1:30 pm equator-crossing time on the ascending node and has been routinely acquiring data since March 2005. The payload (Deschamps et al., 1994) consists of a POLDER-like digital camera with a 274x242-pixel CCD detector array, wide-field telecentric optics and a rotating filter wheel enabling measurements in nine spectral channels from blue (0.443 μm) through to near-infrared (1.020 μm) and in several polarization directions at 0.490 μm, 0.670 μm and 0.865 μm. The bandwidth is between 20 nm and 40 nm depending on the spectral band. As it acquires a sequence of images every 20 seconds, the instrument can view ground targets from different angles, +/− 51° along track and +/− 43° across track. Because of limited on-board fuel budget, the PARASOL orbit was decreased by 4 km at the end of 2009. The drift of the orbit since 2009 has resulted in a crossing time of around 2:30pm at the end of 2011.

8.5.2 Algorithm description

Over ocean, the inversion scheme uses the total normalized radiances in the 670 nm and 865 nm channels only, to avoid the contamination by the ocean reflectance that is very variable at 555 nm and in the blue region. The polarized Stokes parameters at 865 and 670 nm are also used for deriving the best aerosol model. The rough ocean is modeled with the Cox and Munk (1954) equations assuming a wind speed of 5 m/s for considering the surface–atmosphere multiple interactions. The actual wind speed provided by the ECMWF weather forecast model is used in the glint mask and for computing the foam reflectance. The foam coverage is calculated according to the Koepke's model (1984) with a reflectance of 0.22 at both wavelengths. The underwater contribution is taken equal to 0.001 and 0.000 at 670 and 865 nm respectively. Let us add that PARASOL has always at least one viewing direction out of the glint making aerosol τ retrievals possible everywhere over water.

The algorithm is based on a comparison between spectral, directional and polarized measurements and LUT built for a set of aerosol models (size distribution, refractive index, optical thickness) and for geometrical and other atmospheric conditions as close as possible to the actual situation. The aerosol size distribution is bimodal with an accumulation mode between 0.1–0.2 μm and a spherical coarse mode around 2.50 μm (Herman et al., 2005). Each mode of spherical particles follows a lognormal distribution, and several real refractive indices for each mode are considered. m = 1.35, 1.45 and 1.60 for the accumulation mode and m = 1.33, 1.35 and 1.37 for the coarse mode. Particles are considered non-absorbing in both modes. Since PARASOL measurements are sensitive to the shape of aerosols, a model of nonspherical dust is included in the LUT. The phase matrix is taken from Volten et al. (2001) and the model assumes no spectral dependence. Concerning its physical properties, the effective radius is taken equal to 2.50 μm and there is no retrieval of the refractive index of the coarse mode in presence of dust. The dust can be mixed with the spherical coarse mode through 5 values: 0.00, 0.25, 0.50, 0.75 and 1.00. The LUTs are built using the Successive Order of Scattering (SOS) method (Chapter 3) including polarization computations (Lenoble at al., 2007). The radiances as well as the normalized Stokes parameters Q and U are pre-computed following Eq. 8.4.

Limitations:

• For low aerosol content , the aerosol model is fixed and only the aerosol optical thickness is derived.

• The refractive indices and modal radii of the small and large spherical particles are not derived everywhere since the retrieval requires a scattering angle coverage larger than 125°–155°, which depends on the viewing conditions for a given pixel.

Over land, aerosol remote sensing from solar radiance measurements is more difficult than over ocean because the reflectance from the surface is generally much larger than the reflectance from aerosols at the top of the atmosphere, except over dark surfaces such as vegetation in the blue part of the solar spectrum (see Section 8.3). The use of polarized radiances is very attractive in such conditions. Airborne experiments have shown that the relative contribution of the surface compared to the atmosphere, is less important in polarized light than in total light (Deuzé at al., 2001), which makes the polarized satellite radiances more subject to the presence of aerosols than the total radiances. In addition, contrary to the total radiances, polarized radiances reflected by terrestrial surfaces are fairly spectrally independent and uniform. The surface contribution in the POLDER algorithm is then handled in the following way: (i) two surface models are considered, bare soils and vegetated areas, and (ii) the contribution is then estimated from a relationship using empirical coefficients adjusted for the different classes of land surfaces according to the main IGBP biotypes and the NDVI (Nadal and Bréon, 1999). Nevertheless, there is a drawback using polarization that comes from its deficiency to detect all types of particles. Large particles do not polarize sunlight and, for instance, desert dust is not accessible for a quantitative inversion. Aerosol polarization mainly comes from the small spherical particles with radii less than about 0.5 μm. So, the present algorithm over land, based on a best fit between polarized measurements at 670 and 865 nm and LUTs, retrieves aerosols within the accumulation mode only. So, particles within the coarse mode are not considered in the LUT. They are computed for 10 effective radius (from 0.075 to 0.225 μm) of the fine mode with a complex refractive index equal to 1.47–0.01i, which corresponds to a mean value for aerosols resulting from biomass burning or pollution events (Dubovik et al., 2002).

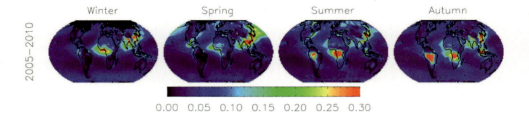

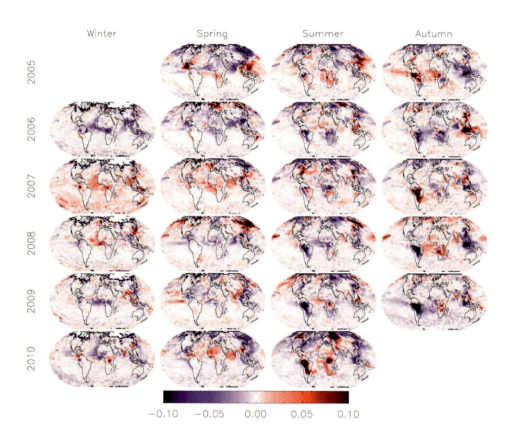

Figure 8.5 Top panel: AOD at 550 nm resulting from the accumulation mode for the 4 seasons averaged over 6 years. Bottom panel: AOD anomalies for the six years from 2005 to 2010. Blues indicate that the AOD of the year was lower than the long term mean and reds indicate higher values.

Limitations:

• Surface models are less accurate over deserts near the horizon.

• The coarse mode can contribute to the polarization for very intense events that are difficult to identify.

Figure 8.5 illustrates the capabilities of the sensor to derive the aerosol optical depth of the accumulation mode over land and ocean. It is a very important parameter since aerosol types dominated by the fine mode are mostly the result of combustion activity and consequently anthropogenic (Kaufman et al., 2005). In Figure 8.5, aerosol optical thickness anomalies for six years are compared with the 2005–2010 mean values (top panel) for the four seasons. The top panel illustrates the large seasonal variability of the source regions. In the autumn season, high values of the AOD are reported over Brazil, South Africa and Indonesia due to fires. High values are observed in the Indo-Gangetic Plain of India and eastern China during all seasons. In summer, North America and Europe display moderate aerosol loading. Biomass burning occurs again in Africa, north of the equator, in winter, and in Central America during the spring. The bottom panel in Figure 8.5 shows the yearly evolution with important variations in specific regions and seasons. The causes of the fluctuations are multiple: (i) changes in intensity and number of fires or controls of pollutant emissions; (ii) changes in meteorological fields that affect aerosol transport; or (iii) changes in scavenging processes, precipitations or dry deposition.

8.5.3 Advanced POLDER retrieval

A new algorithm for enhanced retrievals of aerosol properties from POLDER/PARASOL observations is under development As illustrated in Kokhanovsky et al. (2010) using synthetic data, PARASOL-like measurements combined with a sophisticated inversion algorithm (Dubovik et al., 2011) are the most accurate means for retrieving the aerosol properties when compared with other algorithms and sensors.

Indeed, PARASOL observations form the most comprehensive data set currently available from space, which provides an opportunity for a deeper utilization of statistical optimization principles in satellite data inversion. The retrieval is designed as a statistically optimized multi-variable fitting of the complete PARASOL observation set, including both measurements of total radiances and polarized at all available spectral channels (Dubovik et al., 2011). Based on this strategy, the algorithm is driven by a larger number of unknown parameters and aimed at a retrieval of an extended set of parameters affecting measured radiation. For example, this approach allows the retrieval of both the optical properties of the aerosol and the underlying surface from PARASOL observations over land. Also, the algorithm is expected to provide more detailed (compared with the current operational PARASOL algorithm) information about aerosol properties over land including some information about aerosol sizes, shape, absorption and composition (refractive index).

In addition, the algorithm was inspired by the MAIAC inversion for MODIS (Section 8.9.2), and is developed as a simultaneous inversion of a large group of pixels within one or several images. Such a multi-pixel retrieval regime is expected to take advantages of known limitations on spatial and temporal variability in both aerosol and surface properties. Specifically, the pixel-to-pixel or day-to-day variations of the retrieved parameters are forced to be smooth by imposing additional appropriately set *a priori* constraints. We anticipate higher consistency in the retrievals because the retrieval over each single pixel will be benefiting from co-incident aerosol information from neighboring pixels, as well as from the information about surface reflectance (over land) obtained in preceding and consequent observation over the same pixels.

This new inversion procedure has been applied to a limited PARASOL data set. The method is very promising since retrievals of the total AOD over bright surfaces like deserts have been performed and show good agreement with coincident AERONET measurements over Banizoumbou, Niger (Dubovik et al., 2011).

8.6 OMI-Aura

8.6.1 Mission and instrument descriptions

The Ozone Monitoring Instrument (OMI) is one of four sensors on the EOS-Aura platform launched on July 15, 2004 (Schoeberl et al., 2006). The Aura spacecraft is part of the A-train, a set of satellites flying in close proximity. Other A-train aerosol sensors are Aqua-MODIS (aerosol component described above), PARASOL (described above), and CALI-OP (Chapter 10). Aura's equator crossing time changed from 13:45 at launch to 13:38 in May 2008 to reduce the time difference of the observation with other A-train sensors. The OMI project (Levelt et al., 2006) is an international effort that involves the participation of the Netherlands, Finland, and the USA. OMI is a high spectral resolution spectrograph that measures the upwelling radiance at the top of the atmosphere from the ultraviolet to the visible (270–500 nm). It has a 2600 km wide swath and provides daily global coverage at a spatial resolution varying from 13×24 km at nadir to 28×150 km at the extremes of the swath. The OMI sensor was designed to map on a daily basis the global distribution of the total atmospheric ozone content and its vertical distribution. The hyper-spectral nature of OMI observations allows the application of techniques to derive the atmospheric content of other important trace gases such as NO_2, SO_2, HCHO, BrO, etc. (Platt, 1994). OMI observations are also used for the estimation of cloud heights, and for the characterization of atmospheric aerosols, which will be described below.

8.6.2 Algorithms description

Two aerosol inversion schemes are applied to the OMI measurements; the OMI near-UV (OMAERUV) algorithm uses two UV wavelengths to derive aerosol extinction and absorption optical depth. A multi-wavelength algorithm (OMAERO) that uses up to 19 channels in the 330–500 nm spectral range is used to derive aerosol extinction optical depth at several wavelengths.

Near UV aerosol algorithm (OMAERUV)

The OMAERUV algorithm makes use of two major advantages of the near-UV spectral region for deriving aerosol properties. The reflectance of all terrestrial surfaces (not covered with snow) is small in the UV and, therefore, the retrieval of aerosol properties is possible over a larger variety of land surfaces than in the visible, including the arid and semi-arid regions of the world that appear very bright in the visible and near-IR. The second advantage is the strong interaction between aerosol absorption and molecular scattering from below the aerosol layer that allows one to estimate the aerosol absorption capacity of the atmospheric aerosols (Torres et al., 1998, 2005). The OMAERUV algorithm is an application to the OMI observations of the near-UV technique of aerosol detection and characterization developed based on observations by the TOMS sensor and discussed in detail in Chapter 7.

The OMAERUV aerosol algorithm uses measurements at 354 nm and 388 nm to retrieve aerosol extinction optical depth and single scattering albedo at 388 nm. In addition to these parameters, the UV Aerosol Index (UVAI) is calculated as described in Section 7.5.1 for $\lambda = 354$ nm and $\lambda_0 = 388$ nm.

Aerosol Index

The first step in the derivation of the Aerosol Index is the calculation of the Lambertian reflectivity R_{388} using Eq. 7.6. The quantity R_{388} is usually larger than the true surface reflectivity, R_{388}^{sfc}, due to scattering from clouds and aerosols but it can be smaller if the aerosols are highly absorbing. The term R_{354} is estimated by correcting R_{388} for the spectral dependence of the actual surface reflectivity,

$$R_{354} = R_{388} + [R_{388}^{sfc} - R_{354}^{sfc}], \tag{8.6}$$

using a pre-computed climatological database of surface reflectivity in the 331-380 nm spectral region obtained from TOMS multi-year observations similar to that of Herman and Celarier (1997). The calculated top of the atmosphere 354 nm radiance associated with the adopted surface-atmosphere model representation, as given by

$$L_{354}^{cal} = L_{354}^0 + \frac{R_{354}T_{t\_354}}{1 - S_{354}R_{354}}, \tag{8.7}$$

is then used in the computation of the OMI UV aerosol index, AI,

$$AI = -100 \ \log\left[\frac{L_{354}^M}{L_{354}^{cal}}\right], \tag{8.8}$$

where L_{354}^M is the satellite observed radiance at 354 nm. All parameters on the right side of Eq. (8.7) are defined in Chapter 7.

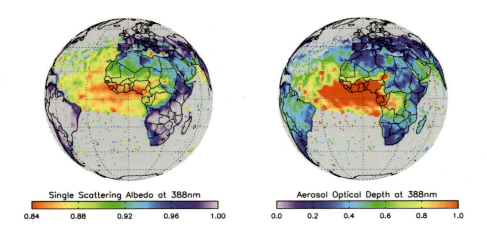

Figure 8.6 OMAERUV January 2007 monthly average aerosol optical depth and single scattering albedo.

Aerosol extinction optical depth and single scattering albedo

Aerosol extinction optical depth and single scattering albedo at 388 nm are derived using a standard inversion algorithm that uses pre-computed reflectances for a set of assumed aerosol models. The retrieval algorithm is in many ways similar to the one applied to TOMS observations (Torres et al., 2002). Three major aerosol types are considered: desert dust, carbonaceous aerosols associated with biomass burning, and weakly absorbing sulfate-based aerosols. The carbonaceous aerosol model accounts for the presence of black and organic carbon as the main light absorbing components (Jethva and Torres, 2011). Each aerosol type is represented by seven aerosol models of varying single scattering albedo, for a total of 21 microphysical models (Torres et al., 2007). In the current algorithm version, the selection of the aerosol type is carried out based on spectral considerations and the use of ancillary information available from other sensors. The separation of absorbing and non-absorbing aerosols is done by means of the AI concept. Carbonaceous aerosols are differentiated from desert dust aerosols making use of the observed correlation between the AI and the CO column amount product of the Aqua-AIRS sensor.

For a chosen aerosol type, represented by a subset of seven aerosol models, the extinction optical depth and single scattering albedo are retrieved by examining the variability of the 354–388 nm radiance ratio spectral contrast and the 388 nm reflectance. Since the retrieved parameters are sensitive to the aerosol layer height, the altitude of the absorbing aerosol layer must be prescribed. OMAERUV uses a global monthly climatological data set of aerosol height derived from CALIOP observations.

Given the relatively large size of the OMI pixels (compared to sensors specifically designed for aerosol retrieval, such as MODIS), OMI pixels are often cloud contaminated. The effect of sub-pixel cloud contamination is the overestimation of the extinction optical depth and underestimation of the single scattering co-albedo (Torres et al, 1998) and, therefore, a partial cancellation of sub-pixel cloud contamination errors takes place in the calculation of the aerosol absorption optical depth (AAOD). Thus, AAOD results are still reliable even in the presence of small amounts of cloud contamination. OMAERUV retrieval results have been evaluated using airborne (Livingston et al., 2009), and ground-based (Torres et al., 2007) sun-photometer measurements and by comparison to MODIS and MISR observations (Ahn et al., 2008).

Figure 8.6 shows the January 2007 monthly averages of retrieved aerosol single scattering albedo and extinction optical depth over Africa and the Atlantic Ocean. Aerosol optical depth values larger than unity from biomass burning activity cover the Tropical Atlantic. Much lower values are observed over the Saharan desert and the North Atlantic where desert dust aerosols are present. Single scattering albedo values as low as 0.82 are associated with the smoke layer whereas values in the vicinity of 0.92 are observed for dust aerosols.

Multi-wavelength aerosol algorithm (OMAERO)

The OMI multi-wavelength algorithm (OMAERO) is used to derive aerosol characteristics from OMI spectral reflectance measurements of cloud free scenes at up to 19 wavelengths in the spectral range between 330 nm and 500 nm (Torres et al., 2007). The sensitivity to the layer height and single scattering albedo is related to the relatively strong contribution of Rayleigh scattering to the measured reflectance in the UV (Torres et al, 1998). The absorption band of the O_2-O_2 collision complex at 477 nm is used in OMAERO to enhance the sensitivity to the aerosol layer height (Veihelmann et al., 2007) in a way similar to its use in the retrieval of cloud height (Acarreta et al., 2004). Spectral surface albedo in the UV-VIS spectral region is taken from a climatological data set produced from three years of OMI observations (Kleipool et al., 2008).

Aerosol parameters are determined by minimizing the merit function Ψ_m,

$$\Psi_m = \sum_{i=1}^{N} \left(\frac{L^*(\lambda_i) - L_m(\tau(\lambda_{ref}), \lambda_i)}{\varepsilon(\lambda_i)} \right)^2, \tag{8.9}$$

where $L^*(\lambda_i)$ is the measured reflectance, $L_m(\tau(\lambda_{ref}), \lambda_i)$ is the reflectance for the aerosol model m as a function of the aerosol optical thickness τ at the reference wavelength λ_{ref}, and $\varepsilon(\lambda_i)$ is the error in the measured reflectance. The merit function is a sum over N wavelength bands. An optimal τ is determined for each model by a non-linear fitting routine using a modified Levenberg-Marquardt method (Moré, 1978). The best fitting aerosol model with the smallest Ψ_m is selected for the present ground pixel. The retrieved AOD value and precision of the best fitting aerosol model are provided in the OMAERO product. Also, the values of single-scattering albedo, size distribution and aerosol height that are associated with the best fitting aerosol model are provided.

8.7 GOME-SCIAMACHY

8.7.1 Instrument description

GOME (Global Ozone Monitoring Experiment) and SCIAMACHY (Scanning Imaging Absorption Spectrometer for Atmospheric Chartography) are satellite spectrometers measuring reflected sunlight. Although these instruments are intended for measuring trace gases like O_3, NO_2, CO, and CH_4, and have limited spatial resolution, some of their characteristics, like contiguous spectral coverage from UV to near-IR, inclusion of O_2 A-band, and polarization channels, give them unique aerosol observational capabilities. Furthermore, the combination of trace gas and aerosol measurements may lead to a better understanding of the formation of secondary aerosols.

SCIAMACHY flies on the European Space Agency's (ESA's) Envisat satellite, launched in March 2002. SCIAMACHY is a grating spectrometer with eight spectral channels covering the range 240–2380 nm. The spectral resolution varies from 0.2 nm in the UV to 1.5 nm in the near-IR (Bovensmann et al., 1999). SCIAMACHY has three viewing modes: nadir, limb and occultation (Sun and Moon). The ground pixel size in nadir mode is 30×60 km² for most wavelengths. The full spectrum is measured at 30×240 km² resolution. In limb mode the spatial resolution is 3×240 km². Both in nadir and limb mode the swath is 960 km wide. The nadir and limb modes alternate each other along the orbit, and therefore full global coverage is achieved in six days.

GOME was developed as a simpler version of SCIAMACHY, and was launched in April 1995 on the ERS-2 satellite of ESA. GOME has four spectral channels covering the range 240–790 nm with spectral resolution 0.2-0.4 nm, and has nadir mode only (Burrows et al., 1999a). Its pixel size is 40×320 km², and its swath is 960 km. Global coverage is achieved in three days. Due to its solid state recorder failure, GOME lost global coverage in June 2003; after that the coverage has been limited to 30–40%. The instrument and satellite were switched off in July 2011.

GOME-2 is the successor of GOME and flies on EUMETSAT's Metop-A satellite, launched in October 2006 (Callies et al., 2000). GOME-2 has 40×80 km² pixels and a 1920 km wide swath, providing global coverage in 1.5 days. GOME-2 has the same spectrometer capabilities as GOME, but it has enhanced polarization detectors (PMDs): its 15 spectral PMD channels, in the range 330–750 nm, for two perpendicular polarization directions enable measurement of both Stokes parameters I and Q with spatial resolution of 40×10 km².

The in-flight radiometric and spectral calibrations of GOME and SCIAMACHY consist of daily Sun observations (via a diffuser) and on-board lamp monitoring (Lichtenberg et al., 2006).

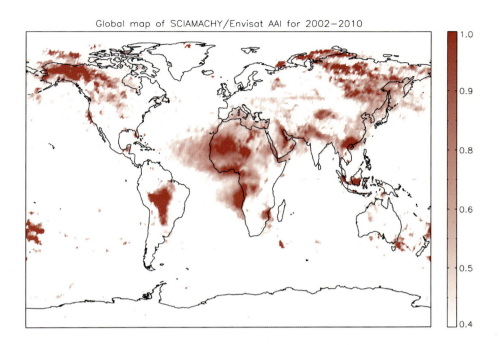

Figure 8.7 Global map of the Aerosol Index from SCIAMACHY, averaged over the period 2002-2010. The major UV-absorbing aerosol events are clearly visible, for example Saharan desert dust, Asian desert dust, biomass burning smoke in South-West Africa and South America, biomass burning smoke in South-East Asia, and smoke from wild fires in Siberia, Canada and Alaska. (Figure from L. G. Tilstra, KNMI).

8.7.2 Aerosol algorithm description and limitations

The main aerosol product from GOME is the UV absorbing aerosol index, AI (Chapter 7, Eq. 7.9). In the GOME application λ_0 (reference wavelength) is 380 nm and the λ is 340 nm.

The GOME AI was first used by Gleason et al. (1998) for a case study. The sensitivities of the AI to aerosol optical thickness, single scattering albedo, layer height, and other parameters are well documented (Torres et al., 1998; De Graaf et al., 2005). A global GOME AI product has been created.

Aerosol optical depth retrievals from GOME, using a standard multispectral least-squares fitting technique, have been performed by several groups (Torricella et al., 1999;

Veefkind et al., 2000; Kusmierczyk-Michulec and De Leeuw, 2005), but with limited success. The main problem is the large pixel size of GOME, which causes a high chance of cloud contamination. Sub-pixel cloud detection is not possible without additional information. Therefore, Holzer-Popp et al. (2002) used a combination of GOME with the ATSR-2 imager, but also in that case the number of useful pixels is very small. The use of the reflectance of the O_2 A-band measured by GOME for aerosol height retrieval was investigated by Koppers et al. (1997).

In the case of SCIAMACHY cloud contamination and uncertain surface albedo prevent accurate retrievals of τ (Von Hoyningen-Huene et al., 2007). However, the AI detects aerosols also in cloudy circumstances, whereas the surface albedo is low in the UV. Therefore, the AI is the most used aerosol product from SCIAMACHY (De Graaf and Stammes, 2005; Tilstra and Stammes, 2007). Figure 8.7 shows the global map of AI from SCIAMACHY, averaged over the period 2002–2010. The wide spectral range of SCIAMACHY has been used to analyze the complete reflectance spectrum of absorbing desert dust and biomass burning aerosols (De Graaf et al., 2007, 2012). Usually only positive values of the AI are being considered, which indicate absorbing aerosols. Negative values of the AI – i.e. scenes being bluer than aerosol-free scenes – from SCIAMACHY have been studied by Penning de Vries et al. (2009), showing a correlation with scattering aerosols.

The AI is the standard aerosol product for GOME-2, and as in the case of SCIAMACHY, the wavelength pair is 340/380 nm (Tilstra et al., 2010). As an improvement to GOME and SCIAMACHY, the Stokes parameter Q measured by GOME-2 will enable better characterization of aerosol microphysics, like particle size and refractive index. The planned retrieval method is based on fitting the modeled I and Q values to measured quantities in an optimal-estimation sense (Hasekamp and Landgraf, 2005a).

8.8 APS

8.8.1 Instrument description

The Aerosol Polarimetry Sensor (APS) was launched aboard the Glory satellite on March 4, 2011 but did not reach orbit due to the failure of the payload fairing to detach from the spacecraft. At the time of writing, it is unclear whether or not NASA will build and fly a copy of the APS instrument. Since APS data may never be available for the interested reader to evaluate, we will describe both APS and also an airborne instrument, the Research Scanning Polarimeter (RSP). The RSP instruments are airborne simulators of the APS, using the same measurement approach, and have been used to demonstrate the capability of that measurement approach to deliver reliable remote sensing observations with a polarimetric accuracy of ~0.1%. The RSP instruments have been flown in field experiments since 2001 and there are currently over 1000 hours of measurements available to the public that sample North and Central America from near the equator to close to the Pole.

The key measurement requirements for the accurate retrieval of a broad range of aerosol and cloud properties from photopolarimetric data are high accuracy, a broad spectral range, and observations from multiple angles, including a method for reliable and stable calibration of the measurements. The polarimetric accuracy of the APS sensor is based on the use of Wollaston prisms to separate orthogonal polarization states from exactly the same scene at the same time and allows the required polarimetric accuracy of 0.2% to be attained even for scenes with sub-pixel heterogeneity. To measure the Stokes parameters that define the state of linear polarization (I, Q, and U), the APS employs a pair of telescopes. One telescope in the pair measures I and Q while the second telescope in the pair has the polarization azimuths of its measurements rotated 45° with respect to the first telescope so that it is measuring I and U.

In order to provide a broad spectral range, the APS uses three pairs of telescopes and in each telescope pair the light is split into three spectral bands for a total of nine spectral bands, in all of which the Stokes parameters I, Q and U are measured. The spectral bands are separated from one another using dichroic beam splitters and interference filters are then used to define each spectral channel. The APS measures wavelengths of 413, 555 and 866 nm in the first telescope pair, 444 (469 for RSP), 674 and 911 (962 for RSP) nm in the second telescope pair, and 1376 (1884 for RSP), 1603 and 2260 nm in the third telescope pair. Blue enhanced silicon detectors are used in the first two telescope pairs, but in order to provide sufficient sensitivity and low noise for observing aerosols over the oceans the detectors in the third telescope pair are made from HgCdTe and are cooled to 160° K. All spectral channels but the 911 (962) and 1376 (1884) nm ones are free of strong gaseous absorption. The 1376 (1884) nm exception is centered on a major water vapor absorption band and is specifically intended for detection and characterization of thin cirrus clouds. The 1884 nm band used by the RSP is better for cirrus cloud detection over snow and ice than the 1376 nm band, that is preferred for the APS, because of stronger ice absorption that makes the surface darker. However, for a satellite instrument a cirrus detection band at 1376 nm is better because of its capability to detect and characterize stratospheric aerosols with the accuracy required to simultaneously sense aerosols in the troposphere (Gao and Kaufman, 1992). The 910 (962) nm band provides a self-contained capability to determine column water vapor amount that is used to correct the longest wavelength bands (1603, 2260 nm) for contamination by water vapor. The locations of the other APS (or RSP) spectral channels are similar to those used in MODIS, MISR or POLDER and provide a wide spectral range (410–2260 nm) while minimizing the effects of absorbing gases.

The critical ability to view a scene from multiple angles is provided by scanning the 8 mrad APS instantaneous field of view (IFOV) along the spacecraft ground track with a rotation rate of 40.7 revolutions per minute (rpm) with angular samples acquired every 8 mrad. This yields ~220 useable view angles per scene over ±60° from nadir that are acquired over a 5-minute period as the spacecraft flies over the scene. The polarization-compensated scanner assembly includes a pair of matched mirrors operating in a "crossed" configuration such that polarized light that is incident "parallel" to the first mirror is "perpendicular" to the second mirror and vice versa. This type of scanner assembly has been demonstrated to yield instrumental polarization less than 0.05% and also allows a set of calibrators to be viewed on the side of the scan rotation opposite to the Earth. The APS calibrators provide comprehensive tracking of polarimetric calibration throughout each orbit, while radiometric stability was to be tracked monthly using lunar views. (The RSP is similar in many respects and in order to view a scene from multiple angles is oriented to scan along the aircraft groundtrack. It has a rotation rate of 71.3 rpm, a 14 mrad IFOV and acquires samples every 14 mrad such that 150 views are obtained over ±60° with respect to nadir. To collect observations of a given scene at the ground over that view angle range takes about 5 minutes for an aircraft at 8.5 km altitude flying at 100 ms^{-1} similar to the time required to aggregate an APS scene.)

8.8.2 Aerosol retrieval algorithm description and limitations

The APS and RSP aerosol retrieval algorithms are essentially the same for clear-sky scenes, the only noteworthy difference being that calculations for an aircraft embedded in the atmosphere are somewhat more complicated than those for a satellite sensor. Since the APS sensor is not flying, we will therefore describe the algorithms currently being used in the analysis of RSP observations. The first step in preparing the observations for analysis is to correct them for ozone and nitrogen dioxide absorption which is assumed to lie above the tropospheric aerosols in the stratosphere and can therefore be corrected using Beer's law for the absorption in the path down to the aerosol layer and back to the observation point. The column water vapor is estimated from the reflectance ratio of a non-absorbing band (866 nm) and a band that has moderate water vapor absorption (911 nm) (Gao and Goetz, 1990) and is used to calculate an effective absorption optical depth for the 1603 and 2260 nm bands. Such an approximation is quite accurate because water vapor absorption is generated by the far wings of the absorption lines and absorption by carbon dioxide (1603 nm) and methane (2260 nm) in the lower atmosphere, where the interaction of absorption with aerosol scattering is important, is also dominated by line wings. The rapid attenuation of radiation in the upper atmosphere in the line centers that are contained in these spectral bands is included as a solar zenith dependent transmission correction (Cairns et al., 2003). Once the observations have been corrected for gaseous absorption effects the aerosol retrieval, whether over land or ocean, is based on an iterative adjustment of all the parameters required to define the surface and a bimodal aerosol model that minimizes the difference between the model and the measurements (Rodgers, 2000; Waquet et al., 2009b; Cairns et al., 2009; Dubovik et al., 2011; Hasekamp et al., 2011). The retrieved aerosol model consists of the effective radius, variance and spectral refractive index of each size mode and its optical depth with the accumulation, or fine, mode being restricted to effective radii smaller than 0.5 μm and the coarse mode consisting of aerosols with effective radii larger than 0.5 μm. If the residual fit between the observations and models is larger than expected based on a χ^2-test then a second retrieval is attempted using a nonspherical model for the coarse mode. At present we use a nonspherical model consisting of an equi-probable mix of prolate and oblate spheroids with aspect ratios from 0.3 to 3. We do not at present allow for a nonspherical accumulation mode because the differences between small spherical and nonspherical particles are relatively small and not readily detectable.

Over oceans the chlorophyll concentration ([Chl]) and windspeed define the ocean surface reflectance model, as discussed above, and are the surface parameters that are retrieved. The only difference between the ocean model used in the analysis of RSP measurements and the models of ocean reflectance described elsewhere in this chapter is the inclusion of polarization in the scattering of light within the ocean body and the calculation of the polarization of light emerging from the ocean. This is needed so that we have a complete and consistent model of the contribution of the ocean to the observed unpolarized and polarized radiances (Chowdhary et al., 2006). The benefit of having such a model of the ocean body reflectance that only depends on [Chl] and predicts the oceanic contribution across the entire solar spectrum is that all of the APS unpolarized and polarized radiance observations from 410 to 2260 nm can be used in the aerosol retrieval. The value of being able to use the spectral bands across the visible part of the spectrum in aerosol retrievals is that the spectral shape of the ocean body contribution and the effects of aerosol absorption are very different and this difference allows an accurate determination of the single scattering albedo of the aerosols (Chowdhary et al., 2005). Size distribution and spectral optical depth retrievals over the ocean have also been validated during the Chesapeake Lighthouse Aircraft Measurements for Satellites (CLAMS) and the Megacity Initiative: Local and Global Research Observations (MILAGRO) field experiments (Chowdhary et al., 2005, 2011).

Over land the surface polarized reflectance is assumed to be grey (i.e., the same in all spectral bands) and has the same form, given by Nadal and Breon (1999) that is used in POLDER aerosols retrievals. The primary difference between the APS/RSP aerosol retrievals over land and the POLDER approach is that the pre-factor in the surface polarized reflectance model is estimated as part of the APS/RSP retrieval process. This surface parameter and the bimodal aerosol model parameters are the retrieval products over land. At present, only the polarized reflectance in all spectral bands is used in the aerosol retrievals over land. The advantage of only using the polarized reflectance is that the same method can be applied to aerosols over snow as well as other surfaces. This is possible because the polarized reflectance of snow is also grey and, although it has a smaller magnitude than is typical of other land surfaces, has a similar dependence on scattering angle to other surfaces types. Examples of retrievals over a range of conditions from very low aerosol optical depths to very high aerosol optical depths have demonstrated the capabilities of polarization observations over land including the capability to retrieve the aerosol single scattering albedo (Waquet et al., 2009b; Knobelspiesse et al., 2011) which complements the capability to determine aerosol absorption in the UV that was presented in Section 8.5.

One of the main limitations of the existing analysis approach over both ocean and land surfaces is that the iterative estimate is relatively slow because the calculation of how the polarized radiation field varies as a function of all the aerosol and surface parameters is performed using direct numerical differentiation. Hasekamp and Landgraf (2005b) have shown that perturbations to the radiation field as a function of surface and aerosol parameters can be calculated analytically and we are currently testing an implementation of that approach using our own radiative transfer code. Our own code uses the doubling/adding method, which has the advantage for the analysis of multiangle observations that the effect of any possible perturbations on all view angles can be evaluated using a single radiative transfer calculation. In addition, comprehensive look-up tables are also being tested as a quicker method for initializing the iterative search.

An issue regarding the aerosol retrievals over land is the generality and accuracy of the existing surface polarization models (Maignan et al., 2009; Litvinov et al., 2011) such as that of Nadal and Breon (1999) and may be a particular problem at high scattering angle for snow surfaces. Although we have found our equi-probable shape distribution approach to modeling dust and soil to work for far travelled dust off the coast of Virginia and soils blown off the coast of Mexico other models of dust may be required such as that used in the AERONET retrievals (Dubovik et al., 2006). Lastly, although the use of only polarized reflectance to perform aerosol retrievals over land allows for a relatively simple algorithm and the multiangle polarization observations in long wavelength bands at 1603 and 2260 nm do provide some sensitivity to coarse mode aerosols, the inclusion of unpolarized reflectance observations in the retrieval of aerosols over land from POLDER observations has been demonstrated by Dubovik et al. (2011) to have significant benefits and we plan on implementing a similar retrieval approach in future analyses of the RSP data sets.

8.9 Advances in remote sensing of aerosol using modern sensors

The previous sections have presented a survey of many sensors and algorithms used for aerosol retrievals during the 2000s and beyond. All of the algorithms presented in previous sections are more or less "standard algorithms", supported by various space agencies to provide aerosol information to the public. The algorithms were developed over a number of years and applied to the sensors at launch. There have been changes since launch, and some of those changes have been major. For example, the MODIS Deep Blue algorithm was implemented five years after Terra launch. However, change comes slowly. While the standard algorithms continue to produce a continuous time series of important aerosol products, research into how to extract even more information by applying new techniques is taking place. In this section, we introduce some recent advances in aerosol remote sensing applied to the observations from the passive sensors described above. Some of these techniques may eventually be included in the operational algorithms.

8.9.1 Aerosol retrieval over clouds

One characteristic that all sensors and retrieval methods described above have in common is that they retrieve aerosol only in cloud-free scenes. This severely limits true global coverage of aerosol properties and misses the ability to quantify possible positive aerosol forcing when dark aerosol is lofted above clouds (Chand et al., 2009). Two techniques for retrieving aerosol optical depth above clouds have been developed over the last few years. The aerosol load above clouds can be retrieved from observations of polarized radiances as described by Waquet et al. (2009a) and Knobelspiesse et al. (2011), and from radiance measurements in the near UV (Torres et al., 2011).

The Polarization method

One of the observations of cloud top pressure (CTP) from PARASOL, a standard PARASOL data product, makes use of the polarized reflectance. Since the polarized reflectance at $0.49\,\mu$m increases with the "thickness" of the molecular atmosphere, an apparent cloud top pressure called "Rayleigh CTP" (Goloub et al., 1994) can be derived from the value of the polarized reflectance. The presence of an aerosol layer mixed with molecules above a cloud changes the polarization at $0.49\,\mu$m and makes the cloud higher than it would be in the absence of aerosols. So, in such conditions, inconsistencies occur with the second PARASOL CTP based on differential absorption measurements in two oxygen A-band channels at 0.76

μm (Vanbauce et al., 2003) or with the MODIS "IR CTP" in the thermal infrared channels when available (Menzel et al, 2008) that are not affected. Let us state that discrepancies are anticipated between the three CTPs (Sneep et al., 2008) since the retrievals are based on different physical processes that are not exactly sensitive to the same altitude range, but when the "Rayleigh CTP" over some regions is smaller than the two others, it is in conflict with our understanding.

Such observations (Waquet et al., 2009b) were noticed from PARASOL in areas located in regions and times associated with the presence of aerosols in the fine mode, i.e. with pollution and/or smoke generated mainly by anthropogenic activities. Several cases have been investigated over the subtropical South Atlantic ocean in August/September when fires in Southwest Africa are frequent and meteorological processes favor biomass burning aerosol transport to the West. Waquet et al. (2009b) analyzed specific collocated POLDER/PARA-SOL, MODIS and CALIOP observations in 2006, from 14 to 18 August and found an "additional" polarization signal over the 90°–120° scattering angle range. This additional polarization leads the "Rayleigh" algorithm to an underestimate of the CTP confirmed by the CALIOP data. The excess of polarization is due to the presence of an aerosol plume, as also confirmed by CALIOP observations. Modeling the radiative transfer through the cloud-aerosol atmosphere can then retrieve the aerosol optical thickness ($\tau_{aerosol}$). Presence of dust cannot be detected with a similar approach since the dust polarization is not significant, but a method using the polarized rainbow is an alternative under development (Waquet el al., 2010).

Similar observations of aerosols above cloud were made by the RSP as part of the Megacity Initiative: Local and Global Research Observations (MILAGRO) field campaign on March 13, 2006. In this case, the presence of aerosols was identified as a result of the inconsistency between cloud top pressure estimates using polarized reflectance measurements in spectral bands at 0.41 and 0.469 μm. The aerosol layer was lofted above a low altitude marine stratocumulus cloud close to shore in the Gulf of Mexico. The analysis approach used with these aircraft observations is to first determine the cloud droplet size distribution using the angular location of the cloud bow and other features in the polarized reflectance. This cloud retrieval is then used in a multiple scattering radiative transfer model optimization to determine the aerosol optical properties and the cloud droplet size distribution using only the polarized reflectance observations over the spectral range from 0.41 to 2.260 μm. This approach is taken because the cloud optical depth does not then need to be estimated since the polarized signal generated by a cloud saturates at an optical depth of 2–3. The retrieved aerosol variables are the aerosol optical depth, the fine mode aerosol size distribution and its complex refractive index (Knobelspiesse et al., 2011). Subsequent unpublished analysis demonstrates that the imaginary refractive index can be retrieved far more accurately by including total reflectance observations in the optimization process and also retrieving the cloud optical depth, similar to what is discussed in greater detail in the next subsection.

The near UV method

This retrieval technique uses observations of backscattered near-UV reflectance observed by the OMI sensor on the Aura satellite. The retrieval requires both the OMI-derived absorbing Aerosol Index and the observed 0.388 μm reflectance, which are fed into an inver-

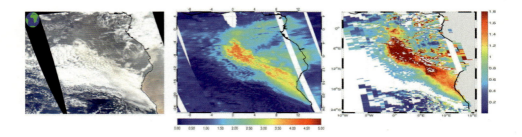

Figure 8.8 Stratocumulus cloud as seen by Aqua-MODIS (left), OMI observed Aerosol Index (center), and derived aerosol optical depth above the cloud (right).

sion procedure that simultaneously retrieves the optical depth of both the cloud and the aerosol above the cloud. Sensitivity analysis studies (Torres et al., 2011) indicate that the magnitude of the Aerosol Index associated with absorbing aerosols above clouds depends on cloud and aerosol optical depth, wavelength-dependent aerosol single scattering albedo and, to a lesser extent, on aerosol layer-cloud separation. Thus, to simultaneously retrieve $\tau_{aerosol}$ and τ_{cloud}, knowledge of other cloud and aerosol parameters is needed. An error analysis was carried out to estimate the uncertainty in the retrieved values of aerosol and cloud optical depth associated with the uncertainty in the assumed values of single scattering albedo and aerosol cloud separation. Results indicate that the combined uncertainty of ±0.03 in single scattering albedo and ±2km in aerosol-cloud separation yields a $\tau_{aerosol}$ error between −26% and 54% for typical cloud and aerosol layer optical depths of 10 and 0.5, respectively. Retrieval errors decrease with increasing cloud optical depth. Errors in retrieved τ_{cloud} are smaller than 20% in most cases which are comparable to the reported uncertainty of the MODIS product for optical depths larger than about 10 (Platnick et al., 2003).

An example of the application of the UV method of above-cloud-aerosol optical depth retrieval is shown in Figure 8.8. The left panel shows the MODIS-Aqua true color image composite of a horizontally extended stratocumulus cloud over the South Atlantic Ocean off the coasts of Namibia and Angola on August 4, 2007. The center panel illustrates the OMI Aerosol Index indicating the unmistakable presence of an absorbing aerosol layer overlying the stratocumulus cloud. The right panel depicts the aerosol optical depth field derived by the UV technique.

8.9.2 The MAIAC retrieval

The standard MODIS aerosol products described above, both Dark Target and Deep Blue, are applied to the data collected on each orbit, soon after that orbit's data are downloaded from satellite to processing center. The retrieval is completely independent of observations taken yesterday or tomorrow, except in terms of using long-term means or climatology as ancillary data.

The Multiangle Implementation of Atmospheric Correction (MAIAC) algorithm (Lyapustin et al., 2011a) was designed for simultaneous retrievals of aerosol information and surface bidirectional reflectance over land. To achieve this goal, it uses a time series analysis of images from multiple days and simultaneous processing for groups of 25×25 1 km^2 pixels, called blocks. In this way, the processing accumulates enough measurements to exceed the number of unknowns – the necessary condition for solution of the inverse problem. This method relies on different space-time scales of variability of aerosol and land reflectance: namely, the land surface is variable in space but changes little over short time intervals, while the aerosol is homogeneous across a spatial domain of ~25 km but can change rapidly from one day to the next.

MAIAC stores in memory from 16 days of measurements over the equator to five days at the Poles, and implements a sliding window algorithm. The surface BRDF is character-ized by the linear Ross-Thick Li-Sparse BRF model (Lucht et al., 2000) depending on three parameters, which are derived for each pixel from up to 16 observations at different view angles acquired on different days. The multi-day block-level processing is used to derive spectral regression coefficient (SRC) for each pixel which relates BRF in the Blue and shortwave infrared (SWIR, 2.13 μm) bands. The inversion is simplified by an assumption of spectral invariance of the BRF shape (Veefkind et al., 1998; Diner et al., 2005), which works well over dark and moderately bright surfaces. Once SRC is known, the BRF in the Blue band (B3) is computed at 1 km resolution as a product of SRC and BRF in the SWIR band. This allows MAIAC to derive aerosol information at 1 km resolution using the latest MODIS measurements.

The MAIAC aerosol algorithm applies a look-up table approach (LUT) (Lyapustin et al., 2011b) to computing top-of-atmosphere reflectances in reflective MODIS channels. Similar to the standard MODIS land algorithm, the aerosol models are prescribed by region and season. Unlike the standard algorithm, the models are tuned on a local level in order to achieve a good agreement of spectral MAIAC τ with AERONET measurements. Also unlike the standard over land algorithm, the prescribed fine and coarse models are mono-modal lognormals, not multimodal models, and MAIAC combines the designated fine and coarse modes using the η parameter much like the standard MODIS over ocean technique (Section 8.2.2 and Equation 8.4),. MAIAC obtains an assessment of physical size param-eter information over land, unlike the standard over land retrieval, because the temporal information fixes the surface contribution and reduces the overall uncertainty; however, the accuracy of retrieving Ångström parameter by MAIAC remains low (Lyapustin et al., 2011a).

MAIAC's assumptions are met over dark and moderately bright surfaces. Over deserts, an alternative retrieval path is implemented, including an additional dust aerosol model in the LUT and an empirical correction factor for the angular dependence of SRC which becomes prominent over bright surfaces. Results over bright surfaces are promising, but still in development.

MAIAC retrieval of aerosol optical depth compare as well with AERONET observations as the standard MODIS land algorithm over dark and vegetative surfaces (Lyapustin et al., 2011a), but at 1 km spatial resolution instead of 10 km. There is also evidence that MAIAC produces a more accurate retrieval in some regions that have been difficult for standard retrieval, including western North America. Recently, MAIAC was applied to MODIS observations over the European Alps in order to obtain a finer resolution product that could resolve aerosol gradients and particulate pollution in narrow Alpine valleys. After some specialized filters were applied to the results to account for residual snow and cloud contamination, the results proved to be useful (Emili et al., 2011).

From a historical perspective, MAIAC's development was strongly influenced by the extensive heritage of the MISR and MODIS retrieval algorithms. Influences include employing a rigorous radiative transfer model with a non-Lambertian surface for simultaneous aerosol/surface retrievals (Diner et al., 2001), the concept of using the image spatial structure for aerosol retrievals over land (Martonchik et al., 1998) and the Contrast Reduction method by Tanré et al. (1988) that showed that consecutive images of the same surface area, acquired on different days, can be used to evaluate aerosol differences between days (Section 7.4.3).

8.9.3 Single scattering albedo using critical reflectance methods

Aerosol absorption, as denoted by the single scattering albedo, is an important particle property to retrieve. MISR is able to use its multiangle capability to retrieve ϖ under the right conditions, and OMI is sensitive to aerosol absorption in the UV range of the spectrum where aerosol absorption introduces a deviation from expected Rayleigh scattering. MODIS produces a ϖ through its Deep Blue pathway, based on similar principles as OMI. The PARASOL advanced retrieval will obtain this parameter using multiangle and polarization information. There is an alternative method to obtain aerosol absorption information using MODIS. This method does not require polarization, multiangle looks or Deep Blue channels and can be applied to any wavelength as long as there is sufficient aerosol loading in that wavelength. This is the critical reflectance method.

When a scattering aerosol overlays a dark surface, the reflectance measured at the top of the atmosphere will be greater than without the aerosol because the particles are scattering sunlight back to space. The particles prevent the surface below from absorbing some of that sunlight. The more aerosol added to the scene, the more light scattered, the higher the measured reflectance at top-of-atmosphere. If the aerosol were an absorbing aerosol with low single scattering albedo and the underlying surface was bright, then the reflectance measured at the top of the atmosphere would be less than without the aerosol. In this case, the particles are absorbing the sunlight that would otherwise be reflected by the bright surface back to space. The more aerosol in this case, the more absorbed light, the lower the reflectance at top-of-atmosphere. An example of the interplay between aerosol absorption and underlying surface is illustrated in Figure 1.1, where the underlying surfaces are dark ocean and bright cloud. Thus, between these two situations there exists a combination of surface brightness and aerosol absorption properties in which adding more aerosol makes no change to the measured reflectance at top-of-atmosphere. This point is called the "critical reflectance" (Fraser and Kaufman, 1985). If the surface reflectance is constant, then the critical reflectance is determined solely by the aerosol properties and is most sensitive to the absorbing properties of the particles. Therefore, critical reflectance provides a remote sensing technique to derive aerosol single scattering albedo (Kaufman, 1987; Kaufman et al., 1990; 2001; Zhu et al., 2011).

The practical crtical reflectance remote sensing algorithm requires two images of the same scene with nearly the same solar and viewing geometries. If the aerosol loading did not change from image to image and the surface reflectance also did not change, a scatter plot of top-of-atmosphere reflectances of one image against the other would have all pixels aligned along the 1:1 line where $x=y$. However, if the aerosol loading did change while the surface remained constant, then plotting the hazier day against the clean day would create a scatter plot in which the pixels would align linearly, but not on the 1:1 line. For darker surfaces and a scattering aerosol, the hazier pixels would have higher values than the corresponding pixels of the clean day. For brighter surfaces and a more absorbing aerosol, the hazier day would darken the top-of-atmosphere reflectance and appear below the 1:1 line on the scatter plot. The point at which the line created by the hazier pixels intersected the 1:1 line would be the critical reflectance. Single scattering albedo (ϖ) could be derived from look-up Tables (LUTs) that relate ϖ to critical reflectance for a particular set of scattering properties (size distribution and real part of the refractive index) for the type of aerosol in question (Zhu et al., 2011; Wells et al., 2012).

The critical reflectance method has been applied to MODIS reflectances to derive spectral ϖ in smoke (Zhu et al., 2011) and dust (Wells et al., 2012), at moderately high spatial resolution. There are limitations to the method that have been explored in sensitivity studies, but the final product shows agreement with AERONET retrievals of ϖ to within 0.02 (Zhu et al., 2011; Wells et al., 2012).

8.9.4 Aerosol plume heights

One additional capability made possible by the multiangle MISR imaging is the retrieval of wildfire smoke, volcanic effluent, and desert dust plume heights (Kahn et al., 2007). The technique, based on interpreting the parallax in terms of elevation above the geoid, requires features in the aerosol plume to be matched in the multiangle views (Moroney et al., 2002; Muller et al., 2002); as such, this technique can be applied primarily within a few hundred kilometres of aerosol sources, where aerosol plumes are well defined. However, height maps of entire near-source regions are routinely produced, and the stereo imagery provides global coverage approximately once per week, so climatologies of smoke, dust, and volcanic plumes are contained in the 11-year MISR data record.

MISR aerosol plume heights are complementary to the CALIPSO active sensor aerosol profile retrievals (Chapter 10), which provide only very narrow-swath (~70 m) "curtain" coverage and rarely, if ever, capture active sources, but are sensitive to aerosol layers as optically thin as 0.005 (Winker et al., 2009). Thus MISR data can be used to constrain aerosol injection heights and near-source plume evolution, whereas CALIPSO lidar can provide aerosol elevation downwind, when the plumes have dissipated into much more extensive, sub-visible layers, the combination offering powerful constraints on aerosol transport modelling (Kahn et al., 2008).

The MISR Stereo Height algorithm also includes a wind correction, which makes use of cloud motions observed with the steeply viewing MISR cameras; the heights are then retrieved by the automatic algorithm using the near-nadir cameras, to limit the relative distortion and improve stereo matching (Moroney et al., 2002). An interactive computer tool has also been developed which provides more detailed height and wind information on individual plumes than is available from the operational algorithm (Nelson et al., 2008). A database of plumes digitized with this tool can be found at: http://misr.jpl.nasa. gov/getData/accessData/MisrMinxPlumes/index.cfm. Early work with the MISR plume height products has demonstrated that smoke from wildfires can be injected above the atmospheric boundary layer, which means it will stay aloft longer, travel farther, and have broader environmental impact than is often assumed in aerosol transport model simulations (Kahn et al., 2007; Mims et al., 2009; Val Martin et al., 2010).

8.9.5 Advances using combination of sensors or sensors and models

One of the most important innovations in aerosol remote sensing occurring during the 2000s is the placement of multiple instruments on the same platform or formation flying satellites so that the strengths of different instruments could complement each other for more comprehensive understanding of the aerosol system. The Terra satellite includes both MODIS and MISR so that they view some of the same scenes simultaneously. The A-Train is a constellation of satellites that follow the same orbit and allow views of the same scene within minutes of each other. Of the sensors described above, Aqua-MODIS, Aura-OMI and PARASOL-POLDER have all flown in formation as part of the A-Train. Also included in the A-Train is the active sensor, CALIPSO-CALIOP, described in Chapter 10.

Combining information from different sensors can take various forms. Most common are the studies that use different products from different sensors to answer scientific questions. For example, cloud-aerosol studies might use MODIS aerosol optical depth and particle size parameter, MODIS cloud optical properties, CALIOP vertical profiles of aerosol and clouds, and other sensors in the A-Train providing information on chemistry, clouds and precipitation (Yuan et al., 2011, among many). In a similar way, MISR smoke plume heights and aerosol optical depth have been used in conjunction with MODIS fire radiative power to enhance our understanding of plume rise from fires (Val Martin et al., 2010). Kalashnikova and Kahn published in 2008 an excellent exploration of the complementary contribution that could be obtained by combining the information from two sensors. In this work, MISR images were used to fill the gaps left by sun glint in the MODIS over ocean retrieval to better track the evolution of desert dust across the Atlantic ocean. Use of the MISR data increased the MODIS coverage by 50%. In addition, the paper notes that MODIS provides the necessary coverage to track plumes that are difficult for MISR's narrower swath to resolve, but that MISR provides measures of particle properties that are essential to identify dust and quantify the evolution of particle size and ϖ as the plume is transported. The two sensors together provide a much clearer picture of dust transport than would be obtained from either one working alone.

A step beyond using standard archived products together to answer specific science questions and provide complementary views of the aerosol system is to use information from one sensor to constrain assumptions within the retrieval of another sensor. Joint retrievals can include passing cloud mask information from a moderate spatial resolution instrument such as MODIS or MISR to a coarser resolution instrument such as OMI or POLDER, or using retrieved particle property information from MISR or POLDER back to MODIS or OMI to constrain choice of aerosol model in the retrieval's LUT. An example of a joint retrieval is found in Satheesh et al. (2009) where MODIS retrieved τ is used to constrain the assumptions in the OMI retrieval, allowing for a more accurate retrieval of ϖ. The section above entitled "Near UV aerosol algorithm (OMAERUV)" describes the OMAERUV retrieval of τ and ϖ. The sensitivity of the retrieval to aerosol layer height is noted, which is constrained operationally by assuming climatological values for aerosol height. In the joint OMI-MODIS retrieval, MODIS provides τ, reducing the number of unknowns in the OMI retrieval, which then returns ϖ and a retrieved (not climatological) aerosol layer height.

Finally, much clarity of the aerosol system can be obtained by combining satellite re-mote sensing products with numerical models, either by constraining model output that encourages modification of model assumptions and parameterizations, or by active assimi-lation of aerosol products resulting in hybrid model-observation depictions of the aerosol system. Such assimilated data sets are becoming more commonplace and are shown to be an improvement to model-only generated aerosol fields (Zhang et al., 2008; Benedetti et al., 2009). These assimilated fields remove some of the biases in the satellite products and fill in holes in the retrieved fields due to clouds, sun glint, inappropriate surfaces, etc.

8.10 Summary

The 21st century has seen an explosion of space-based passive aerosol remote sensing using sensors that make use of all aspects of the radiation field: direct solar attenuation through the atmosphere, wide spectral range, multiangle views, fine spectral resolution in the ultraviolet spectrum and polarization. This has spawned a wide variety of responses as different development teams have worked to extract as much information as possible from the measurements of each particular sensor. Each sensor has strengths that allow for specific capabilities in terms of aerosol retrievals and each sensor has limitations. Table 8.2 summarizes these strengths and limitations.

Ideally, the perfect passive aerosol satellite sensor would have broad swath coverage, pixel spatial resolution of less than 1 km, a spectral range stretching from the UV to the mid-infrared, multiple angle views of the same scene and highly accurate polarization in all channels. Such an instrument does not exist now, but a combination of information from different sensors does increase capability, and also aids in the important effort of character-izing uncertainty in the products and keeping an eye on calibration drifts.

The 21st century continues to be a golden era for passive shortwave aerosol remote sensing. The constellation of satellite sensors observing Earth is supplying unprecedented aerosol information. This fleet of instruments provides an important global perspective that complements the more local focus of ground-based instruments. Still, these instruments observe aerosol only during daylight and provide only total column measures through the lower atmosphere. The complete picture of the aerosol system requires IR and lidar techniques that offer nighttime views and vertical profiling.

Instrument	Swath width (km)	Pixel resolution – nadir (km)	Wavelength range (μm)	Number of wavelengths	Angular range	Number of angles per scene	Polarization
SAGE(s)			0.38–1.55	4 to 9			
MODIS	2330	0.5x0.5	0.41–2.13	8 for aerosol	±70°	1	none
MISR	380	0.275 x 0.275 to 1.1 x 1.1	0.446–0.866	4	70.5° aft to 70.5° forward	9	depolarized
POLDER	1600	5.3 x 6.2	0.443–1.020	7 for aerosol	±43° across track ±51° along track	14–16	in 3 channels, I,Q,U
OMI	2600	13 x 24	0.27–0.50	Hyperspectral	±70°	1	none
GOME	960	40x320	0.24–0.79	Hyperspectral	±30°	1	limited
SCIAMACHY	960	30x60	0.24–2.3	Hyperspectral	±30°	1	limited
GOME-2	1920	40x80	0.24–0.79	Hyperspectral	±55°	1	Q/I
APS	6	6x6	0.41–2.26	7 for aerosol	±60°	220	all channels I,Q, U

Table 8.2 Comparison of instrument capability for selected passive shortwave sensors producing aerosol products in the 2000s.

9 Longwave passive remote sensing

Clémence Pierangelo
Contributions from Alain Chédin, Michel Legrand

9.1 Introduction

The longwave spectral domain dealt with here, also called the "thermal infrared" domain, roughly covers the $3-15\,\mu$m spectral range. The main radiation source is not the sun but the Earth system, that is, the Earth's surface (land and ocean) and the atmosphere. The infrared emission from bodies is directly linked to their temperature as described by Plank's law: the hotter the bodies are, the higher their emission. The Earth and its atmosphere are heated by the fraction of sunlight they absorb. Their increase in temperature results in increased infrared emission to space, thus ensuring the energy balance of the Earth system.

The exact ranges of spectral domains are the subject of frequent discussion. To avoid any confusion, we hereby give the definitions that apply in this chapter:

- The word "longwave" applies to the thermal radiation emitted by the Earth system. Thus, it goes practically from 3 to 25 μm.

- The word "shortwave" applies to the radiation emitted by the sun. For satellite nadir observation, it roughly goes up to 4 μm. Beyond this wavelength, the shortwave contribution turns to be negligible in comparison with the longwave one. But for satellite limb observations or solar occultation, the shortwave radiation might be not negligible at higher wavelengths[1].

---

1 The infrared emission spectrum of the Earth goes up to 1000 μm. Between 25 and 1000 μm, the far-infrared is not useful for remote sensing applications as the atmosphere becomes opaque. Beyond 1 mm, the radiation emitted by the Earth is in the microwave domain and does not interact with aerosols as particle size is too low, with respect to the wavelength.

However, not all the longwave domain is relevant for remote sensing applications: the totally opaque spectral ranges due to strong gaseous absorption bands must be avoided. Therefore, only the 3–15 μm spectral range is of interest here. In this spectral domain both "longwave" and "shortwave" radiation (by day only for the latter) are to be considered:

- The "medium infrared" range, roughly corresponding to the 3–4 μm range, is characterized by the contribution of both shortwave and longwave radiations.

- The "thermal infrared" range, corresponding to the 4–15 μm range and including the 10-μm atmospheric window, is the domain of the longwave radiation.

Several interactions can occur between the electromagnetic field, the Earth's surface, and the molecules and particles of the atmosphere. The infrared radiation emitted by the Earth's surface and by the atmosphere can be absorbed, scattered, and emitted, by the aerosol. Compared to shorter wavelengths, the specificity of the longwave domain regarding aerosols lies in the emission process.

Moreover, in the thermal infrared, many atmospheric gases (CO_2, H_2O, ozone, etc.) have strong absorption lines, caused by the rotation and vibration of atoms inside the molecules. Most of the spectrum in the infrared is then affected by gaseous absorption and emission; consequently, this feature must be taken into account when building up a retrieval of aerosol properties.

The heating and cooling effects of the aerosols result from their combined interaction with longwave and shortwave radiations. In the longwave, a non-negligible part of the radiation upwelling from the surface is absorbed and scattered by the particles. The radiation is then re-emitted, partly downwards, and also back-scattered towards the Earth's surface. Thus, the aerosols in the thermal infrared produce a greenhouse effect in the lower atmosphere and a surface heating. This effect is the opposite of the direct radiative effect observed in the shortwave resulting in a surface cooling. During daylight, the shortwave effect is prevailing and the aerosol presence results in a surface cooling, whilst at night it results in a surface heating. A short review of the literature shows that the direct radiative forcing in the longwave domain is far from negligible (Markowicz et al., 2003; Vogelman et al., 2003). However, the domain remains largely unexplored and is poorly understood. Coordinated research strategies for its study are thus recommended (Yu et al., 2006).

The longwave radiation primarily depends on the characteristics of the Earth's surface (temperature and emissivity), on the thermodynamic state of the atmosphere (temperature and humidity profiles), and on its chemical and mineral composition, including the aerosols These coupled dependencies are one of the main difficulties when using thermal radiation for aerosol remote sensing. However, longwave remote sensing of particles offers some unique possibilities, such as night-time aerosol observations, the retrieval of the aerosol layer altitude or dust detection over deserts. Besides, aerosol effects on the infrared radiance must be taken into account for other retrievals using the infrared spectrum. For instance, the retrievals of the temperature and moisture profiles, used for weather forecasts, are affected by the presence of dust, as shown by Weaver et al. (2003).

Note about the units: when dealing with high resolution spectra, often recorded by a Fourier Transform spectrometer, the gaseous spectroscopy is of first significance. Thus, because of the way this instrument concept records the spectrum and because of the spectroscopy habits, the wavenumber (in cm^{-1}) is a more natural unit than the wavelength (in μm). On the other hand, when dealing with aerosol optical properties, the wavelength is more appropriate because these properties depend on the ratio of the particle size to the wavelength. Consequently, we keep both units hereafter, and a conversion scale is given at the end of the chapter.

In this chapter, we first focus on the aerosol optical properties in the longwave domain, and on their significant consequences on the remote sensing validity domain and strategy. Then, the radiative transfer equation in the thermal infrared is specifically studied, including the scattering contribution by aerosols. Next, we review the techniques and algorithms currently used for aerosol detection and remote sensing in the longwave, including the pre-processing steps of cloud clearing. Finally, some illustrations of experimental results are given, obtained both from satellite sensors (sounders and imagers) and from ground-based instruments.

9.2 Aerosol optical properties in the longwave domain

As the definition and the modeling of aerosol optical properties (absorption and scattering) have previously been described in this book (see Chapter 2), we focus here on the specificity of the longwave domain.

9.2.1 On the relative impact of aerosol species in the infrared

Before studying the effect of the microphysical properties of aerosols on their infrared optical properties in more detail, we first examine the relative impact of the main aerosol species. For the optical remote sensing of aerosols, the variable of prime interest is optical depth, as it combines both the information on the columnar amount of aerosols (number of particles, mass, etc.) and the optical extinction efficiency factor (i.e. the strength with which the aerosol interacts with the radiation, by scattering or absorbing the photons). The optical depth at a given wavelength is usually the quantity observed by a passive remote sensor recording the radiation at this given wavelength. However, the optical depth is strongly dependent on the wavelength. This is why the primary quantity that may be retrieved from measurements in the longwave domain is the AOD at infrared wavelengths (and not at some visible wavelength). Nevertheless, most aerosol observations are performed in the visible or near-infrared domain: generally, the 0.55 μm wavelength is taken as a reference. This is why we might estimate the relative impact of aerosol species in the infrared by comparing their infrared AOD to their visible AOD, or, equivalently, their infrared extinction coefficient to their visible extinction coefficient.

The ratio of infrared to visible optical depths (normalized at 0.55 μm), at wavelengths from 4 to 12 μm, for several aerosol species shows a high variability of aerosol type (see Figure 9.1: aerosol types are taken from the OPAC – Optical Properties of Aerosol and Clouds database from Hess et al. (1998)). Consequently, this high variability implies that the longwave and shortwave remote sensing observations are not sensitive to the same aerosol types. One of the main reasons why some aerosols are optically thick and others have

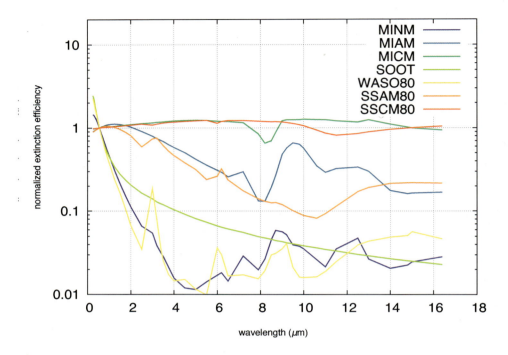

Figure 9.1 Normalized extinction coefficient as a function of wavelength for 7 types of aerosol component, taken from the OPAC database (Hess et al., 1998) (MINM=Mineral Nucleus Mode, MIAM=Mineral Accumulation Mode, MICM=Mineral Coarse Mode, WASO80=Water Soluble, 80% humidity, SOOT=soot, SSAM80= Sea-Salt Accumulation Mode 80% humidity, SSCM80=Sea-Salt Coarse Mode 80% humidity). The normalization is performed at 0.55 μm.

almost no effect in the infrared is their size distribution, as will be explained in more detail in Section 9.2.3. As an illustration, it can be seen on Figure 9.1 that aerosols in the coarse mode have the highest impact in the infrared (e.g. mineral aerosols or sea-salt in coarse mode). Amongst tropospheric aerosols, it is dust that has the main impact in the infrared. Indeed, aerosol types which are more typical of pollution or biomass burning (such as water-soluble aerosol or soot on Figure 9.1) are usually smaller in size. The optical depth of sea-salt in coarse mode is not negligible either. However, sea-salt particles are usually confined to the bottom of the planetary boundary layer, which prevents them from having an impact on radiance collected at satellite level, as will be explained in Section 9.3.6. Mineral dust aerosol is firstly mainly composed of particles in coarse size mode, potentially promoting a high infrared optical depth, and secondly can reach high altitudes, for example in the so-called Saharan-boundary layer (Chiapello et al., 1995). Subsequently, the remote sensing of aerosols in the longwave domain will mostly focus on retrievals of mineral dust properties.

Another serious consequence of these considerations is that there is no point in trying to retrieve visible optical depth from infrared observations, unless some other *a priori* information is available. Indeed, there is no unique relationship between the directly observable (the infrared optical depth) and the visible optical depth, but many possibilities depending on the aerosol type and size mode. This also implies that the validation of infrared optical depth with visible observations is difficult and limited.

Let us also mention volcanic aerosol. There are two types of volcanic aerosol: ash particles in the troposphere and sulfuric acid droplets in the stratosphere. Volcanic ash is emitted in great quantities during extreme events and has a lifetime of a few days in the troposphere. It contains particles in the 10 μm size range which have an effect on infrared observations. The stratospheric aerosol consists of sulfuric acid droplets (H_2SO_4) transformed from SO_2 emitted by volcanic eruptions (e.g. El Chichon or Pinatubo eruptions) into aerosol through gas-particle conversions. Although their infrared optical depth is below 0.1, they can be detected by infrared sensors, thanks to their global distribution, their long life in the stratosphere (several months or even years) and their high altitude (see Section 9.5.1 for an illustration).

9.2.2 Effect of refractive indices (or composition)

The chemical or mineralogical composition of aerosol is represented in its optical properties through the real and imaginary parts of its refractive index at every wavelength. The composition of aerosol is highly variable, depending on the source location, the emission process and the atmospheric conditions. Aerosol particles, and especially mineral aerosols, are generally composed of a mixture of mineral and/or carbonaceous elements that can be internally mixed or more often externally mixed (Chou et al., 2008). Therefore, the refractive index is one of the lesser-known quantities when computing the optical properties of aerosols in the infrared. Indeed, most studies have focused on the visible refractive index (e.g. McConnell et al, 2010).

Two different approaches are possible to obtain refractive indices of aerosols. The first one consists of taking values measured from real atmospheric aerosol particles, whereas the second consists of combining values from elementary components (or minerals for dust aerosols) in order to obtain the refractive indices of the mixture (following the approach of Sokolik and Toon, 1999, for example). The advantages of the second method are the availability of data (numerous measurements from soil samples, however not globally distributed), and the possibility of relating various minerals and particle size modes, while the difficulty lies in the modelling of the mixture.

Amongst one of the first inventories of refractive indices of aggregated particles, Volz (1973) published a dataset of refractive indices in the infrared for various kinds of aerosol: Saharan dust, volcanic dust, coal-fire fly ash and ammonium sulphate droplets. Concerning mineral dust, Sokolik et al. (1998) list the measurements performed in the infrared range (see their Table 1) and previously published in Volz (1972, 1973), Fischer (1976), Patterson (1981), Fouquart et al. (1987) or Sokolik and Golitsyn (1993). Each dataset comes from aerosols collected on a regional scale only and cannot thus be applied worldwide. For example, no measurements from Asian dust are available so far. Sokolik et al. (1998) note that the only measurements of dust particles far from the aerosol sources are the measure-

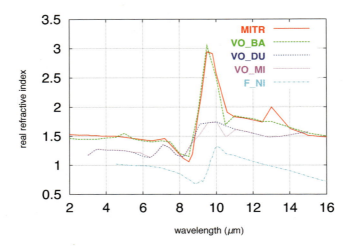

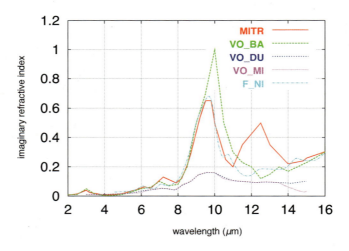

Figure 9.2 Refractive index of 5 mineral dust aerosol data set: MITR=Mineral Transported aerosol from Carlson and Benjamin (1980) and Sokolik et al. (1993), VO_BA=Saharan dust aerosol at Barbados (Volz, 1973), VO_DU=continental dust in Germany (Volz, 1973), VO_MI=mineral aerosol in Germany (Volz, 1973), F_NI=Saharan dust aerosol above Niger (Fouquart, 1987) from 4.5 to 16 μm. Real part (top) and absolute value of imaginary part (bottom).

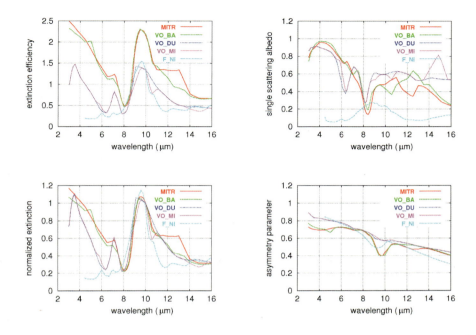

Figure 9.3 Effect of the refractive indices of mineral dust on the optical properties for 5 data sets for a Mie calculation. See legend in Figure 9.2 for the meaning of the data set.

ments from Volz (1973). The remote sensing of aerosols from space requires knowledge of aerosols on a global scale, especially concerning those far from their sources. This is why, following Sokolik et al. (1998), we recommend the indices from Volz (1973) for remote sensing of Saharan aerosols far from their sources. The "Mineral Transported" model from the OPAC database mentioned above is similar to, indeed is mainly based on, the Volz (1973) indices slightly modified by Carlson and Benjamin (1980). As an example Figure 9.2 illustrates the great variability of dust refractive indices as a function of wavelength. These spectral features are strongly dependent on the minerals that constitute the aerosol (see below). The variability between aerosol sources is also very great, in particular between desert sources and continental sources, and this greatly affects the optical properties important for radiative transfer (see Figure 9. 3), in agreement with Highwood et al. (2003). The composition of aerosols also changes during transport, as can be seen by comparing refractive indices for Saharan dust over Niger or Barbados.

Since the variability of soil minerals is very high, Claquin et al. (1999) have chosen a mineralogical set adapted to aerosol remote sensing, based on the abundance, optical properties and chemical properties of minerals. The study of refractive indices cannot be performed without taking into account the size distribution of particles, as minerals do not all appear in the same particle size range (e.g. Claquin et al, 1999). Their final set consists of eight major minerals: illite, kaolinite and smectite in the clay fraction, and calcite, quartz, feldspar, hematite and gypsum in the silt fraction. The refractive indices of illite, kaolinite, smectite and feldspar (which are all aluminosilicates) are not significantly different, and finally, the variability of refractive indices of dust species is summarized by quartz, clay (and feldspar), calcite and hematite. Similarly, Sokolik and Toon (1999) examined the refractive indices of seven minerals (illite, kaolinite, montmorillonite, hematite, quartz, calcite and gypsum), and they ended up with three main classes relevant to the modeling of aerosol optical properties: clays, quartz and hematite. Others, such as Chou et al. (2008), made in-situ observations that could sum up the variability of dust to aluminosilicates (illite and kaolinite), quartz and iron oxide (i.e. hematite and goethite). The mineralogical species to be considered must also be adapted to the spectral domain of the observation instrument. For instance, Turner (2008) worked on the 8–13 μm domain and, according to the signatures of minerals in this spectral domain, he considered three main minerals that have distinctly

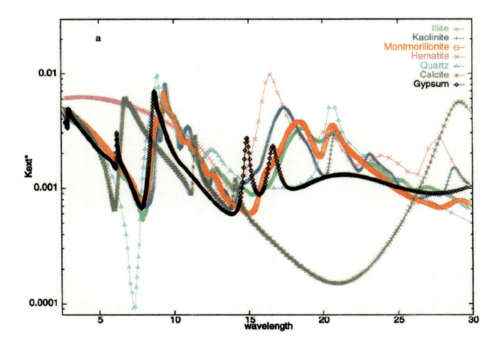

Figure 9.4 Spectral extinction coefficient for individual minerals (normalized for 1 particle per cm⁻³), computed for a lognormal size distribution with mode radius=0.5 μm and width=2; from Sokolik and Toon (1999).

different absorption bands: kaolinite, gypsum and quartz. The exact minerals in hand must also be adapted to the region of observations (e.g. Hansell et al., 2008).

The description of the main spectral features of each mineral is outside the scope of this chapter and we therefore refer readers interested in this subject to the paper from Sokolik and Toon (1999) and its references. To sum up, the spectral peaks in the extinction efficiency are caused by high imaginary indices, related to absorption of light by stretching vibrations of molecules. For instance, quartz has a major absorption line at 9.2 μm due to Si-O-Si asymmetric stretching vibrations. More generally, silicates show this spectral feature between 8 and 12 μm as the vibration line can be shifted in presence of other ions in the crystal lattice (Al, water, hydroxyl, etc.), whereas calcite exhibits a strong line at 7 μm and weak lines between 11.4 and 14.3 μm, and gypsum shows spectral lines near 8.7 and 16 μm (Sokolik et al., 1998). The great variability of spectral features for pure minerals is shown in Figure 9.4, taken from Sokolik and Toon (1999). Obviously, the extinction maximum between 9 and 10 μm typical of quartz and clays is also visible in the spectrum obtained from aggregated aerosols, as plotted in Figure 9.3.

An updated list of refractive index measurements available in the infrared domain for 7 minerals is given in Balkanski et al. (2007), together with the spectral domain covered by each measurement. It can be noted that very few refractive index spectra cover both the visible and infrared domains. This explains why the extrapolation of infrared optical depth or radiative forcing from only visible measurements requires many assumptions and results in great uncertainties.

9.2.3 Effect of size distribution

The particle size distribution (PSD) of aerosols, and particularly of dust, is often bimodal, as shown by Dubovik et al. (2002). So, as the interaction between light and a particle reaches its maximum when wavelength and size are of the same order of magnitude (see Chapter 2), it is mainly the coarse mode that has an impact on the thermal infrared radiation.

Let us consider the example of a typical size distribution retrieved from AERONET over Capo Verde in April 2003 (http://aeronet.gsfc.nasa.gov/). It can be modeled by a bimodal lognormal volumic size distribution (see Chapter 1) characterized by its modal volumic radii r_{vi} and widths σ_{vi} for mode i:

$$r_{vf}=0.138 \ \mu m, \ln(\sigma_{vf})=0.508;$$
$$r_{vc}=2.00 \ \mu m, \ln(\sigma_{vc})=0.608;$$

where the subscripts v, f and c mean "volumic", "fine mode", and "coarse mode".

The volume of dust in the fine (respectively coarse) mode is V_f, (respectively V_c). The size distribution retrieved by the AERONET sun-photometer also provides the ratio $V_c/V_f=2.71$.

A Lorenz-Mie calculation based on these data and the refractive indices of mineral aerosols in the OPAC database, using the code available at ftp://ftp.giss.nasa.gov/pub/crmim/spher.f (Mishchenko et al., 2002), gives the extinction coefficients for the fine and coarse modes in Table 9.1. The extinction efficiency, or efficiency factor for extinction, Q_{ext}, is defined here as the ratio of the mean extinction cross-section per particle by the mean surface area per particle, consistently with Chapter 2.

For dust with a typical bimodal distribution, the fine mode contribution to the total optical depth at 10 μm is usually of the order of 10%. This differs from the visible domain where the fine mode is predominant. We can therefore assert one of the main conclusions of this chapter: even for the same aerosol species, the infrared and visible AOD cannot be compared directly; their ratio obviously depends on the relative abundance between fine and coarse particles. However, the comparison of both products contains some valuable information on dust size distribution. A second consequence is that making the approximation of a monomodal lognormal particle size distribution (PSD) is acceptable, since these simulations demonstrate that the contribution of the accumulation mode to the total optical depth is of the order of 10%. Indeed, infrared radiation mostly interacts with coarse mode particles.

This makes it possible to reduce the number of parameters of the PSD to just one: the coarse mode effective radius (see Chapter 1), which is the most significant size parameter in terms of radiative impact. Indeed, the effective radius is well suited to describe the effect of size on optical properties (Mishchenko et al., 2002) while the PSD width (its geometric standard deviation) has a weak effect with regard to its range of variation. For instance, the impact of the geometric standard deviation of the PSD (in the range 1.6 to 2.5) on the normalized extinction coefficient between 3.8 and 15 μm is below 10%. The geometric standard deviation of the distribution is hereafter fixed at 2.2, in agreement with in-situ measurements reported by Reid et al. (2003).

The impact of the effective radius on optical properties (extinction, single scattering albedo, asymmetry parameter) is computed with the above-mentioned Lorentz-Mie code (Mishchenko et al., 2002) and plotted from 3 to 16 μm on Figure 9.5. At first glance, the impact of the effective radius on extinction efficiency seems huge. However, this effect depends slightly on the wavelength, as can be seen on the plot of the extinction efficiency normalized at 10 μm. It is worth pointing out that the radiation at satellite level does not solely depend on the extinction efficiency Q (or extinction cross-section C), but rather on the vertically integrated extinction coefficient (i.e. the optical depth), that is the product of the extinction cross-section to the particle number. An aerosol layer with extinction cross-section C and particle number N or an aerosol layer with extinction cross-section C/2 and particle number 2N are indeed strict equivalents with regard to the measurement. The particle number or concentration is usually totally unknown, because it is the optical depth that is directly measured. We will show through the radiative transfer equation in Section 9.3 that a single channel retrieval algorithm cannot distinguish between N and C without *a priori* information. A multispectral retrieval algorithm requires knowledge of optical depth at every channel frequency $\tau_{v0}, \tau_{v1}, \tau_{v2}$, etc. Once a reference frequency v_0 has been chosen, this is equivalent to knowledge of $\tau_{v0}, C_{v1}/C_{v0}, C_{v2}/C_{v0}$, etc. Consequently, for a given optical depth at a given wavelength, only the spectral variations of the extinction cross-section can modify the effect of aerosol on the observation. We call "normalized extinction" the ratio between extinction cross-section at wavelength λ and at a reference wavelength λ_0, usually 10 μm here. This is why we focus on the normalized extinction (and not the extinction efficiency) when evaluating the impact of a microphysical property on optical properties. Going back to Figure 9.5, the effective radius moderately modifies the

Wavelength (μm)	Fine mode		Coarse mode	
	Extinction efficiency	AOD contribution	Extinction efficiency	AOD contribution
0.55	1.957E-2	0.793	6.7662	0.207
10	6.749E-4	0.137	5.6373	0.863

Table 9.1 Contributions of the fine and coarse modes of a bimodal size distribution of mineral dust aerosol on the total optical depth, at 0.55 μm and 10 μm

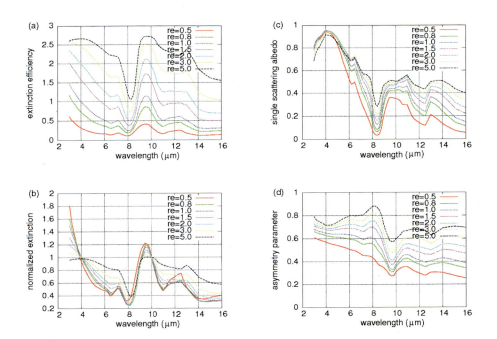

Figure 9.5 Effect of particle size distribution on dust optical properties. The extinction efficiency (a), the extinction normalized at 10 μm (b), the single scattering albedo (c) and the asymmetry parameter (d) are plotted from 3 to 16 μm for 7 effective radii "re", from 0.5 μm to 5.0 μm.

normalized extinction. Considering realistic effective radii for the coarse mode, between 1 and 3 μm, the spectral impact on extinction reaches 30% at its maximum. This agrees with Sokolik et al. (1998) simulations showing that, for many dust samples, the shape of extinction coefficient at infrared wavelengths remains almost unchanged when the mean dust radius increases. The impact on the single scattering albedo and the asymmetry parameter (which are less significant than extinction efficiency for longwave radiative transfer and aerosol retrievals) is of the same order of magnitude.

As we expect that readers are more familiar with aerosol remote sensing in the short-wave, we show on Figure 9.6 the variations of extinction from 0.5 to 15 μm for three effective radii. The situation differs greatly in the shortwave and in the longwave:

- from 3 to 15 μm, as mentioned previously, the three curves are very similar (the normalized extinctions are very close), which means that all the wavelengths are affected in the same way by a change in the particle size

- below 3 μm, the curves of extinction efficiency are not proportional but cross each other, which means that the relative sensitivity of wavelengths (or instrument bands) to the aerosol content is strongly dependent on particle size.

In short, the impact of PSD is crucial for aerosol remote sensing in the solar domain when using several channels, but it is less significant for the longwave domain. A final comment on Figure 9.6 is that the ratio between infrared and visible extinction (or AOD) increases with the effective radius, from 0.2 when re=1μm to 0.8 when re=3μm. This great variability explains why it is not possible to retrieve or extrapolate the infrared AOD from the visible AOD if the fine and coarse mode of the PSD are not perfectly known. This last comment implies that retrieval of infrared AOD from infrared observations is indispensable in order to estimate the aerosol forcing on the longwave radiation.

9.2.4 Effect of particle shape

With the exception of liquid aerosol particles as sulphate stratospheric aerosol that are obviously spherical, modeling the shape of aerosol particles is rather complex: observations show that dust exhibits a high variability of shapes (e.g. E.A. Reid et al., 2003 ; Chou et al., 2008). Until recently, this difficulty has led most authors to consider that dust particles were spherical when simulating their infrared optical properties.

However, over recent years, much laboratory work has been conducted to evaluate the effect of asphericity at infrared wavelengths (Hudson et al., 2007, 2008; Mogili et al., 2007 ; Kleiber et al., 2009). The comparison between laboratory measurements on elementary mineral aerosols tends to show that the Mie theory (i.e. the assumption of spherical homogeneous particles) fails to reproduce the extinction spectrum of minerals. Hudson et al. (2007) and Mogili et al. (2007) also showed that, surprisingly, simpler disc models (Bohren and Huffman, 1983) may work well in reproducing the position of the resonance band of clays or quartz which is shifted towards lower wavelengths when using the Mie theory. However, the results are less clear for the coarse mode particles where the Mie theory behaves better, and some uncertainty remains about the transposition of the laboratory work to the real atmospheric aerosol. As discussed in Kleiber et al. (2009), it is also possible that refractive indices directly measured in the collected atmospheric aerosol (and not on individual minerals in bulk crystal samples) indirectly take into account the shape of the particle as this information could be incorporated in the refractive indices.

Amongst the available algorithms for computing optical properties for nonspherical aerosol particles, one of the most widely used is the T-matrix code by Mishchenko et al. (2002). The particles can be modeled as oblate or prolate spheroids (see Chapter 2), characterized by their aspect ratio. The aspect ratio of desert dust particles is usually in the [1.7–2.0] range (Mishchenko et al., 1997), as confirmed by numerous analyses of scanning electron microscope (SEM) images. However, shape observations by Nadeau (1987), for example, are consistent with the sheet-like mineral structure of silicate clays. Similarly, the best fits of measurement spectra with T-Matrix results assuming spheroid particles require highly eccentric oblate spheroids (for silicate clays) or a very broad distribution including both extreme oblate and prolate spheroids (Kleiber et al., 2009).

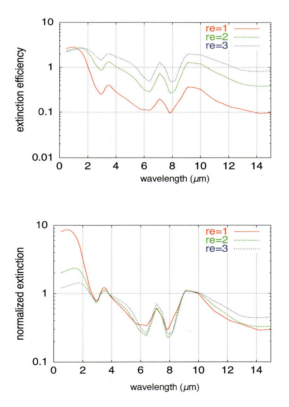

Figure 9.6 Effect of effective radius "re" on the extinction efficiency (top) and the extinction normalized at 10 μm (bottom) from 0.5 to 15 μm: comparison of the spectral behavior in the shortwave and longwave domains.

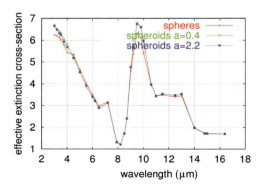

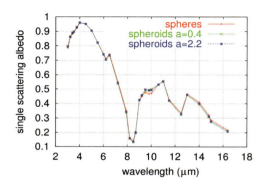

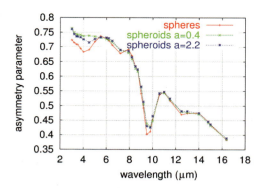

Figure 9.7 Effect of the aspect ratio of mineral dust particles on the optical properties from 3 to 16 μm.

Subsequently, the aim of the current paragraph is not a quantitative evaluation of optical properties for a realistic aerosol population but an estimation of the expected error when assuming spherical aerosol. We are thus only considering here an ensemble of randomly-oriented oblate or prolate spheroids with an aspect ratio varying from 0.4 to 2.2. This is quite an extreme range considering aspect ratios usually reported from SEM images. If we take a single value for the aspect ratio rather than a distribution, we might overestimate the impact of asphericity: Mishchenko et al. (1997) have shown that averaging over various aspect ratios tends to reduce the difference observed with spherical particles.

Simulations are performed using the T-matrix algorithm described in the book by Mishchenko et al. (2002) and available at http://www.giss.nasa.gov/staff/mmishchenko/t_matrix.html. The optical properties for three particle shapes, oblate spheroids (aspect ratio a=0.4), prolate spheroids (a=2.2) and spheres, are plotted on Figure 9.7. The differences between spheres and spheroids are minor: some discrepancy occurs around 4 and 9 μm only. The difference is maximal at precisely 3.75 and 9.5 μm. At these wavelengths, the optical properties C_{ext}, ϖ and g generally show a minimum for spheres and increase when the aspect ratio moves away from 1. The phase function for spheroids differs from the phase function for spheres only at 3.5 μm. By comparison, for shorter wavelengths (e.g. 0.55 μm), the sensitivity to shape is high for the phase function at high scattering angles (see Chapter 2). We must also underline here a difference between the longwave and the shortwave domains: the change in the phase function at high angles (120 to 180°) has very low impact on the satellite-based infrared observations. In fact, as the main source of the radiation, the Earth's surface, is "behind" the aerosol layer, the signature of aerosols from space in the longwave is primarily due to their extinction (while in the shortwave it is primarily due to the scattering of photons towards the instrument).

9.2.5 Summary

This section is summarized in Table 9.2 comparing the conclusions from this chapter with what occurs in the shortwave domain.

9.3 The radiative transfer equation in the thermal infrared range

Chapter 3 dealt with the radiative transfer equation in the shortwave domain, i.e. considering the sun as the only source of radiation. In this chapter, we establish the radiative transfer equation considering the Earth as the main source of radiation. The observed radiation is emitted by the Earth's surface and/or by the atmospheric layers. It is partly absorbed by the molecules and aerosols, and/or scattered by the aerosols: it thus contains the signature of the atmosphere.

First, let us introduce some specific vocabulary, avoiding equations.

The observation of the radiation in the infrared can be performed by a radiometer or by a spectrometer, located either on the ground, or onboard an aircraft, a balloon or a satellite.

Whereas a radiometer collects the photons to measure filtered radiance in a few wide spectral bands, a spectrometer measures the spectral radiance in many continuous, very narrow, spectral channels. A typical radiometer can have up to 20 spectral bands (or channels), whereas a typical spectrometer has several thousands channels (note that the word "band" is not used for spectrometers). The Atmospheric InfraRed Sounder AIRS (Aumann et al., 2003) or the Infrared Atmospheric Sounding Interferometer IASI (Chalon et al., 2001) are two examples of new-generation spectrometers.

Mineral aerosol microphysics	Thermal infrared	Visible and solar infrared
Refractive indices (composition)	+++ great impact -high spectral variability of indices (absorption peaks) -high inter-model variability - few bulk aerosol measurements - difficulty in handling the mixture of pure minerals (high geographical variability)	++ moderate impact - spectral dependency of indices relatively low and well-known - more observations
Size distribution	++ moderate impact - effect of the PSD width is extremely weak - coarse mode prevails - effect on normalized extinction, single scattering albedo and asymmetry parameter to the order of 30% at most - large impact on extinction efficiency if not normalized, but no effect for remote sensing applications	+++ great impact - relative sensitivity at different wavelengths depends on the PSD - accumulation and coarse mode must both be considered
Shape	+ (++?) *a priori* low impact - impact lower than 10% only around specific wavelengths for relatively moderate aspect ratios - laboratory measurements in progress tend to show that very eccentric aspect ratio should be considered	++ moderate impact - phase function at high angles is modified, which directly affects the observation from a satellite

Table 9.2 Influence of microphysical properties of dust aerosol on its optical properties and consequences for remote sensing applications. Comparison between shortwave and longwave domains.

The **spectral radiance** is the power received by the sensor facing a given direction, per surface unit of the receiver, per unit solid angle of its field of view, and per wavelength or wavenumber unit[2]. Thus, it can be expressed in $W/(m^2.sr.\mu m)$ or $W/(m^2.sr.cm^{-1})$. However, it is often more convenient to consider an equivalent quantity: the **brightness temperature** (BT), in kelvin (K). The BT is defined as the temperature of a black body emitting the same spectral radiance at the same wavelength (by extension, at the central wavelength of the spectral channel being considered). Whereas the radiance varies by several orders of magnitude from the longer thermal infrared wavelengths to the shorter wavelengths, the brightness temperature scales the whole spectrum at comparable levels. Another reason why BT is convenient is that the brightness temperature measured from space is usually not far from the skin temperature of the surface (in atmospheric windows) or the thermo-dynamic temperature of the atmospheric level observed (in absorbing regions).

An important category of infrared radiometers or spectrometers on board satellites are **vertical sounders**, or just "sounders", which means that they provide information at several vertical levels in the atmosphere. Vertical sounders on board satellites have been used for decades to provide vertical profiles of temperature and water vapor for weather forecasting (McMillin et al., 1973). This ability to separate the radiation from several altitudes using the spectral information mostly relies on the absorption and emission by the molecules.

As it is critical for readers to figure out the "weighting function" of a spectral channel, indicating which slice of the atmosphere is preferentially observed, we will first try to define it on the basis of physical considerations, before introducing the formal equations in the next paragraph. We consider here an instrument on board a satellite observing the Earth downward.

Due to the **absorption** of the radiation by atmospheric gases (water vapor, ozone, carbon dioxide, etc.), the **transmission** of the upwelling radiation between an altitude level (or equivalently a pressure level) in the atmosphere and the satellite decreases when the altitude level decreases. For instance, if the transmission between the surface and the satellite is 0.5, only 50% of the radiation emitted by the surface can reach the satellite. The transmission between an elevated layer and the satellite is thus higher than 0.5 (e.g. the transmission could be 0.7 at 800 hPa). At the top of the atmosphere, which is usually fixed at a few pascals (between 50 to 100 km), the transmission reaches its maximum value of 1. Indeed, the radiation emitted by the top of the atmosphere in the direction of the satellite does not meet any absorbing molecule and reaches the satellite level without attenuation.

The **thermal emission** of an atmospheric layer (i.e. of the molecules and particles contained in this layer) is the physical mechanism counteracting the absorption; the higher the absorption of a medium, the higher its emission, following Kirchhoff's law. Out of two layers at the same temperature, the one that absorbs the most is also the one that emits the most. At the same time, considering two similarly absorbent layers at different temperatures, the warmer one will emit more radiation than the colder one.

---

2 For broad-band radiometers, the measured signal is sometimes given in term of integrated radiance, in $W/(m^2.sr)$. The integrated radiance, which is instrument-dependent, is equal to the product of the spectral radiance with the instrument filter – or instrument spectral response function (ISRF) – integrated over the full range of observed wavelengths. See Section 9.3.1 and Eq. (9.17)

When absorption is strong enough, the surface and the lower part of the atmosphere might not be observed from space (then the total transmission is zero, see Figure 9.8, for example at 6 μm). On the other hand, the upper atmosphere cannot be observed either, because emission is negligible due to the scarcity of particles and molecules. This involves that only radiation emitted by a given intermediate range of vertical levels might be observed. It is the **weighting function** that gives the exact contribution of each atmospheric level to radiation going up to space. The weighting function usually shows a typical bell-shaped curve, see Figure 9.9 for a schematic illustration. As absorption depends on the wavelength, each spectral channel has its own weighting function (see Figure 9.10).

When absorption strength is moderate, the total transmission is no longer zero (see Figure 9.8, for example at 12 μm) and there is some contribution of the surface and lower part of atmosphere to the radiance outgoing to space. Then the weighting function curve is cut at the surface (see Figure 9.9).

Lastly, the special case with weak absorption corresponds to the so-called "window channels" (see Figure 9.8, for example at 4 μm) and a weighting function limited at the surface and quickly vanishing further up in the atmosphere.

Now that the reader is more familiar with the notions of sounding and weighting functions, let us build the radiative transfer equation (RTE) in the infrared, first without aerosols (Section 9.3.1) and then with non scattering and scattering aerosols (Sections 9.3.3 and 9.3.4). This is only a very short introduction to radiative transfer theory, and a far more detailed presentation can be found in the books from Liou (1980), Lenoble (1993) or Thomas and Stamnes (1999).

9.3.1 The RTE in an aerosol-free atmosphere

The hypotheses considered here are, for the sake of simplicity:

- a plane parallel atmospheric model;
- no refraction in the atmosphere;
- no sun (night observations only or wavelengths higher than 4 μm);
- no scattering (Rayleigh scattering by molecules is negligible in the longwave domain);
- local thermal equilibrium prevailing everywhere.

Modeling a spherical atmosphere or the refraction effects is not difficult, but beyond the scope of this book. The effect of sun radiation and scattering is introduced in Section 9.3.4. Firstly we develop a monochromatic case, i.e. considering a single wavelength, followed by the explanation of the convolution by the instrument spectral response.

Let us consider an infinitely small path in the atmosphere. The Beer-Lambert equation, already set out in this book (Chapter 2) introduces the extinction coefficient σ_{ext} as the proportion of energy lost per elementary length unit ds at a given wavelength

$$\frac{dI_v}{ds} = -\sigma_{ext}(v)I_v,\tag{9.1}$$

where I_v is the spectral radiance at frequency v.

Without scattering, the extinction is caused by absorption only and thus

$$\frac{dI_v}{ds} = -\sigma_{abs}(v)I_v. \tag{9.2}$$

However, the loss of energy by absorption is not the only process to consider. The atmospheric medium considered here also emits radiation, as any body at temperature T>0. Both processes, absorption and emission, are to be considered and thus the radiative transfer equation (RTE) in its differential form is

$$\frac{dI_v}{ds} = \sigma_{emis}(v)J_v - \sigma_{abs}(v)I_v, \tag{9.3}$$

where σ_{emis} is the emission coefficient of the body, and J_v is the so-called source function.

In order to give the analytical expression of the source function, let us first introduce the concept of Local Thermodynamic Equilibrium (LTE). In the atmosphere, up to altitudes of roughly 60 to 70 km, the molecule density is such that the "thermal" emission (caused by collisions between molecules) is predominant and spontaneous emission of photons by molecules is negligible. In other words, the repartition of molecules in the energy levels is driven only by temperature: this is LTE. At higher altitudes where there is no LTE, the density of molecules is so low that this effect can be dismissed for a nadir sounder on board

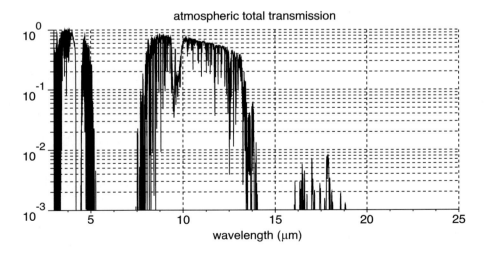

Figure 9.8 Total transmittance of the atmosphere from 3 to 25 μm, for a tropical situation without aerosol.

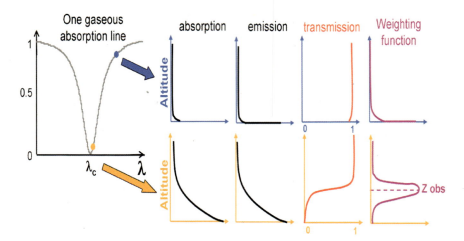

Figure 9.9 Simplistic view of the weighting function. If the gaseous absorption is low (blue dot), then, according to Kirchhoff's law, the emission is low and the atmospheric transmittance is almost equal to 1 down to the Earth's surface. The weighting function is zero except close to the surface (window channel). If the absorption is strong (orange dot), then transmission at the surface is zero and the weighting function peaks at an observation altitude Zobs (sounding channel).

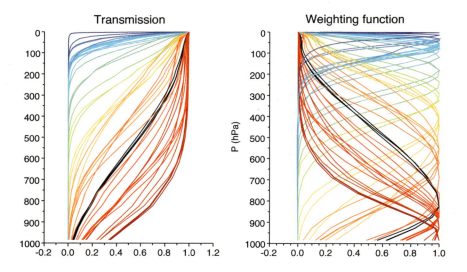

Figure 9.10 Examples of transmission profiles (left) and normalized weighting function (right) for the IASI instrument, for a subset of one over 10 channels in the spectral range [660–780] cm^{-1}. The color code indicates the altitude of the peak of the weighting function.

a satellite (looking down to the Earth's surface). Non-LTE must be taken into account for limb view sounders (looking to space through the atmosphere) (see Goody and Young's book, 1989), but this type of observation geometry prevents the observation of the lower atmosphere and is consequently out of the scope of tropospheric aerosol remote sensing.

Assuming LTE, the source function of each atmospheric layer depends on its temperature only and is given by Planck's law

$$J_v = B_v(T) = \frac{2hv^3}{c^2 \left[\exp\left(\dfrac{hv}{kT}\right) - 1 \right]}. \tag{9.4}$$

This function, established by Max Planck in 1900, describes the spectral radiance of the radiation emitted at frequency v by a blackbody at temperature T. It contains the following constants:

• c, velocity of light in the vacuum;

• h, Planck's constant;

• k, Bolzmann's constant.

Let us now introduce Kirchhoff's law of thermal radiation. The law states that, at thermal equilibrium, the emissivity of a body equals its absorptivity. Applying Kirchhoff's law locally in a given atmospheric layer gives the equality of absorption and emission coefficients

$$\sigma_{emis}(v) = \sigma_{abs}(v)$$

Thus, Eq. (9.3) can be written

$$\frac{dI_v}{dl} = \sigma_{abs}(v)(B_v(T) - I_v). \tag{9.5}$$

This equation is the differential RTE at local thermal equilibrium in a non-scattering medium. Another form of this equation may be preferred, using either the optical depth or the transmittance instead of the absorption coefficient.

The transmittance of a medium, between z and z' is defined as

$$t_v = \exp\left(-\int_z^{z'} \sigma_{abs} \, dl\right) = \exp(-\tau), \tag{9.6}$$

where we introduce τ, the optical depth of the layer between z and z'. This is equivalent to

$$\frac{dt}{t} = -\sigma_{abs} \, dl. \tag{9.7}$$

Using the above expression in Eq. (9.5), we obtain

$$\frac{dI_v(t)}{dt} = -\frac{1}{t}\left[I_v(t) - B_v(T(t))\right].$$

(9.8)

We can also use the definition of the optical depth from Eq. (9.6) which results in

$$\frac{dI_v(\tau)}{d\tau} = -\left[I_v(\tau) - B_v(T(\tau))\right].$$

(9.9)

For atmospheric studies, the radiative transfer equation is of more use in its integral form than in its differential form. So, we have to integrate the RTE between the Earth's surface and the satellite. We first consider a nadir view (i.e. the angle between the local vertical at the Earth's surface oriented downward and the satellite viewing direction is 0°). The integrating path is along the vertical axis, going from 0 (at the surface) to the top of the atmosphere.

The integration gives the spectral radiance at the top of the atmosphere, assuming a surface emissivity of ε_{surf} (between 0 and 1)

$$I_v = \varepsilon_{surf} B_v\left(T_{surf}\right)\exp(-\tau_{surf}) + \int_0^\tau B_v\left(T(\tau)\right)\exp(-\tau)d\tau + \left(1 - \varepsilon_{surf}\right)\exp(-\tau_{surf})\int_\tau^0 B_v\left(T(\tau)\right)\exp(-\tau)d\tau.$$

(9.10)

Three terms appear in Eq. (9.10): the first is the radiation emitted by the surface and transmitted to the satellite; the second is the so-called "upward" radiation emitted by the atmospheric layers; the third is the "downward" radiation emitted by the atmosphere and reflected by the surface. Obviously, the sum of the surface reflection coefficient and emissivity equals 1. Note that the surface emissivity is sometimes close to 1 (e.g. 0.99 over the ocean, around 10 μm), so the third term can be disregarded in the RTE in these cases.

The same equation as a function of transmittance t is

$$I_v = \varepsilon_{surf} B_v\left(T_{surf}\right)t_{surf} + \int_0^1 B_v(T(\tau))dt + \left(1 - \varepsilon_{surf}\right)t_{surf}\int_1^0 B_v(T(\tau))dt ,$$

(9.11)

and, after replacing the integration variable dt by $d(ln(P))$ with P the pressure at a vertical level, it becomes

$$I_v = \varepsilon_{surf} B_v\left(T_{surf}\right)t_{surf} + \int_{\ln(Psurf)}^{-\infty} B_v(T(P))\frac{\partial t(P)}{\partial \ln(P)}d\ln(P) + \left(1-\varepsilon_{surf}\right)t_{surf} \int_{-\infty}^{\ln(Psurf)} B_v(T(P))\frac{\partial t(P)}{\partial \ln(P)}d\ln(P).$$

$$(9.12)$$

This provides the opportunity to introduce the weighting function

$$W(P) = \frac{\partial t(P)}{\partial \ln(P)} \; . \tag{9.13}$$

The weighting function is thus defined as the derivative of the transmission with respect to the pressure logarithm (or altitude). Its name comes from its role in the RTE: if we consider, for example, the upward term, the radiation emitted by each layer $B(T(P))$ is weighted by the weighting function W. The maximum contribution comes from the pressure levels where W is maximum, which corresponds to an inflection point on the transmission profile. A sketch of the weighting function is given in Figure 9.10.

When the local zenith angle θ (angle at the ground surface between the local vertical and the satellite direction) is not equal to 0, then we introduce the secant of this angle, using $\mu = cos(\theta)$

$$\sec(\theta) = \frac{1}{\mu}\, , \tag{9.14}$$

and Eq. (9.5) becomes

$$\mu\frac{dI_v}{dl} = \sigma_{abs}(v)\left(B_v(T) - I_v\right), \tag{9.15}$$

and the transmission becomes

$$t_v = \exp\left(-\int_z^{z'}\frac{1}{\mu}\,\sigma_{abs}\,dl\right); \tag{9.16}$$

so that Eq. (9.13) is unchanged (this is an advantage of working with transmission rather than optical depth).

Readers will have noticed that the previous equations are all monochromatic. Indeed, in the infrared domain, the gaseous absorption lines are so numerous and thin that an accurate simulation requires a line-by-line approach. A real instrument never senses a single wavelength but over a wavelength range constituting a channel, the radiative contribution of each wavelength being weighted by the so-called instrument spectral response function (ISRF), or instrument line shape (ILS), or sometimes also referred to as the convolution function. Thus, the last step of the simulation of the radiative transfer usually consists in the convolution by the ISRF f_i of each channel i through the normalization

$$I_i = \frac{\int f_i(v)I(v)dv}{\int f_i(v)dv}. \tag{9.17}$$

9.3.2 The data required as input for a gaseous atmosphere

A short analysis of the integrated RTE gives us the main variables to take into account.

The first one is the temperature, not only of the surface (skin temperature), but also at every level in the atmosphere. Indeed, the radiation emitted by the surface and by each atmospheric layer is directly dependent on the temperature through Planck's function. A second surface parameter of importance is the emissivity (related to the materials at the surface). The surface pressure (related to the altitude and the meteorological situation) also needs to be taken into account.

A third variable is the transmittance profile, which is equivalent to the absorption or optical depth profiles. The absorption of photons is caused by rotation and vibration transitions of atmospheric gas molecules. In a given layer of pressure P, the total optical depth of absorption is the sum of the optical depths for all species with absorption lines at the given wavelength

$$\tau(P,v) = \sum_j \tau_j(P,v). \tag{9.18}$$

The optical depth for an individual gas at a given wavelength is the product of its absorption efficiency (for a single molecule) by the number of molecules

$$\tau(P,v) = \sum_j \sigma_{abs\,j}(P,T(P),v)N_j(P). \tag{9.19}$$

The absorption efficiency of a molecule depends on the frequency, but also on the thermodynamic conditions (mainly the temperature and pressure). Indeed, absorption lines corresponding to transitions between two states of energy, their intensity, width and shape depend on partial pressure of the gas in question and other gases, and the temperature. The spectral width of a line, for example, is due to collisions between molecules (which obviously depend on pressure) and to the Doppler effect. However, the accurate description of these spectroscopic effects and line modeling is beyond the scope of this book, and can be found in Bernath (2005).

Consequently, the concentration profile ($N_j(P)$) must be known for each gas that has a non-negligible absorption. Also, gaseous spectroscopy (for water vapor, ozone, carbon dioxide...) is a necessary input. All the parameters needed to compute absorption line shape and intensity can be found in the GEISA (Jacquinet-Husson et al., 2008) or HITRAN (Rothman et al., 2009) spectroscopic databases, both available on the Internet: http://ether.ipsl.jussieu.fr/ (GEISA) and http://www.cfa.harvard.edu/hitran/ (HITRAN). These databases compile results from the work of teams at many spectroscopic laboratories. For instance, the GEISA-09 sub-database of line transition parameters involves 50 molecules (111 isotopic species) and contains 3,807,997 entries.

All the variable parameters (thermodynamic profile, i.e. temperature and humidity profiles, other gas concentration profiles and Earth's surface temperature, emissivity and pressure) must be known with sufficient accuracy for remote sensing applications.

The surface temperature and the surface emissivity over land might come from the instrument observations themselves (Péquignot et al., 2008, Chédin et al., 2004). The surface emissivity might also come from a climatological data set (e.g. the ASTER spectral library, available at http://speclib.jpl.nasa.gov/), thus offering the possibility to easily retrieve the surface temperature from a window channel. However, the surface emissivity can be highly variable in space (low emissivity over deserts or bare soils, high emissivity over forest or ocean) and in time (seasonal cycle related to the growth of vegetation). This difficulty explains why retrieval algorithms in the infrared are usually first developed over oceans, where the emissivity is well known, almost spatially homogeneous and almost spectrally constant (close to 1 in the thermal infrared). The surface pressure is usually known with sufficient accuracy by just considering the altitude of the ground pixel.

The temperature and humidity profiles can be obtained from in-situ radiosoundings, Numerical Weather Prediction (NWP) models, or sounder retrievals. However, when using sounder retrievals for aerosol remote sensing applications, great care must be taken not to choose channels sensitive to aerosols when retrieving the thermodynamic profile. Not only the profile of water vapor but also the profile of other gases must be provided: ozone, for instance has strong absorption lines between 9 and 10 μm, in the middle of the window 8−12 μm which is of particular interest for aerosols. Other gases absorbing in the infrared range are CO_2, CH_4, CO, etc. If the gas variability with time and space is low, or if the channel chosen has a weak sensitivity, then a constant profile can be assumed (for instance, for CO_2). Otherwise, highly variable gas absorption lines are usually avoided for aerosol remote sensing applications. Water vapor strong lines are to be avoided as much as possible. However, theses lines are so numerous, and there is a continuum of water vapor absorption, so the impact of water vapor can not be dismissed. Henceforth, we shall refer to the vertical profiles of temperature and all the absorbing gases (essentially water vapor) in the spectral domain of interest as the "atmospheric situation".

We should also mention climatologic databases containing a huge sample of atmospheric situations (e.g. the Thermodynamic Initial Guess Retrieval – TIGR- database from Chédin et al. (1985), and Chevallier et al. (1998)) that can be used for sensitivity studies or for the look-up tables approach of aerosol remote sensing.

We illustrate in Figure 9.11 the sensitivity of the radiative transfer to atmospheric and surface parameters with some typical examples of infrared spectra (without scattering).

9.3.3 The particular case of RTE in an atmosphere containing non-scattering aerosol

When the scattering of the particles can be neglected, the RTE for a gaseous atmosphere remains valid, with the addition of aerosol absorption optical depth to molecular optical depths. The transmittance is now the product of particles and molecules transmittances

$$t_v = \exp\left(-\int_z^{z'}\left(\sigma_{molecules} + \sigma_{particles}\right)dl\right) = t_v^{molecules}\, t_v^{particles}. \tag{9.20}$$

The transmission from the Earth's surface to the satellite must also be modified in the same manner. This particular case applies mainly for stratospheric volcanic aerosols due to their refractive indices.

Although very simple, this formulation of the RTE is quite useful as it gives us the contribution of aerosol absorption and emission to the radiance. If the single scattering albedo is not too high, the main effect of aerosols on the brightness temperature spectrum is a cooling effect.

Indeed, the total transmittance decreases, and the Earth's surface contribution (first term in Eq. (9.10), which is the dominating term for channels of interest to aerosol remote sensing, decreases as well. The second term in Eq. (9.10), of atmospheric contribution, is also

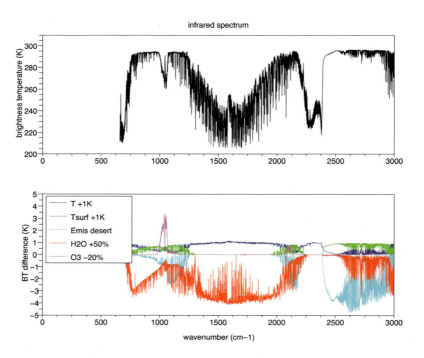

Figure 9.11 Top: infrared spectrum convolved with the IASI instrument response function, in brightness temperature (night-time, over ocean, no aerosol); bottom: the infrared spectrum sensitivity to a change in the atmosphere or the surface (blue: temperature profile, green: surface temperature, cyan: surface emissivity, red: water vapor, pink: ozone profile).

modified: the transmittance at levels inside the aerosol layer and below, is lower. Thus, the weighting function is increased at all the levels where aerosol is present. This means that the relative contribution of the black-body radiation emitted at these levels to the total observed radiance increases. If the aerosol is located at altitudes where the temperature is lower than the mean BT seen by the channel, which is usually the case for channels peaking close to the surface, then the second term in Eq. (9.10) also gives a decrease of the BT in presence of absorbing aerosols.

9.3.4 The RTE with scattering processes

As previously shown (considering the values of single scattering albedo of Figure 9.3), scattering cannot be ignored in the presence of aerosols such as desert dust. Thus, the RTE must be modified by introducing scattering.

Let us assume that the electromagnetic wave propagates in the direction (θ, ϕ) and let $\mu = \cos(\theta)$. In a scattering medium of single scattering albedo ϖ and phase function p, the differential form of the RTE in Eq. (9.8) becomes

$$\frac{dI_\nu\left(t, \mu, \varphi\right)}{dt} = -\frac{1}{t} I_\nu\left(t, \mu, \varphi\right) + \frac{1-\varpi}{t} B_\nu\left(T(t)\right) + \frac{\varpi}{(4\pi)t} \int_0^{2\pi} \int_{-1}^{1} p(t; \mu, \varphi; \mu', \varphi') I(t, \mu', \varphi) d\mu' d\varphi'.$$

(9.21)

The Planck emission of the medium which is no longer a black body is reduced by the scattering (second term), but the scattering itself contributes to the radiation in the direction (μ, ϕ), depending on the phase function (third term in Eq. (9.21)).

If solar radiation from direction (θ_s, ϕ_s) cannot be dismissed (e.g. for shorter wavelength up to 4 μm during daytime), then a fourth term must be added to the previous equation

$$J_{sun} = \frac{\varpi}{(4\pi)t} I_{sun} p(t; \mu, \varphi; \mu_s, \varphi_s) t^{\mu/\mu_s},$$

(9.22)

where I_{sun} is the solar radiance at the top of the atmosphere and $\mu_s = 1/\cos(\theta_s)$ is negative.

The RTE is now an integro-differential equation with no analytical solution. Several numerical techniques have been developed to solve this problem, for example the Discrete Ordinate algorithm (DISORT), the method of the successive orders of scattering, or the two-stream approximation (see Chapter 2 or below). For a detailed description of these techniques, readers may refer to the books from Thomas and Stamnes (1999) or Hanel et al. (2003).

In some cases, a simplification might occur. Let us separate the radiance I into I+ and I-, with I+ the upwelling radiance and I- the downwelling radiance. These radiances and the phase function might be decomposed in Fourier series over $\phi-\phi_0$ where ϕ_0 is the azimuthal solar angle

$$I_\nu^{\pm}\left(t, \mu, \varphi\right) = \sum_{i=0}^{m} I_{\nu,m}^{\pm}\left(t, \mu\right) \cos\left(m(\varphi - \varphi_0)\right).$$

(9.23)

When applying the RTE to the mth Fourier component, we obtain a more useful form for numerical integration

$$\frac{dI^{\pm}_{v,m}(t,\mu)}{dt} =$$

$$-\frac{1}{t}I^{\pm}_{v,m}(t,\mu) + \frac{1-\varpi}{t}B_v\left(T(t)\right) + \frac{\varpi}{(4\pi)t}\int_{-1}^{1}P_m(t;\mu;\mu')I^{\pm}_{v,m}(t,\mu)d\mu' + \frac{\varpi}{(4\pi)t}I_{sun}P_m(t;\mu;\mu_0)I^{\pm}_{v,m}(t,\mu).$$

$$(9.24)$$

Further simplification can arise, for instance if the solar radiation can be ignored. Figure 9.12 shows that the effect of solar radiation might be ignored, either by night or for wavelengths higher than 4.2 μm for a satellite instrument viewing at the Earth's surface. In that case, if the phase function has a revolution symmetry, which is what is usually observed for aerosols, then the azimuthal dependency is removed and only for m=0 is the Fourier component not null.

9.3.5 Solving the RTE

A numerical solution of the RTE equation can be computed with a code coupling a line-by-line model (required by the very thin spectral lines of molecules and the very high spectral resolution of infrared sounders) and a scattering algorithm, such as DISORT (Discrete ordinate algorithm, first described by Chandrasekar (1960)) or SOS (successive orders of scattering) algorithm (e.g. Hansen and Travis, 1974; Heilliette et al., 2004) or the adding-doubling algorithm (Van de Hulst, 1963; Liou, 2002). The DISORT algorithm has been coded by Stamnes et al. (1988) and is available as a Fortran subroutine.

In a coupled code (for example Dubuisson et al. (1996) or the 4A/OP code originally from Scott and Chédin (1981), and described in Chaumat et al. (2009)(see also http://ara.lmd.polytechnique.fr), the input requires the usual parameters of the line-by-line absorption code as described in Section 9.3.2, and additional parameters needed for the simulation of scattering. The DISORT or SOS scattering codes require the gaseous optical depth profile (that is, an output of the line-by-line absorption code) and aerosol parameters. The scattering parameter (extinction, single scattering albedo and phase function) are needed at each vertical level, but we usually assume that only the aerosol content varies with altitude. Note also that the Henyey-Greenstein approximation is often used instead of the full phase function. The Henyey-Greenstein approximation consists of replacing the phase function by a function of the asymmetry parameter g only. The approximated phase function defined from g differs from the true one at high scattering angles (i.e. for backscattering), but this has a negligible impact on the infrared radiance.

Such a coupled code is very accurate and is needed as a reference code. However, in the line-by-line case, computation is still time-consuming, so various techniques have been suggested to speed up the calculation, by taking into account the instrument spectral response function: k-distribution (e.g. S. Yang et al., 2000 ; Lacis and Oinas, 1991), spectral averaging (e.g. in the 4A/OP code).

Two situations in particular can benefit from faster algorithms. First, for window channels and dry atmospheres, when there is no (or almost no) gaseous absorption, the radiative transfer equation can be written much more simply, and can be solved, for instance, with the two-stream algorithm (Meador and Weaver, 1980). Second, for broad-band channels, band models, which are much faster than line-by-line models, can be used, for example the MODTRAN code (Berk et al., 1989) or Streamer (Key and Schweiger, 1988).

9.3.6 Aerosol effect on the longwave spectrum

The main difficulty for aerosol remote sensing in the longwave is that the sensitivity of brightness temperature (BT) to atmospheric temperature, water vapor and surface state is very high, and the aerosol has only a second-order impact, as illustrated in Figure 9.13. Indeed, a change from a typical tropical atmosphere to a mid-latitude atmosphere decreases the BT by several tens of kelvins, whereas the effect of the aerosol is usually in the range of a few kelvins.

A second difficulty is that the impact of aerosols on the observed radiance is also strongly dependent on the underlying atmospheric situation. For example, an aerosol of fixed optical depth, single scattering albedo and phase function profile might decrease the brightness

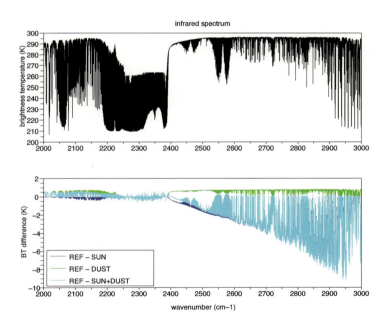

Figure 9.12 Top: simulated infrared spectrum in the medium infrared domain (2000 to 3000 cm⁻¹); the effect of solar radiation must be taken into account in this spectral domain; bottom plot: difference in BT (K) between the reference spectrum (no sun, no aerosol), and the same situation with sun only (blue), dust only (green) or sun and dust (cyan).

temperature by 6 K for a given atmospheric situation, and by only 2 K for another one with a different temperature and water vapor content, all other things being equal. As shown in Figure 9.14, this variability is observed both for window channels (AIRS channel #315 has a surface transmission greater than 0.9), mainly caused by a change in surface temperature, and for sounding channels (AIRS channel #134 has a surface transmission of about 0.55).

To sum up:

- The absolute value of BT depends on the surface and atmospheric thermodynamic situation.

- The relative effect of the aerosol on BT strongly depends on the surface and atmospheric thermodynamic situation.

We focus now on the effect of optical properties (single scattering albedo, asymmetry parameter, AOD) on the observed brightness temperature. Observing simulation results by night in Figure 9.15, it can be noted that:

- The effect of the single scattering albedo is complex: in this example, the sign of the albedo impact can be either positive (e.g. channel at 3.82 μm) or negative (e.g. channel at 11.48 μm) depending on the wavelength. This can be explained by a trade-off between two effects: in the RTE with scattering Eq. (9.21), the emitted radiation (second term), is weighted by $(1-\varpi)$, whereas the scattered radiation (third term) is weighted by ϖ.

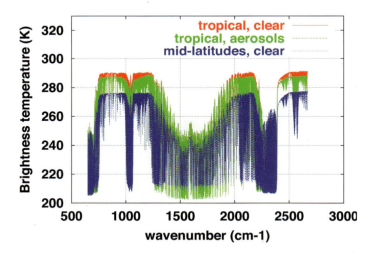

Figure 9.13 Comparison between the latitudinal effect (humidity and temperature profile) and the aerosol effect (Mineral Transported model from OPAC data-base) on the longwave spectrum: simulations for night-time, over oceans.

- The BT increases with the asymmetry parameter g: if g is close to 1, then forward scattering (in the direction of the satellite) is favored, while radiation lost by scattering is reduced (no solar radiation is assumed here).

- The BT decreases when the AOD increases, all other parameters being unchanged. This can be easily understood if the aerosol emission is lower than its extinction, which is generally the case for a temperature profile decreasing with altitude: when the aerosol extinction increases, the number of photons emitted by the surface and the atmosphere and reaching the satellite decreases.

These results are obtained for a single thermodynamic atmospheric situation (i.e. a single temperature and humidity profile). However, some cases might occur where the behavior changes. For instance, in case of very strong temperature inversion in the lower layers (a situation sometimes occurring by night), the emission from the aerosols might then prevail over the absorption and thus the BT might increase with AOD. Also, if the weighting function of the channel peaks higher in the atmosphere (strong absorption by gases), then the relative impact of the aerosol would change, but it would remain very low in the absolute, which is not a useful configuration for aerosol remote sensing.

These simulations show that the radiative effect of atmospheric aerosol on atmospheric spectrum is difficult to predict without a proper radiative transfer model, when both scattering and absorption occur. However, BT varies smoothly with optical properties, thus enabling some linear interpolations in order to avoid long computations.

We now focus on a specificity of infrared observation: the role of vertical distribution of aerosols on the observed signal. Not only can the quantity of aerosol at each level (i.e. the AOD within each atmospheric layer) change, but the single scattering albedo and asymmetry parameter can also vary, for instance if the size distribution or the composition of particles is modulated with the altitude. However, as there are very few experimental data of such variations, we are restricting these considerations here to the effect of the extinction vertical profile, assuming albedo and phase function constant over the whole atmospheric column.

We deal firstly with tropospheric aerosols, such as dust, located in one single atmospheric layer of the model. Note that radiative transfer codes usually work with the pressure as a vertical coordinate (and not the altitude). However, it is rather simple to convert a pressure grid to an altitude vertical grid (if a moderate accuracy can be accepted) and so we will generally present results using the vertical altitude. Figure 9.15 gives the sensitivity of BT to the layer altitude at two wavelengths (3.82 μm and 11.48 μm). The higher the layer, the greater its impact on the BT. Indeed, in the troposphere, the higher the aerosol layer, the lower its temperature, which reinforces the aerosol signature through the contrast between the radiation emitted by the surface and the radiation emitted by the aerosol. Secondly, when the aerosol is close to the surface, the absorption of radiation by gases between the aerosol layer and the satellite partly (or totally) hides the aerosol signature. This is why a change of altitude of the aerosol layer does not have a major impact on a window channel (at 3.82 μm here, the surface transmittance is greater than 0.9, whereas at 11.48 μm it is only 0.25). So, not only is there an impact from aerosol altitude on the signal observed, but also this impact depends on the wavelength (through the effect of gas absorption). Another

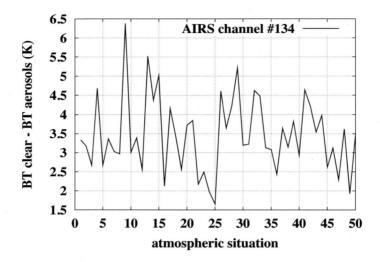

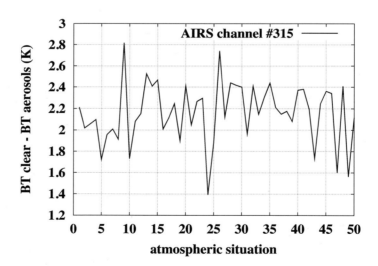

Figure 9.14 Examples of calculated impact of an aerosol layer on the brightness temperature measured by 2 AIRS channels: channel #134 at 843 cm^{-1} (top) and channel #315 at 2616 cm^{-1} (bottom). Nighttime simulations over oceans. The impact strongly depends on the thermodynamic atmospheric situation (i.e. T, H$_2$O and O$_3$ vertical profiles, taken in the TIGR database).

way to analyze this effect is to consider the aerosol layer temperature: as the temperature decrease in the troposphere is nearly uniform, the information about aerosol plume altitude or the information about its temperature are roughly equivalent. These considerations make aerosol remote sensing rather difficult in the longwave domain, but they do provide the unique possibility of measuring their altitude from space (using passive remote sensing).

In the case of aerosol profiles with complex structures (for example, two overlapping layers), it can be shown that the altitude retrieved corresponds to the median altitude in terms of infrared extinction. In other words, half of the AOD comes from aerosol below the retrieved altitude, and half comes from above. Except for perfect window channels with a transmittance of 1 (which do not really exist...), there is a particular case where the altitude can be disregarded: stratospheric aerosols observed with channels sounding the low troposphere. Indeed, for these low peaking channels, the transmittance at aerosol altitude is close to 1 and the pressure dependency in the RTE is removed.

Conversely, the infrared radiation is not as sensitive to the particle size or shape (all other things being equal, especially the infrared AOD at 10 μm), as can be seen on Figures 9.16 and 9.17. Of course, if the shortwave AOD, or the number density or mass density was kept constant, the size effect would be larger. This is why it is appropriate to retrieve the aerosol infrared optical depth from longwave observations. Figures 9.16 and 9.17 correspond to simulations performed with a relatively high amount of dust, with the AOD at 10 μm being 0.6. Thus, we conclude that the effect of size (less than a kelvin) and shape (a few tenths of a kelvin), can be dismissed on a first approach. This also means that the retrieval of size or shape requires firstly that the aerosol infrared AOD and altitude be well known.

9.4 Inversion principles

The previous paragraph described the direct problem: the modeling of the aerosol effect on measurements. We now focus on the inverse problem: the retrieval of some aerosol properties from observations. Even though there might be many more spectral measurements than unknowns (in the case of high resolution spectrometers), this inverse problem is ill-posed, as only one or a few quantities of interest to aerosols can be retrieved, and not the full description of the vertical profile of all the optical or microphysical properties of the aerosol. So, a careful sensitivity study, using radiative transfer simulations, can select the observable parameters and is a prerequisite to any inversion algorithm design. Indeed, an ill-posed problem can be solved only by introducing *a priori* knowledge, for example using parameterization (i.e. reduction of the number of unknowns through the description of the aerosol by a limited number of parameters) or using regularization (*a priori* of smoothness, for instance).

We have seen that many geophysical parameters not related to the aerosol, like the surface or atmospheric temperature, influence the radiative transfer. Therefore, these parameters must be either known *a priori* (from NWP analysis, or from another instrument), or retrieved (in a first-step process or simultaneously with the aerosol inversion), or by-passed by determining a combination of channels removing their influence.

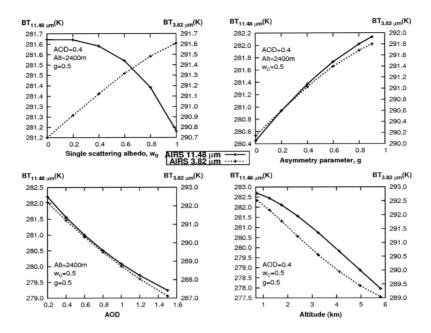

Figure 9.15 Simulated effect of aerosol optical properties and altitude on brightness temperature for 2 AIRS channels corresponding to the wavelengths 11.48 μm (thick line) and 3.82 μm (dotted line).

The purpose of this paragraph is to describe the great diversity of retrieval algorithms that can be used in the longwave domain for aerosol remote sensing. However, readers must bear in mind that the algorithm strongly depends on the instrumental data type (broadband radiometer or spectrometer, on ground or on board a satellite) but also on the purpose of the retrieval. For instance, an aerosol retrieval designed for near real-time monitoring in the framework of air quality must be fast enough. For a regional case study, some parameters in the algorithm might be tuned "by hand", whereas global retrievals for climate studies must be resistant to any local change.

9.4.1 Useful spectral domain

In order to have an impact on the observed radiance, the transmission from the aerosol layer to the satellite must be sufficiently high. This is why most channels are selected within, or close to, the atmospheric windows: around 4 μm, 8 μm and 10.5–12 μm. The water vapor absorption lines in fact totally preclude the use of the 5–8 μm band, not only because of the low transmission but also because the water vapor variability is extremely

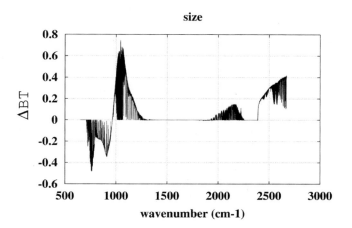

Figure 9.16 Simulated effect of the effective radius on the infrared spectrum: difference in brightness temperature between a spectrum computed with an effective radius of 3 μm and a radius of 1 μm. Night-time simulation over ocean, for a high mineral dust load.

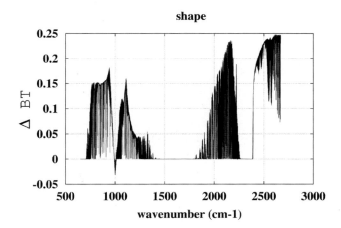

Figure 9.17 Simulated effect of the particle shape on the infrared spectrum: difference in brightness temperature between a spectrum computed with a spheroid – aspect ratio of 2 – and a sphere. Night-time simulation over ocean, for a high mineral dust load.

high. To a lesser extent, the same reason holds for the ozone band centered at 9.6 μm, which must be avoided.

The 8–12 μm band is appropriate for dust (which has a strong absorption band around 9 μm, see Section 9.2.2) but also for volcanic ash or volcanic sulphuric acid droplets.

Combining observations from the medium infrared band (3–4 μm) and the thermal infrared band (8–12 μm) is useful to discriminate the aerosol AOD from its height (or equivalently the aerosol layer temperature), as the respective sensitivities differ: Figure 9.18 shows that the sensitivity to a change in the aerosol altitude is stronger in the thermal infrared domain than in the medium infrared domain.

For recent infrared sounders, a channel selection is usually first needed to limit the computation time, because of the very high number of spectral channels. The number of channels to be used depends on the computation time and the number of observations to be retrieved, and can vary from 2 or 3 for global processing of measurements over long periods to over 100 for case studies (e.g. Carn et al., 2002). The channel selection for aerosol retrieval generally relies on statistics obtained from a high number of radiative transfer simulated spectra or Jacobians (the partial derivatives of the radiance with respect to a change in the aerosol optical depth or altitude). Sensitivity to aerosol must be as high as possible, together with a sensitivity as low as possible to variable gas (water vapor and ozone mainly, but also carbon monoxide or methane). The channels which are less affected by the aerosol (for instance at a wavelength where the aerosol extinction efficiency is relatively low) but which have similar weighting functions are also useful to constrain the emission from Earth's surface and atmosphere. Selecting channels with different transmissions of the gaseous atmosphere brings vertical information too: a proper coverage of the lower tropospheric column is ensured by selecting channels with transmittance varying in the range [0,1], because the lower the transmittance, the higher the altitude of the channel weighting function peak (see Section 9.3).

9.4.2 Preliminary steps

Building an aerosol-dedicated cloud mask

Distinguishing aerosols from clouds is always a tricky problem for aerosol remote sensing (King et al., 1998). Henceforth, we shall refer to a scene without aerosols or clouds as "clear" or "aerosol-free", and a scene with a water or ice cloud as "cloudy" and a scene with aerosol particles as "aerosol". Note that the cloudy scenes might also contain aerosols but are disregarded anyway, even if Hong et al. (2006) have shown that it should be possible to distinguish co-existing cirrus-dust scenes from those associated only with cirrus clouds or dust alone.

Usually an aerosol-dedicated cloud mask must be developed, because standard cloud masks used for meteorological applications need a high confidence in clear sky detection and are consequently designed to filter both water/ice clouds and aerosols together (e.g. EUMETSAT, 2009). Very thick clouds are usually easy to detect (they correspond to very cold BT) and the principal difficulty lies in the detection of thin cirrus clouds and low altitude clouds. The accurate description of clouds' edges might be another concern for imaging instrument with high spatial resolution. The detection by daytime of low altitude

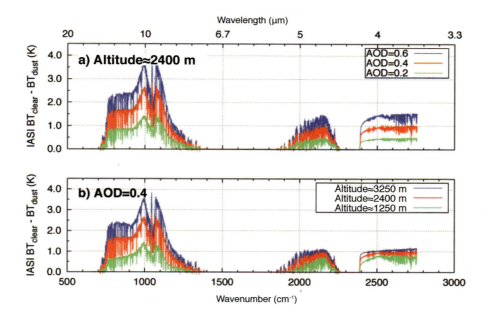

Figure 9.18 Simulated impact of dust AOD (top) and altitude (bottom) on IASI brightness temperature spectra over oceans, by night (courtesy of S. Peyridieu).

clouds, which are usually bright over dark surfaces in the shortwave, may benefit from visible reflectance observations, if available.

Amongst cloud masks used for aerosol remote sensing applications, there are threshold tests (usually, a BT difference between 2 channels that must be higher or lower than a threshold), spatial heterogeneity tests or tests using exogenous data (from other instruments).

The aerosol signature on the infrared spectrum shows a typical "V-shape" observed by many authors, e.g. Sokolik (2002) for dust, or Karagulian et al. (2010) for ash. This is also sometimes referred to as "negative slope" in the 800–1000 cm^{-1}, in contrast to the positive slope observed for cirrus clouds (see Figure 9.19). This effect makes it possible to distinguish clouds (especially thin cirrus) from aerosols using a threshold test. Radiative transfer simulations can help to find couples of channels which are well correlated over a wide range of atmospheric situations except cloudy ones. If the difference between these channel BTs reaches a given threshold, the pixel is flagged as cloudy. In order to prevent a pixel being flagged as cloudy when dusty, the dust effect must be different from the cloud effect. For instance, if the presence of clouds decreases the difference between two channels, the presence of dust should increase this difference or keep it constant. This technique has been used for AIRS by Pierangelo et al. (2004b).

An example of exogenous data is the use of microwave sounder measurements (for example, the AMSU instrument, on board the same satellites as AIRS or IASI). Microwave radiances are not sensitive to most clouds and can be used to predict an infrared radiance or brightness temperature. When the predicted infrared BT is too far from its measured value, the pixel is considered as cloudy. Another example: for the IASI infrared sounder, it is possible to use the cloud mask from the AVHRR imager.

The spatial heterogeneity tests rely on the spatial difference between aerosols (usually forming homogeneous plumes or "clouds") and water or ice clouds which are more heterogeneous. For such a test, one considers that the standard deviation between neighbor pixels must be lower than a given threshold in order for these pixels to be considered cloud-free.

Several approaches are generally combined together. For example, Hansell et al. (2007) combined, for the MODIS instrument infrared channels, the simple difference in the threshold technique with a double difference ("slope method") and analytical expression including a BT difference and visible reflectance ratios.

Removing the "calculated minus observed" biases

Radiative transfer models often show a bias with observed measurements. This systematic bias can come from the model (e.g. a spectroscopy error) or from the instrument (e.g. calibration uncertainty or poor knowledge of the response function). Since the retrieval often relies on simulated data, their application to real data implies that any brightness temperature systematic biases between simulations and observations have been eliminated. This correction is of particular importance for climate applications. This "calculated minus observation", also called "calc-obs", bias is routinely monitored in operational weather forecast centers. For aerosol remote sensing applications, these biases might be evaluated by comparing simulations and observations for a set of collocated satellite observations and radiosondes (e.g. ERA 40) or NWP analyses, for example from the European Center for Medium-Range Weather Forecasts, for clear sky situations. By adding the average bias (calculated − observed), the observed BT are scaled to the radiative transfer model BT reference frame. The biases usually distinguish night and day, land and sea, and sometimes latitude bands. This procedure also prevents the background aerosol, whose loading shows a weak temporal variability, from contaminating the retrieval.

9.4.3 Detecting aerosols

The signature of a strong aerosol event (desert dust, volcanic ash, etc.) on an atmospheric spectrum is usually easily detected "by the naked eye" on a full infrared spectrum, thanks to the previously mentioned "V-shape". However, this signature greatly depends on the atmospheric and surface states. Contrary, therefore, to retrieving dust properties, i.e. producing a well-defined physical parameter (e.g. the AOD at a given wavelength), detecting dust can be performed via simple algorithms, like two channel BT differences. The combination of more channels is also possible, but the principle remains similar: some of the channels are not sensitive to aerosols and are used to estimate the "base-line" while the other channels have decreasing BT when the aerosol load increases. The selection of the channels can be performed through radiative transfer simulations or by comparing spectra with or without an aerosol plume. These algorithms are widely used, both for radiometers

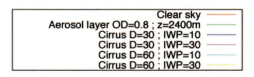

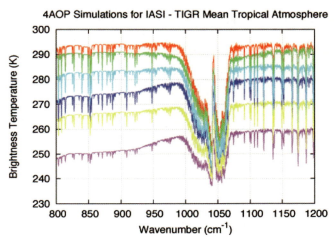

Figure 9.19 Comparison of the effect of aerosol and cirrus clouds on the infrared spectrum from 800 to 1200 cm⁻¹ (8.3 to 12.5 μm). The diameter D and Ice Water Path (IWP) of cirrus models are given in the legend at the top. Simulations over ocean.

and spectrometers (e.g. Ackerman, 1989; Wald et al., 1998; De Souza-Machado et al., 2006; Karagulian et al., 2010, etc.). Their advantages are their simplicity and a very fast computation time, suitable for near-real time. These algorithms are very useful to track aerosol plumes, for the monitoring of air quality for example. However, they do not give access to a geophysical quantity (such as mass or AOD), and they can be applied only for case studies as they might suffer from regional biases. For instance, the BT difference can also be sensitive to the Earth's surface emissivity or temperature. In order to bypass these difficulties, more sophisticated linear regressions are possible, when introducing for example exogenous data for the surface temperature.

As an illustration, Figure 9.20 shows that the difference between HIRS (the High Resolution Infrared Sounder on board the NOAA satellites) channel 8 (11.1 μm) and channel 10 (8.3 μm) depends almost linearly on the surface temperature. So, Pierangelo et al. (2005b) introduced the regression residual r

$$r = BT(10) - BT(8) - \left(aT_{surf} + b \right), \qquad (9.25)$$

where the linear regression coefficient a and b are determined on a clear-sky data set (either from simulations or from observations). The bottom plot of the same figure shows that this regression residual is almost insensitive to the clear-sky atmospheric variability (including water vapor content). Consequently, any departure of r from 0 comes from the presence of an aerosol, which was not included in the regression data set. Since the effect of aerosols on the BT is higher at 8.3 μm than 11.1 μm, the value of r increases with the aerosol load. However, the exact relationship between the residual r and the AOD strongly depends on the aerosol altitude, this is why this algorithm is suitable for detection only but not for the retrieval of optical depth[3].

Another example is the detection of dust over deserts with geostationary satellites: the high number of observations for a given area can be used by algorithms working with a clear and dust-free reference image. For each pixel, the reference BT at a fixed time is the highest BT observed during a period typically lasting half a month. The time period must be long enough to allow clear-sky day(s) and short enough to consider the emissivity of the Earth's surface as locally unchanged. The drop of observed BT with respect to the reference is associated with the presence of dust, both because of its direct effect on the radiation and because of the decrease in surface temperature mainly due to the shortwave irradiance decrease. Legrand et al. (1985) introduced this technique and later Legrand et al. (2001) built an Infrared Difference Dust Index (IDDI) over the Sahara using Meteosat observations.

Dust presence results in a change of outgoing radiance at TOA, arising from the difference between the skin surface temperature and the temperature of the dust layer above giving rise to a *thermal contrast method* (according to King et al. (1999)). In the daytime, the land surface is hotter than the dust, especially in arid regions. The hotter the surface, the stronger the thermal contrast and the radiance deficit at TOA. So, dust detection is performed in the middle of the day (at 12:00 UTC for Meteosat observing Africa).

A simple analytic expression of the dust-induced radiance deficit at TOA is possible with the following assumptions: (i) absorbing gases are not considered; (ii) the surface is assumed black at temperature T_s with no impact of dust on it; (iii) dust is assumed non-scattering; (iv) the dusty layer is assumed isothermal at T_d, with a transmittance t; (v) the spectral integration throughout the channel bandpass width is ignored. The radiance measured at TOA is then

$$L_{sat} = tB(T_s) + (1-t)B(T_d),$$
(9.26)

with B the black body (Planck) function for Meteosat channel. The TOA radiance deficit is thus the difference between the TOA radiances without and with dust

$$\Delta L_{sat} = (1-t)[B(T_s) - B(T_d)].$$
(9.27)

---

3   A second linear regression using HIRS channel 18 (at 4.0 μm) and channel 10 shows a different sensitivity to altitude and AOD. Combining both regressions gives the possibility to disentangle AOD and altitude.

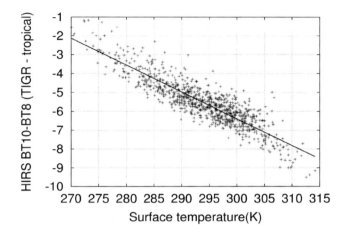

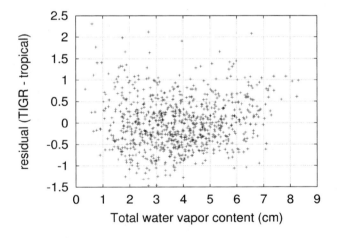

Figure 9.20 Top: difference in calculated brightness temperature between HIRS channel 10 and channel 8 as a function of sea surface temperature for a data set of clear-sky tropical atmospheres from the TIGR database. Bottom: residual of the linear regression between the two channels BT and the surface temperature as a function of total water vapor content.

The transmittance t depends on the amount of dust and its optical properties through the TIR AOD, while the thermal contrast $(B(T_s) - B(T_d))$ depends on the skin surface temperature and on dusty air temperature.

For an arid surface, solar energy is absorbed during daylight and mostly confined within a thin superficial layer. This enhances the skin surface temperature T_s and creates a strong superadiabatic lapse rate in the first meters over the surface of up to 60 K/m, interpreted as a temperature discontinuity at the surface (15–20 K at noon, from measurements near Niamey; Frangi et al., 1992). A consequence is that the IDDI – i.e. the TOA radiance or BT deficit – will be high where the discontinuity of temperature at the surface is large, i.e. over the dry bare ground of the arid and semi-arid regions, these very regions where sources are located and from where dust is emitted. There, the satellite signal will be large, even if dust lies in the surface boundary layer (as it usually does). On the contrary, vegetated surfaces and water bodies will experience little or no discontinuity (due to evaporation and evapotranspiration processes), resulting in a weaker sensitivity to dust (in the latter circumstances dust can be detected at high altitude only).

Illustrations of such detection algorithms are given in Section 9.5.

9.4.4 Direct retrieval

Herein we refer to the inversion from the infrared measurement of a geophysical quantity related to the aerosol (typically its optical depth, but it could also be its total mass, size distribution, altitude, etc.) as "retrieval". We refer to an inversion process where no precomputed brightness temperature data are used and no radiative transfer code is used during the retrieval as "direct retrieval".

The retrieval can rely on a direct relationship between an aerosol property (e.g. AOD) and a function of brightness temperatures. Such a relationship can be established empirically or on the basis of extensive simulations. This approach is widely used for broadband instruments. As an example, the dust loading is retrieved from SEVIRI using a linear combination of channel BTs (Brindley and Russel, 2006).

9.4.5 Physical retrieval

A second family of retrieval algorithms is based on a fit between observed radiances and calculated radiances given by a radiative transfer code. A radiative transfer computation is thus needed for each single processed spectrum. The fit is performed, for instance, by finding the aerosol characteristics and sometimes other parameters such as the temperature profile or Earth's surface state, that minimize a distance, generally a χ^2 type distance. The elements to fit must include all the parameters that are not supposed to be known. For measurements from a spectrometer, this technique has the advantage of using the full spectrum. The Bayesian approach described in the book from Rodgers (2000), known as "optimal estimation", is widely employed for vertical profile retrieval of temperature or gases.

The state vector x, containing all the unknowns, is related to the measurement vector y through the equation

$$y = F(x) + \varepsilon, \tag{9.28}$$

where ε is the measurement noise.

The inverse problem is solved by iterative linearization

$$y = Kx + \varepsilon, \tag{9.29}$$

where K is the Jacobian matrix that contains the partial derivatives of y with respect to the elements of x.

Since this inverse problem is under-constrained, it has to be regularized. This can be done using various methods, such as via Twomey-Thikonov regularization, or Bayesian regularization. Bayesian regularization introduces the *a priori* state vector x_a, the *a priori* covariance matrix of the state vector, S_a, and the noise covariance matrix, S_ε and it can be shown that the most probable state vector, i.e. the best estimate of x, is given by

$$x = x_a + \left(K^T S_\varepsilon^{-1} K + S_a^{-1}\right)^{-1} K^T S_\varepsilon^{-1} \left(y - F(x_a)\right). \tag{9.30}$$

Since the problem is usually moderately nonlinear, it is in fact solved by successive iterations.

The optimal estimation approach has been tried for aerosol retrievals (Clarisse et al., 2010) for intense events where it may be preferable to study the gas and aerosol chemistry together. Although this technique theoretically brings the lowest uncertainty on the retrieved quantity, it is very time-consuming (an iterative scheme with a call to the RTE at each loop is required, including the computation of partial derivative with respect to aerosol and gases), even when using a simplified scheme for the RTE (Kruglanski et al., 2006). The theoretical bases for this approach require the problem to be sufficiently linear and the distribution of errors and parameters to follow Gaussian distributions. The dependency of the result on the first guess and the *a priori* knowledge might be a drawback of this method.

9.4.6 Look-up table retrieval

As radiative transfer computation is often very time consuming, the inversion of a very high number of observations for global applications often requires a look-up table (LUT) approach. This technique is very similar to the previous one, except that the simulated BTs for the fit are not calculated for each retrieval. They are instead calculated in one go for a large and representative set of situations (surface, atmosphere, aerosol, observation geometry, etc.) and stored in the so-called LUT. The distance to minimize generally looks at observed and simulated BTs but also BT differences between two channels, the differential aspects adding a constraint on the proximity recognition. A typical distance for i selected channels and j independent selected couples is

$$D = \sum_i \frac{\left(BT_i^{calc} - BT_i^{obs}\right)^2}{\sigma_i^2} + \sum_j \frac{\left(\Delta BT_j^{calc} - \Delta BT_j^{obs}\right)^2}{\sigma_j^2}, \tag{9.31}$$

where σ_i^2 is the variance of the calculated BT_i over the whole LUT. This normalization is necessary to put the same weight on each selected channel.

Even though the generation of the LUT is time consuming, its application is straightforward. The retrieval consists of computing the distance D between the observation and all the situations in the LUT, or a subset of situations in the LUT if some *a priori* information is known (like, obviously, geometric knowledge as the viewing angle, or the thermodynamic state of the surface or the atmosphere from a first-step retrieval). Then, the situation of the LUT that minimizes the distance D is chosen and gives the desired aerosol properties. More sophisticated selection of several "minimal" situations can be combined (e.g. Peyridieu et al., 2010a), in order to take into account the noise of the individual measurements, for instance.

As an example, the LUT calculated for the retrieval of dust AOD and altitude from AIRS by Peyridieu et al. (2010a) has been built by computing the brightness temperatures for each of the eight selected channels, for the 567 atmospheric situations from the TIGR dataset, for seven viewing angles (0 to 30°), for nine dust AODs (0.0 to 0.8), and for eight mean altitudes of the layer (750 to 5800 m).

Validating the LUT retrieval is an indispensable step. As the full inverse problem is ill-posed, it is necessary to check that the parameterization applied (i.e. the choice of the quantity to retrieve and the channel selected) is sufficient to avoid multiple solutions or zero solutions. The discretization of the variability of the atmosphere through the choice of the thermodynamic profiles and the aerosol parameters must also be checked. This validation can be performed easily through the application of the retrieval algorithm to each simulation of the LUT, removing, of course, the tested situation itself from the LUT when looking for the minimum distance.

Once the LUT is built, instead of minimizing a distance, it should be theoretically possible to educate a nonlinear regression algorithm, such as a multilayer feed-forward neural network. Neural networks have been used for trace gas retrievals (e.g. Chédin et al., 2003; Turquety et al., 2004; Crevoisier et al., 2004), but, as far as we know, not for aerosol retrieval.

9.4.7 Specificities of a retrieval algorithm over land

The capability of detection of aerosols over land (especially dust or ash in the coarse mode) is an advantage of the longwave, compared with the visible domain. In the shortwave, the contrast between the reflectance of the ground and the reflectance of the aerosol layer is sometimes too low (over high albedo surfaces such as deserts) and prevents both signals from being disentangled. This is one of the great assets of using the longwave domain for aerosol remote sensing. Nevertheless, remote sensing of aerosol from infrared observations above land surfaces remains tricky.

As seen before in Section 9.3, the Earth's surface emission depends on surface temperature and emissivity[4]. Over the oceans, the surface emissivity is well known and almost constant spectrally, close to 0.99 at 10 μm, slightly depending on the incident angle. Some tables provide the spectral and directional dependency of sea water (Masuda et al., 1988).

---

4 Note that, in addition to surface emissivity, for a retrieval over land the ground altitude of the observed point must also be taken into account through the surface pressure but this usually does not present any difficulty.

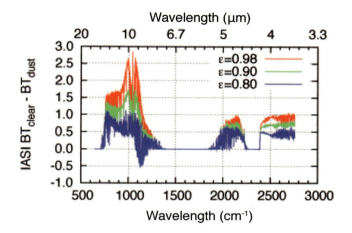

Figure 9.21 Simulated effect of surface emissivity on the BT difference (clear – dust) for IASI channels (AOD=0.4, altitude=2400 m).

However, over continental surfaces, the emissivity is spectrally, spatially and temporally variable. The infrared surface emissivity depends on many parameters: the composition of soil (type of mineral or vegetation), its moisture, and its roughness (Salisbury and d'Aria, 1992). Moreover, the spectral and angular variations of emissivity are significant (e.g. Salisbury and d'Aria, 1992 ; Péquignot et al., 2008). Over bare soils or deserts, the surface emissivity might reach values as low as 0.7 at 8.5 μm. Over semi-arid lands, like savannas, the seasonal cycle of emissivity is significant (Chédin et al, 2004) because of the change in the vegetation influenced by seasonal precipitations. This variability has an influence on the radiation emitted and modulates the aerosol impact (Figure 9.21) as noted by Legrand et al. (2001). This comes directly from the change in the radiation emitted, but also from the energy exchange between the atmosphere and the surface which modulates the surface temperature (Legrand et al., 1992). Note that the ground is always supposed Lambertian with respect to its infrared emissivity. This is a sensible hypothesis, given the large size of satellite sounder footprints.

For aerosol remote sensing over land, there are two main possibilities: the emissivity might be taken from exogenous data, or be by-passed by an adequate choice and combination of channels.

Regarding the first possibility, the knowledge of surface emissivity might come from a prior retrieval with the same infrared observations. The dust detection by a regression algorithm for HIRS channels by Pierangelo et al. (2005b) is extended over land by using the monthly averaged surface emissivity from Chédin et al. (2004). Their emissivity retrieval is based on the assumption that the emissivity varies slowly over time, contrary to surface temperature. De Paepe and Dewitte (2009) used the algorithm described in Minnis et al. (2002) to retrieve the surface emissivity from SEVIRI channels before retrieving mineral dust aerosol. Knowledge of surface emissivity may also come from another instrument with similar channels: Li et al. (2007) used MODIS infrared surface emissivities as input for their Saharan dust retrieval algorithm applied to SEVIRI. In some cases, for example when the knowledge of the surface material is sufficient, surface emissivity might be also taken from a library: let us mention the MODIS/UCSB and ASTER/JPL emissivity libraries which archive very high spectral resolution laboratory measurements of the emissivity of different samples of typical Earth surfaces. They can be downloaded from the Internet (http://www.icess.Ucsb.edu/modis/EMIS/html/em.html and http://speclib.jpl.nasa.gov/ respectively).

The second possibility might consist of a difference in brightness temperatures selecting channels for the baseline close to 12 μm, where the emissivity is almost constant with respect to the soil type (Péquignot et al., 2008). It is also implicitly used by algorithms selecting only channels at wavelengths where the surface emissivity does not vary too much. For that reason, Li et al. (2007) do not use the 8.7 μm channel of SEVIRI.

A third possibility would be to retrieve aerosol and surface properties simultaneously, but this increases the number of parameters to retrieve and would almost certainly lead to difficulties, as emissivity or surface temperature and aerosol effects can compensate each other.

The capability of infrared measurements to remotely sense the aerosol over land surfaces depends on the application: extreme event study or long-term climatologic observation. For a local case study, limited to a small region of desert for example, the emissivity might be indeed adjusted or assumed to be known and/or spatially constant. However, for global retrievals or for long-term climatology, this is far more complicated as the surface emissivity might be the cause of regional or seasonal differences in the retrieved aerosol characteristics. To our knowledge, no truly global data set of aerosol loads obtained from infrared measurements has been built: retrievals either work over the ocean only, or are limited to regional areas with one surface type, like the Sahara desert at the maximum.

9.4.8 Validating and comparing the results

Validating an aerosol product retrieved from a longwave instrument is not an easy task. The optimal way to validate the retrieved AOD would be to compare it with direct measurements, but "direct" observations of TIR optical depth routinely measured from the ground still do not exist. The few direct observations of longwave radiation from the ground (see Section 9.5.2) provide radiances (or BT) containing information on aerosol amounts, "weighted" with the aerosol properties (including aerosol altitude, particle size and composition) and depending on the thermodynamic state of the atmosphere. So AOD retrieval is difficult and will result in significant uncertainty if the aerosol properties and the state of the atmosphere are unknown.

Comparing aerosol products retrieved from infrared measurements with data from the on-ground sun-photometer network AERONET is not straightforward and raises several issues:

   (i) the relationship between the visible or shortwave AOD and the longwave AOD
       depends on the aerosol particles composition, size and shape, which are usually

unknown, and can render the comparison irrelevant: the photometer and infrared sensor are not sensitive to the same particles.

(ii) the aerosol content might change rapidly with time over a given station, which implies that night-time retrievals can not be validated this way.

(iii) AERONET retrievals over some stations are available only 2–5 days each month (because of the cloud cover), and this may not be suitable for a comparison with satellite-based products given for a wider area or averaged over a longer period. The comparison with single satellite pixels also suffers from the time difference, reinforced by the sparseness of the data because of the clouds.

While the comparison of the aerosol optical depth is not always relevant, the effective radius of aerosols retrieved from infrared observations can be interestingly compared with the AERONET retrieval of the coarse mode effective radius, which is computed for both spherical and nonspherical particles. However, some of the previously mentioned difficulties remain (the time difference in particular).

Longwave aerosol products can be compared with other satellite-based products, on the basis of spatio-temporal variability analysis. It has the advantage of a global approach. Even if this type of comparison is not an exact validation, it is nevertheless very useful, not only for increasing the confidence in the new aerosol product, but also for exploiting the synergy between sensors and highlighting some interesting geophysical behaviour. However, as reported by Mishchenko et al. (2007) and Liu and Mishchenko (2008), comparison of aerosol datasets from different satellite-based instruments still represents a challenge.

Infrared sensors that might be exploited for aerosol retrieval are not as widespread as visible sensors. Consequently, the comparison between two infrared AODs is not always possible. Peyridieu et al. (2010b) have compared the Saharan dust AOD over ocean from IASI and AIRS, but this is to our knowledge the only example of direct comparison of two similar infrared products (even if the time pass difference implies that the diurnal cycle of dust is not sampled by the two spectrometers in the same way).

Comparisons of satellite-based IR and visible aerosol products will identify similar patterns of dust clouds, but we expect differences between the two classes of products:

• The visible AOD is sensitive to particles in the accumulation and coarse mode, which implies that visible AOD products detect not only dust but also smaller particles such as biomass burning aerosol, pollution aerosols, etc.

• The infrared and visible wavelengths are sensitive to different particle size, therefore the ratio of IR to visible AOD depends on the PSD, which is modified during transport by gravitational settling (Maring et al., 2003). See for example Figure 9.22.

• The ratio of IR to visible AOD depends on the mineral composition in the case of dust, which in turn depends on the geographic origin (source of the aerosol cloud).

• The time difference is often a critical issue for elementary comparison, as visible retrievals need a high solar zenith angle while infrared retrievals usually prefer night data. This is why comparisons over longer time periods, e.g. monthly means, are preferred.

• The space sampling also differs: infrared sounders, for instance, have round footprints with a diameter of the order of 10 km (3 km for Meteosat Second Generation), while visible imagers have smaller square pixels.

As a consequence, visible and infrared AOD must not be considered as equivalent products which could be converted into each other but rather as fully complementary products whose comparison opens the way to very promising studies.

   Until recently, the IR-retrieved altitude of dust (related to the temperature of the layer) was very difficult to compare, as only a few in-situ or ground-based lidar observations from campaigns were available. It can now be compared with the CALIOP lidar data on board the CALIPSO satellite (Winker et al., 2009), even if the space sampling is extremely different. The lidar has a high horizontal resolution (5 km for CALIOP level 2 product) for detecting aerosol features. However, CALIOP only carries out nadir measurements, and so its coverage is much poorer than a scanning instrument like AIRS due to the distance between two successive orbits (more than 1000 km). This is why comparing CALIOP and AIRS products is not straightforward. Peyridieu et al. (2010a) consequently made comparisons between AIRS and CALIOP on the basis of their respective monthly means. They also have chosen to select cases for which only one aerosol layer is detected and measured by the lidar. This procedure also avoids computing an "average" altitude from various lidar layers with different composition or microphysical properties, which could not be

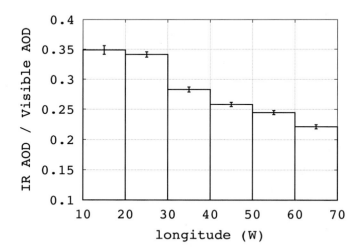

Figure 9.22 Ratio between the AIRS retrieved 10μm AOD and the MODIS retrieved 0.55μm AOD, for July 2003, for 0-30°N, as a function of longitude (from the African coast to the Caribbean sea). From Pierangelo et al. (2004b).

compared with AIRS equivalent altitude, as the infrared effect of these layers might change deeply with composition or properties, especially the median size of particles.

Aerosol products retrieved from infrared observations can also be compared with aerosol characteristics taken from chemistry and transport models (e.g. LMDz-INCA, described in Hauglustaine et al. (2004)), but model representations are still imperfect (vertical transport is still rather coarse, for instance) and the extrapolation of geophysical quantities to optical quantities (e.g. from total mass content to AOD or from mass median diameter to effective radius) usually requires some additional hypotheses.

9.5 Illustrations

9.5.1 Space observations

May we first remind readers that none of the infrared instruments on board satellites were originally designed for aerosol remote sensing. For example, the infrared sounders, HIRS or IASI, are operational instruments dedicated firstly to the retrieval of temperature and humidity profiles in the atmosphere for weather forecast applications. Similarly AVHRR, Meteosat and MSG/SEVIRI, are all meteorological instruments.

Moderate spectral resolution radiometers

Radiometers (or band instruments), as opposed to very high spectral resolution spectrometers, have been flown on board satellites for decades. The NOAA meteorological low Earth orbit satellites have been continuously carrying the infrared sounder HIRS and the imager AVHRR (with both visible and infrared window channels) since 1979. The geostationary Meteosat satellites have been offering observations in the longwave domain since 1981 (re-analysis of the archive by EUMETSAT). The long-term records make the retrieval of aerosol properties from these instruments extremely valuable for climate studies.

The first detection of dust using infrared observations was obtained from these instruments through brightness temperature differences or simple linear regressions. The potential of thermal infrared for monitoring mineral dust events has been investigated focusing on case studies (e.g. Ackerman, 1989; Wald et al., 1998) using data from NOAA/HIRS (High resolution Infrared Radiation Sounder). An 8-year climatology of a dust index over land in the tropical zone has been retrieved from HIRS (Pierangelo et al., 2005b), showing significant seasonal variations of dust sources. As an illustration, Figure 9.23 shows the location of the sources of dust over the Sahara, active for the month of April, from 1988 to 1991.

Following the Pinatubo eruption in June 1991, volcanic aerosols were also widely studied from infrared sounder measurements. Baran et al. (1993) analyzed the effect of Pinatubo aerosols on differences in brightness temperatures from HIRS-2 using in-situ observations of aerosol particle size and number density. They suggested retrieving the aerosol mass loading from this BT difference. Ackerman and Strabala (1994) or Pierangelo et al. (2004a) used three brightness temperatures (at 8, 11, and 12 μm from HIRS-2) to retrieve their vis-

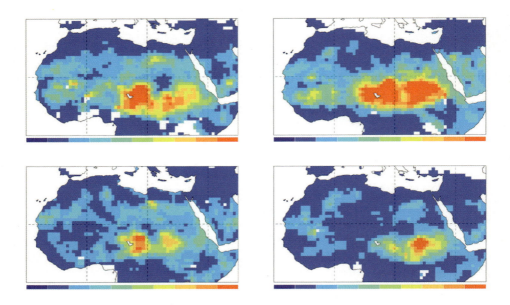

Figure 9.23 Dust index from NOAA/HIRS over the Sahara for the month of April, from 1988 (top left) to 1991 (bottom right). The activity of the sources over Tchad are clearly visible, together with some inter-annual variability. From Pierangelo et al. (2005b).

ible and infrared optical depths. The limitation of these instruments is the small number of broadband channels, which restrains the number of aerosol parameters that can be retrieved and sometimes implies some contamination by trace gases (water vapor or ozone).

Dust detection using geostationary TIR images

The Infrared Difference Dust Index (IDDI) (Legrand et al., 2001) is a Meteosat-derived TIR index dedicated to remote sensing of desert aerosols over land (using the former Meteosat First Generation series with a single IR channel at [10.5–12.5 μm]). The IDDI product proved to be efficacious in studies of dust source location and seasonal activity (Brooks and Legrand, 2000; Léon and Legrand, 2003; Deepshikha et al., 2006a, b). Figure 9.24 shows seasonal IDDI means over Africa calculated throughout the period 1983–98. The structures with high mean IDDI values indicate large atmospheric dust loadings, revealing the active sources of dust emission. This dust index has been used also for the study of the physics of dust emission (Chomette et al, 1999; Marticorena et al., 1997, 1999, 2004), for the description of dust transport (Petit et al., 2005), for the determination of the dust mineral composition with respect to its source of emission (Caquineau et al., 2002) and for meteorological dust forecasting (Hu et al., 2008). Figure 9.25 shows the distribution

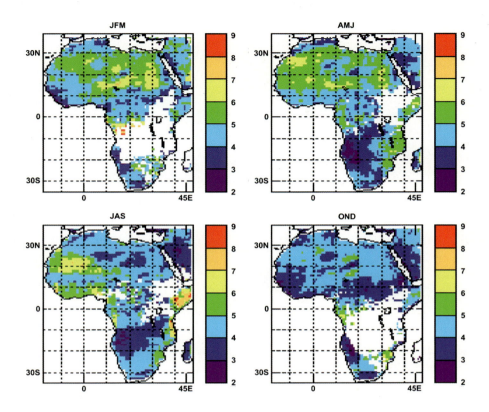

Figure 9.24 Seasonal dust climatology over Africa for the period 1983-1998, represented by 3-month mean fields of IDDI values (in K) on a 0.5°×0.5° grid. White areas are those where cloud presence is frequent.

of dust over China and the corresponding sources, derived from an operational dust detection algorithm generating IDDI and using the data from the meteorological geostationary satellite FY-2C.

Very high spectral resolution sounders: AIRS, IASI

The new generation of sounders (AIRS and IASI) offers a higher number of channels: more aerosol properties can be retrieved from radiance measurements, and the deconvolution of variable gases and aerosol effects is easier, due to the very high spectral resolution. These high performances have also enlarged the number of applications of aerosol products retrieved from infrared observations, as for example the monitoring of aerosols for air quality or the monitoring of volcanic activity for air traffic.

Soon after its launch in 2002, studies demonstrated the ability of AIRS observations to monitor clouds and volcanic aerosol (Ackerman et al., 2004; Carn et al., 2005), or Saharan

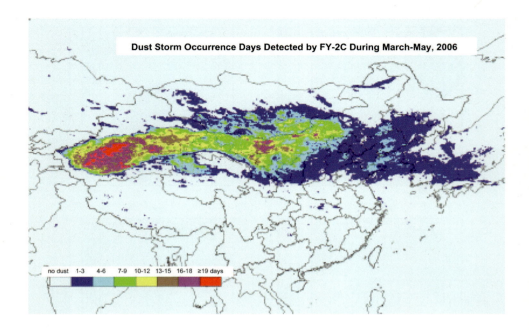

Figure 9.25 Dust occurrence frequency over eastern Asia derived from the geostationary satellite FY-2C (adapted from Hu et al., 2008).

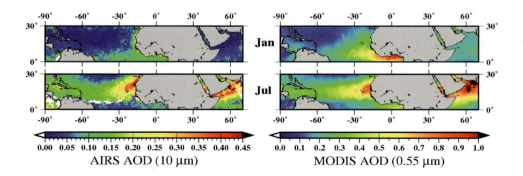

Figure 9.26 6-year (2003–2008) climatology of monthly AOD. Left: 10 μm AOD from AIRS; right: 0.55 μm AOD from MODIS. From Peyridieu et al., 2010a.

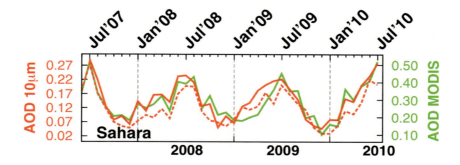

Figure 9.27 Seasonal variability of Saharan dust AOD over the Atlantic Ocean close to the Sahara (10–35°W, 0–28°E): the infrared AOD is retrieved from AIRS (red dotted line) and IASI (red thick line) and compared with MODIS AOD at 0.55 μm (green line). (Courtesy of S. Peyridieu).

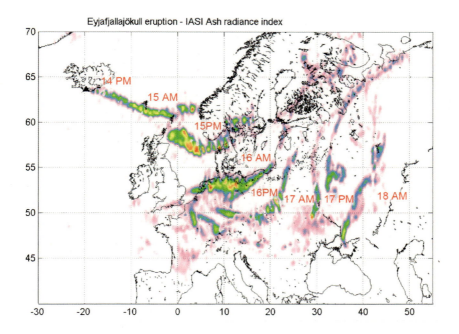

Figure 9.28 Composite image of the ash cloud in the days following the Eyjafjallajökull eruption, in April 2010. The ash radiance index is obtained by a difference between channels. (Courtesy of L. Clarisse.)

dust (Pierangelo et al., 2004b; de Souza-Machado et al., 2006). A tropical climatology of dust (infrared AOD and altitude) has been derived from AIRS for almost eight years (2003–2010) (Peyridieu et al., 2010a). The comparison with MODIS visible AOD gives access to added-value information, like the sedimentation of dust during transport, or the distinction between dust and biomass burning aerosols. It is clearly visible in Figure 9.26 that the aerosol over the Atlantic ocean is mainly composed of dust in summer, while biomass burning aerosols, not seen by AIRS, are predominant in winter. The aerosol layer altitude compares well with the CALIPSO lidar products. The effective radius of dust is also a parameter that can be retrieved in a second-step retrieval (Pierangelo et al., 2005a).

The first results from the IASI instruments, launched in 2006, are very promising. Saharan dust detected with IASI shows a seasonal cycle very similar to the one observed with AIRS, as can be seen on Figure 9.27. The capability of IASI to detect volcanic ash aerosols is reported by Gangale at al. (2010). An illustration for the monitoring of the Eyjafjalla-jökull ash cloud is shown on Figure 9.28.

Limb observations

Limb observation of the infrared emission of the atmosphere can also provide the extinction profile of volcanic aerosols. Note that the lower part of the atmosphere, becoming progressively too opaque, cannot be observed with a limb sounder, which restricts their use to high altitude aerosols (above 7–8 km), i.e. upper tropospheric and stratospheric aerosols. Lambert et al. (1993) studied Pinatubo aerosol using satellite observations in the infrared range from the Improved Stratospheric and Mesospheric Sounder (ISAMS). They retrieved area-weighted mean stratospheric optical thickness at 12.1 μm on a global scale from the zonal mean extinction profiles. The Cryogenic Limb Array Etalon Spectrometer (CLAES) instrument made measurements of the infrared emission of stratospheric aerosols near 12 μm from October 1991 until May 1993. Mergenthaler et al. (1995) obtained the aerosol distribution of absorption cross-section, optical depth and sulfuric acid mass estimates with CLAES measurements. Lambert et al. (1997) combined CLAES and ISAMS to retrieve aerosol composition, volume and area mass density. Echle et al. (1998) also determined optical and microphysical parameters of Pinatubo aerosols using the Michelson Interferometer for Passive Atmospheric Sounding, Balloon-borne version (MIPAS-B).

9.5.2 Ground-based observations

First, let us emphasize that there are no routine measurements or network of instruments dedicated to aerosol remote sensing in the longwave, so far. In the shortwave, ground sun-photometers provide an almost direct measurement of aerosol optical depth, because the radiation source (the sun) is observed through the aerosol. Conversely, the longwave observation of aerosol from the ground is indirect: the source is the aerosol itself and the radiance depends first on the aerosol temperature/altitude. Similarly to space observations, the spectrum recorded from the ground mostly carries the information about the thermodynamic state of the atmosphere: temperature, and also humidity. Consequently, the aerosol optical depth in the longwave retrieved from an instrument on the ground needs exogeneous data and contains a non-negligible error and cannot be used for an accurate validation of space-based retrievals.

Even if there is no routine ground measurement of dust in the TIR domain, the CLIMAT radiometer has been used in field campaigns (Sahel, Sahara, China, Guadeloupe).

The CLIMAT instrument (manufactured by CIMEL Company) is a TIR radiometer developed in the 1990s, to be operated from the ground as well as on board aircraft, firstly for analysis and validation of data acquired by space-borne instruments (Sicard et al., 1999; Legrand et al., 2000; Brogniez et al., 2003). This portable instrument measures TIR filtered sky radiance in a wide channel (8–13 μm) and narrow channels 1 μm wide within this window, for the remote sensing of dust. After cloud-contaminated measurements are eliminated, clear-sky radiances are corrected from gaseous (water vapor) contribution in order to provide a dust radiance component. Figure 9.29 (Legrand and Pancrati, 2006) compares time series of clear-sky radiance derived from CLIMAT channel-W measurements, with aerosol AOD at 670 nm and with the columnar water vapor amount from balloon sounding, obtained during the Sahelian campaign NIGER98 at Banizoumbou (AERONET station) during the dry season of 1998. The CLIMAT radiance and AOD signals show very similar coincident peaks during dust events. Figure 9.30, derived from these results, confirms the expected linear relation and the high correlation between CLIMAT radiance and AOD and shows that correction of water vapor variations effect results in a slightly increased correlation. Figure 9.31 compares the IDDI Meteosat signal for Banizoumbou during NIGER98 with the previous aerosol AOD at 670 nm. Again, the IDDI signal shows coincident peaks with the dust optical depth, like in Figure 9.29 with the CLIMAT signal. This is an example of a succession of dust events characterized by the same radiative properties, which means similar mineral composition, size distribution and altitude of transport. The use of back trajectories explains to a large extent such results, indicating low-level transport in the boundary layer and revealing that dust from the observed events at Banizoumbou originated mostly in the Bodele source. Wind-lifted dust events from this source should only have weak variations of size (confirmed with the Banizoumbou AERONET retrievals of particle size distribution) and in composition. So a fixed mineral composition has been assigned to dust, valid for a Sahelian origin according to Caquineau et al. (2002), namely quartz (19%), kaolinite (72%) and illite (9%) and simulations of the measured radiance have been performed using additional information from AERONET particle size retrievals and Niamey Airport radiosoundings. Finally, in Figure 9.32 we can compare the simulated spectral signature of this Sahelian dust in the CLIMAT channels (i.e. the channel sensitivity to dust) with the corresponding signature derived from CLIMAT measurements, after correction of the water vapor effect. A fair agreement is observed between measurements and simulations. A maximum sensitivity is observed for channel N9 covering the spectral region of the Si-O stretch resonance peaks for quartz (around 9 μm) and clays (around 9.5 μm).

Not only radiometers but also spectrometers are used on ground during campaigns to remotely sense the aerosol. For instance, the Atmospheric Emitted Radiance Interferometer (AERI) has been used during the Aerosol Characterization Experiment Asia (ACE-Asia) and the United Arab Emirates Unified Aerosol Experiment (UAE2). Vogelmann et al. (2003) obtained the longwave radiative effect of aerosol during ACE-Asia by removing a simulated clear-sky spectrum. This radiative effect was then converted to an aerosol IR radiative forcing (flux), which is crucial for climate study. However, it is different from an aerosol retrieval as the forcing includes the net surface radiative effects of aerosol amount, composition, size, height and temperature. Similarly, Markowicz et al. (2003) used the

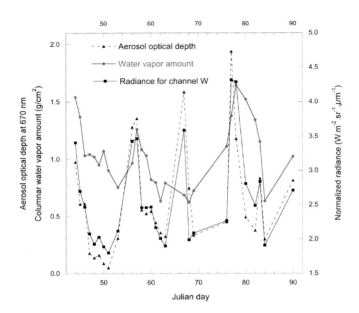

Figure 9.29 For cloud-free days at 12 UTC, 1998, Banizoumbou near Niamey: (i) CLIMAT radiometric measurements (TIR normalized radiance of channel W), (ii) photometric dust optical depth at 670 nm, (iii) water vapor amount from radiosoundings (Niamey Airport).

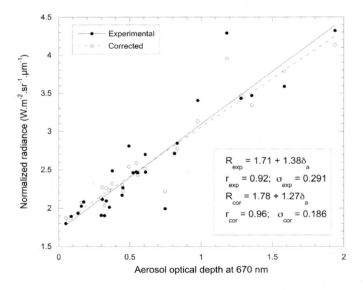

Figure 9.30 Comparison between experimental normalized radiances (for CLIMAT channel W) and the coincident dust optical depth at 670 nm δ_a, before and after correction of water vapor (fixed at 1.03 g/cm^2).

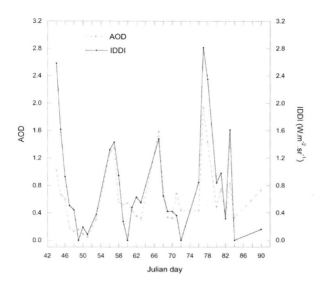

Figure 9.31 Compared temporal series for dust optical depth at 670 nm and IDDI Meteosat in radiance corrected from surface wind effects (Banizoumbou at 12 UTC, same period as in Figure 9.29). The correlation of IDDI with the dust optical depth at 670 nm is r = 0.81. (From Vergé-Dépré et al., 2006.)

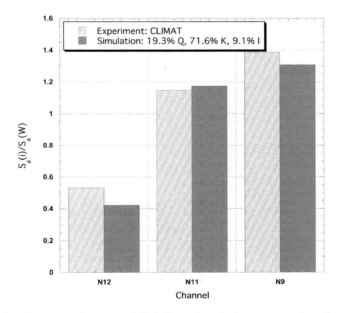

Figure 9.32 Simulated spectral signature of Sahelian mineral dust compared to the CLIMAT-measured signature. These signatures are relative to the wide channel (W) sensitivity to dust (narrow channels are [11.3–12.7 μm] for N12, [10.0–11.1 μm] for N11 and [8.3–9.3 μm] for N9).

infrared spectrometer observations only for computing the aerosol IR forcing and estimated the 10 μm aerosol optical depth from the visible optical depth and an aerosol model. During the UAE2 campaign, aerosol infrared optical depths were retrieved but it was also shown that the characterization of the thermodynamic boundary layer is crucial for accurate dust detection or retrieval (Hansell et al., 2008 ; Nalli et al., 2006). This difficulty is also underlined by FTIR instrument observations during the Lindenberg Aerosol Characterization Experiment (LACE) 1998 campaign (Hollweg et al., 2006). Over all, there is a relatively low number of studies related to aerosol retrieval in the infrared using ground spectrometer, which also emphasizes the remaining difficulties.

9.6 Prospects

The use of the longwave part of the spectrum for aerosol remote sensing is strongly less developed than the shortwave domain. Indeed, the radiative transfer equation in the longwave combines the complexity of absorption and emission by atmospheric gases (water vapor, CO_2, ozone…) and scattering by aerosols. The radiation source is the Earth's surface and atmosphere itself; thus the highly variable surface emissivity and temperature, together with the temperature profile, bring further difficulties in aerosol remote sensing in the infrared. Then, the longwave domain is sensitive to particles in the coarse mode only, such as dust and volcanic aerosol, contrary to the shortwave where particles in the accumulation mode, emitted by biomass burning or industrial activities, can be observed.

However, the longwave domain has unique capabilities, such as night-time observations, or dust monitoring over deserts. The possibility to retrieve the altitude of the aerosol layer or its mineralogical composition is also of great interest. Next, only remote sensing in the longwave can bring accurate observations of infrared aerosol optical depths. Infrared instruments onboard meteorological satellite (e.g. Meteosat or NOAA satellite) give very long time series: extraction of a global climatology of aerosol from their observations provides very valuable information for climate studies. Furthemore, this domain is evolving very quickly, through the availability of new generation instruments, like the very high spectral resolution sounders AIRS or IASI.

The prospects for aerosol remote sensing in the infrared consist of the retrieval of microphysical information, and especially the mineralogical composition (e.g. quartz content in mineral dust particles). The synergy with visible observations also opens the way to innovating products, as both spectral domains are not sensitive to the same species of particles. For example, biomass burning plumes and mineral dust plumes can thus be disentangled.

Regarding the evaluation of aerosol direct radiative forcing, both the shortwave and longwave domain must be taken into account. Thus, the aerosol optical properties retrieved in the longwave are essential to estimate the total radiative impact at top-of-atmosphere and bottom-of-atmosphere. Concerning future applications of infrared retrieved aerosol properties, we foresee that their use in global climate models could constrain further the longwave part of aerosol direct forcing, which is far from being negligible.

Appendix: Conversion between wavenumber and wavelength

The wavenumber σ (in cm^{-1}) and the wavelength λ (in μm) are linked by the following relationship:

$$\sigma = \frac{10000}{\lambda}$$

Wavenumber (cm^{-1})

600	800	1000	1200	1400	1600	1800	2000	2200	2400	2600	2800	3000
16.67	12.5	10	8.33	7.14	6.25	5.56	5	4.55	4.17	3.85	3.57	3.33

Wavelength (μm)

Wavelength (μm)

3	4	5	6	7	8	9	10	11	12	13	14	15
3333	2500	2000	1667	1429	1250	1111	1000	909	833	769	714	667

Wavenumber (cm^{-1})

10 Active lidar remote sensing

M. Patrick McCormick and Kevin R. Leavor

10.1 Introduction

Lidar, an acronym for "**LI**ght **D**etection **A**nd **R**anging", is an active remote sensing technique analogous to radar. Lidar systems use a laser as an active radiation source. The short pulse lengths produced by a laser (approximately 20 ns) and the spectral bandwidth (1 cm^{-1}) allow for highly-resolved ranging measurements with high signal-to-noise. Also, as in radar, lidars could be either monostatic (collocated transmitter and receiver) or bistatic (separated transmitter and receiver). Figure 10.1 illustrates an operational, monostatic lidar. Laser radiation is transmitted and scattered or absorbed by atmospheric constituents, such as clouds, aerosols, or molecules. Photons scattered back to the receiver are then collected, directed to a detector whose signal is analog-to-digitally recorded or counted as a function of altitude or range. The strength of the return signal is related to the physical and optical properties of the scatterers.

Lidar is an active remote sensing instrument with the laser providing the radiation source, and is able to take measurements in the absence of a natural source of radiation. Comparatively, passive remote sensing instruments such as those using the occultation technique, for example, require a natural radiation source such as the sun or stars. Lidar's requirement is that the laser signal is strong enough to overcome any background noise. Active remote sensing instruments typically do not have an ability to self-calibrate, unlike some passive sensors like those that measure unattenuated, exo-atmospheric radiation, just before or after the atmospheric measurements, or those that carry on-board blackbody sources as a means of calibration.

Lidar is one of the only techniques and, if considering only remote sensing techniques, is probably the only technique that can provide routine, height-resolved observations of aerosols and their characteristics in the low-to-middle atmosphere. Lidar measurements from ground-based, aircraft and spacecraft systems are now commonplace. Whereas the ground-based systems are capable of producing long-term observations from a fixed or local location, aircraft up-looking and down-looking lidars can produce data sets over regional areas and spacecraft can produce years of observations on a near global basis, depending on spacecraft orbital characteristics. Networks of ground-based lidar systems, ranging from generalized systems found in the European EARLINET and worldwide GALION networks, to arrays of specialized systems such as Goddard Space Flight Center's MPL-NET consisting of micropulse lidars, provide the capability to study primarily regional characterization of aerosols and their dynamical effects with time and space.

Elastic backscatter lidars were developed soon after the successful optical pumping of ruby material in the Hughes Corporation laboratory, producing stimulated optical emission at 694.3 nm (Maiman, 1960), and the invention of a technique for producing a giant pulse by Q-switching ruby in 1962 (McClung and Hellarth, 1962). The first, atmospheric aerosol lidar measurements were reported by Fiocco and Grams (1964) showing stratospheric aerosol profiling.

The equation governing the strength of the lidar return signal is known as the Lidar Equation, and is given by:

$$P(R) = P_0\eta\left(\frac{A}{R^2}\right)\left(\frac{c\Delta t}{2}\right)O(R)\sigma_b(R)\exp\left[-2\int_0^R\sigma_e(r)dr\right]. \qquad (10.1)$$

$P(R)$ is the power of the return signal from range R^1, P_0 is the emitted laser power, η^2 is the system efficiency, A is the receiver's effective area, c is the speed of light, and Δt is the laser's pulse duration. $O(R)$ is an overlap factor that accounts for the separation of the transmitter and receiver optical axes, the transmitter's output size, shape, and divergence and the receiver's field-of-view and imaging properties. The latter determines whether the entirety of the laser backscatter can be imaged on the detector. $\sigma_b(R)$ is the backscatter coefficient of the atmosphere at range R, and $\sigma_e(r)$ is the atmospheric extinction coefficient at range r.

The expression

$$A\left[\frac{c\Delta t}{2}\right]\left[\frac{O(R)}{R^2}\right]$$

represents the factors related to lidar viewing geometry. Figure 10.2 illustrates the presentation of these terms for a typical lidar geometry. Under ideal conditions, the overlap between the laser's output divergence and receiver's field of view equals unity for concentric laser and receivers. In the event that the laser's return is only partially captured by the receiver, $O(R) < 1$. For a given pulse, the effective pulse length is equal to $c\Delta t/2$. Finally, the solid angle into which photons are scattered is A/R^2.

---

1 Note, throughout Chapter 10 the variable 'R' denotes the range of the lidar beam, while in preceding chapters it represented the reflectance coefficient of a Lambert surface.

2 In Chapter 10, the variable η denotes the system efficiency, while in the preceding chapters it represents the fine mode weighting.

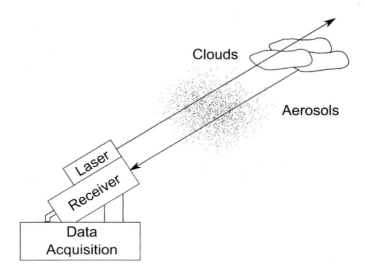

Figure 10.1 An illustration of a monostatic lidar system in operation.

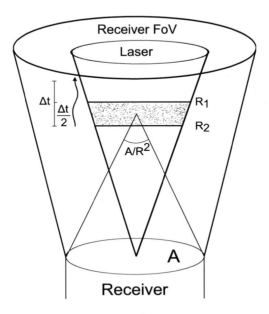

Figure 10.2 Coaxial lidar viewing geometry.

As A and Δt are constant, $O(R)/R^2$ is treated as the geometric factor, denoted

$$G(R) = \frac{O(R)}{R^2}.$$
(10.2)

Furthermore, we may define a system constant

$$C = P_0 \eta \left(\frac{c\Delta t}{2} \right) A.$$
(10.3)

Equation (10.3) is referred to as the lidar constant. This term describes a lidar's operational capabilities, containing the laser's output, the qualities of the receiving optics, and any signal losses or gains that are the result of instrumentation, such as blocking or interference filters.

The backscatter coefficient, $\sigma_b(R)$, is a description of the atmosphere's tendency to scatter light in the π direction during photon collisions (or, for bistatic systems, in the direction of the receiver). σ_b is usually given in units of $\mathrm{km^{-1}sr^{-1}}$.

Finally, $\sigma_e(r)$, the extinction coefficient, is a measure of the loss of photons as the laser pulse travels through the atmosphere, with units commonly expressed in $\mathrm{km^{-1}}$. The integral of the extinction from an origin to range R,

$$\tau = \int_0^R \sigma_e(r)\,dr,$$
(10.4)

is referred to as optical depth. This expression is doubled in the lidar equation to account for attenuation to the scattering particles and back to the receiver. Bistatic systems express this as the sum of integrals along the path from the laser to the scatterers, and then from the scatterers to the receiver, with each path having a different extinction. For monostatic lidars, when exponentiated, Eq. (10.4) represents a lidar transmission term,

$$T(R) = \exp\left[-2\int_0^R \sigma_e(r)\,dr \right],$$
(10.5)

Equation (10.1) may then be expressed in the simplified form (Fernald et al., 1972),

$$P(R) = CG(R)\sigma_b(R)T(R).$$
(10.6)

Another useful expression is the range-corrected lidar signal,

$$X(R) = R^2 P(R).$$
(10.7)

It should be noted that due to the nature of lidar measurements, used to determine the presence of aerosols, clouds, etc., both $\sigma_b(R)$ and $\sigma_e(R)$ are unknown. As a result, external knowledge is often required, either in the form of models, measurements, or assumptions for one of the quantities.

A final expression,

$$L(R) = \frac{\sigma_e(R)}{\sigma_b(R)},$$ (10.8)

known as the lidar ratio, or extinction-to-backscatter ratio, may also be determined. Knowledge of $L(R)$ provides a substitution in Eq. (10.1) for either $\sigma_e(R)$ or $\sigma_b(R)$. A significant source for information on lidar and its uses can be found in Weitkamp (2005).

The following sections detail some of the techniques and lidar measurements involved in the solution of Eq. (10.1) for the extinction and backscatter coefficients. The first section discusses the Fernald inversion method of determining the coefficients from single-wavelength lidar measurements. Although it uses information from only one wavelength, the Fernald method requires assumptions to be made concerning the extinction and scattering properties of the region being measured. This is because each elastic lidar measurement involves two unknown quantities that cannot be determined uniquely from measurements at one wavelength without further constraints. Other lidar techniques, which take advantage of distinctive spectral differences in gaseous absorption (DIAL) and scattering from molecular excitation (Raman) and thermal Doppler broadening (HSRL) effects, are discussed in later sections. These methods require fewer assumptions than the Fernald method and can be used in combination to derive more quantitative information on aerosol physical characteristics.

10.2 Aerosols and elastic backscatter

Since its initial use in atmospheric sciences, lidar has been used extensively for detection of aerosols (Fiocco and Smullin, 1963). Lidar measurements exploit σ_e and σ_b as sums of individual scatterers. As a result, σ_e and σ_b may be written as

$$\sigma_e(R) = \sigma_{e,a}(R) + \sigma_{e,m}(R),$$ (10.9)

$$\sigma_b(R) = \sigma_{b,a}(R) + \sigma_{b,m}(R).$$ (10.10)

The secondary subscripts "a" and "m" represent contributions due to aerosols and molecules, respectively. Typically, absorption due to molecules can be ignored by using laser wavelengths outside wavelengths in gaseous absorption bands. In this case, only contributions due to molecular scattering need be considered.

Elastic backscatter lidar represents the simplest form of lidar measurement. Typically, a single wavelength is used, and photons from the laser are scattered elastically back to the receiver with no energy lost during the collision, assuming no shift in wavelengths or major gaseous absorption events. Equation (10.1) can then be written

$$X(R)=C\left[\sigma_{b,a}(R)+\sigma_{b,m}(R)\right]\exp\left\{-2\int_0^R\left[\sigma_{e,a}+\sigma_{e,m}(r)\right]dr\right\}\quad,\tag{10.11}$$

assuming complete overlap at the scattering layer so that $O(R) = 1$.

Equation (10.8) can be expressed as a molecular or aerosol ratio. The molecular lidar ratio is

$$L_m(R)=\frac{\sigma_{e,m}(R)}{\sigma_{b,m}(R)}=\frac{8\pi}{3}sr.\tag{10.12}$$

Note that the molecular lidar ratio is not range-dependent, as the molecular composition of the atmosphere is homogeneous throughout regions used for typical lidar retrievals. This is especially true of aerosol retrievals.

The corresponding aerosol lidar ratio is, however, range-dependent, as the number density and composition of aerosols typically change from any one position in the atmosphere to another. The aerosol lidar ratio, L_a is given simply as

$$L_a(R)=\frac{\sigma_{e,a}(R)}{\sigma_{b,a}(R)}.\tag{10.13}$$

To further simplify Eq. (10.11), the quantity

$$Y(R)=L_a(R)\left[\sigma_{b,a}(R)+\sigma_{b,m}(R)\right]\tag{10.14}$$

is also defined (Sasano et al., 1985). Note that Eq. (10.14) utilizes the backscatter coefficient in the definition as opposed to extinction, as measurements from lidar are generally of backscatter (or attenuated backscatter, prior to analysis).

Substituting Equations (10.12) and (10.13) into Eq. (10.11) to replace the extinction coefficient terms $\sigma_{e,a}$ and $\sigma_{e,m}$ yields

$$X(R)=C\left[\sigma_{b,a}(R)+\sigma_{b,m}(R)\right]\exp\left\{-2\int_0^R\left[L_a\sigma_{b,a}+L_m\sigma_{b,m}(r)\right]dr\right\}.$$

Adding and subtracting the quantity $L_a\sigma_{b,m}$ to the integral in the exponent, multiplying both sides of the equation by $L_a(R)$, substituting Eq. (10.14), and rearranging gives

$$X(R)L_a(R)\exp\left\{-2\int_0^R\left[L_a(r)-L_m\right]\sigma_{b,m}(r)dr\right\}=CY(R)\exp\left[-2\int_0^R Y(r)dr\right].\tag{10.15}$$

Differentiating the logarithm of both sides of Eq. (10.15) with respect to R produces the differential equation

$$\frac{d\ln\left[X(R)L_a(R)\exp\left\{-2\int_0^R[L_a(r)-L_m]\sigma_{b,m}(r)dr\right\}\right]}{dR} = \frac{1}{Y(R)}\frac{dY(R)}{dR} - 2Y(R). \quad (10.16)$$

Note that Eq. (10.16) is a Bernoulli differential equation which can be arranged into the standard form of

$$\frac{dy}{dx} + P(x)y = Q(x)y^n. \quad (10.17)$$

Equation (10.16) is solved using the boundary condition at reference range R_0

$$Y(R_0) = L_a(R_0)\left[\sigma_{b,a}(R_0) + \sigma_{b,m}(R_0)\right], \quad (10.18)$$

which yields

$$\sigma_{b,a}(R) + \sigma_{b,m}(R) = \frac{X(R)\exp\left\{-2\int_{R_0}^R[L_a(r)-L_m]\sigma_{b,m}(r)dr\right\}}{\dfrac{X(R_0)}{\sigma_{b,a}(R_0) + \sigma_{b,m}(R_0)} - 2\int_{R_0}^R L_a(r)X(r)T(R_0,r)dr}, \quad (10.19)$$

where

$$T(R_0,r) = \exp\left\{-2\int_{R_0}^r[L_a(r')-L_m]\sigma_{b,m}(r')dr'\right\}.$$

Equation (10.19) expresses the aerosol ($\sigma_{b,a}$) and molecular ($\sigma_{b,m}$) backscatter coefficients at range R as a function of the range-corrected lidar return (X), the molecular lidar ratio (L_m), and the assumed aerosol lidar ratio (L_a) integrated from the reference range R_0 to the scattering range. The aerosol lidar ratio is normally assumed to be constant along the laser path. The reference range could be taken as an upper (aerosol-free) region, and the integral evaluated back towards the laser.

This treatment is commonly referred to as the "Fernald method" (Fernald, 1984) or "Klett Method" (Klett, 1981). Molecular scattering could be modeled (for known temperature and pressure) or otherwise derived in order to produce backscatter due to aerosols. Aerosol extinction may then be determined from Eq. (10.13). Figure 10.3 shows measurements of aerosol extinction coefficient over Hampton University with respect to time (left) and a time-averaged profile (right) using the above method. Higher values of aerosol extinction are seen within the Planetary Boundary Layer (PBL), below approximately 3 km, with less loading above in the free troposphere. Higher levels of aerosol extinction are measured higher in the atmosphere, most likely due to clouds.

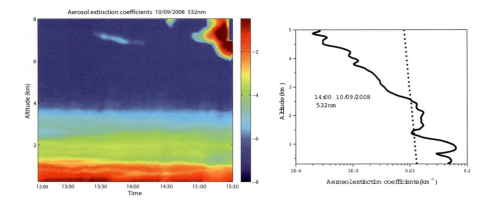

Figure 10.3 Measurements of aerosol extinction coefficient (km⁻¹) over Hampton University on October 9, 2008. Note: The color scale in the left panel is logarithmic such that -3 is equivalent to e^{-3}. The dotted line in the right panel is the corresponding molecular extinction as a function of altitude.

The Fernald method shows that aerosol retrievals are possible with a single wavelength lidar, but assumptions have to be made. For example, one must assume a lidar ratio, typically using a priori knowledge of the types of aerosols to be measured. Furthermore, aerosol retrievals are typically normalized to an aerosol-free region of the atmosphere (e.g., the middle stratosphere or free troposphere). However, in cases of high aerosol loading, such as during fires or volcanic eruptions, a particular lidar system either might not have the capability to produce believable data at high enough altitudes to reach such a region, or the regions normally used have become populated with aerosols.

10.3 DIAL

For completeness, we introduce DIfferential Absorption Lidar (DIAL) which makes use of absorption features unique to specific atmospheric constituents. DIAL is not used for aerosol measurements. Two wavelengths are used, one on-peak, within the absorption band, and one off-peak, outside of the absorption band. Assuming no other atmospheric constituents absorb with any degree of significance in the chosen wavelengths, then differences in on- and off-peak absorption should be due to changes resulting from the species of interest. DIAL measurements, relying on absorption, are generally made of gaseous species, such as ozone. Also, due to the nature of the atmosphere, especially the troposphere, wavelengths in the ultraviolet spectrum are most common, with the possibility for systems ranging from UV to near IR (Gimmestad, 2005).

As before, DIAL begins with Equation (10.1), the lidar equation. However, both the on- and off- peak wavelengths used have a separate lidar equation. The difference in atmospheric extinction, as stated above, is due to the gas species to be measured, and is expressed

$$\Delta \sigma_a = N \Delta C_a,$$ (10.20)

where

$$\Delta C_a = C_a(\lambda_{on}) - C_a(\lambda_{off}),$$ (10.21)

N is molecular number density of the gas species, and $C_a(\lambda)$ is the absorption cross-section at wavelength λ.

Since the lidar constant and overlap function are system-related, a ratio of the on- and off-peak lidar equation yields

$$\frac{P_{on}(R)}{P_{off}(R)} = \frac{\sigma_{b,on}(R)}{\sigma_{b,off}(R)} \exp\left[-2\int_0^R \Delta\sigma_a(r)dr\right].$$ (10.22)

In most cases, the difference between on- and off-peak wavelengths is on the order of a few nm, making it possible to assume that the corresponding backscatter coefficients are equal. With this assumption, solving Eq. (10.22) for $\Delta\sigma_a(R)$ and substituting Eq. (10.20) produces the number density,

$$N(R) = \frac{1}{2\Delta C_a}\left[\frac{d}{dR}\ln\left(\frac{P_{on}(R)}{P_{off}(R)}\right)\right].$$ (10.23)

However, the assumption that the backscatter coefficient at each wavelength is constant can impose significant errors even over small differences in wavelength (Fredriksson and Hertz, 1984). Also, the lidar constant might not be equal at both wavelengths due to differences in optics or the use of multiple lasers. Furthermore, a generalization of extinction due to molecular and other aerosols in the atmosphere should be included for completeness. Extending Eq. (10.22) with these considerations yields

$$\frac{P_{on}(R)}{P_{off}(R)} = C'\frac{\sigma_{b,on}(R)}{\sigma_{b,off}(R)} \exp\left\{-2\int_0^R \left[\Delta\sigma_{a,g}(r) + \Delta\sigma_{a,a}(r) + \Delta\sigma_{a,m}(r)\right]dr\right\},$$ (10.24)

where $C' = C_{on}/C_{off}$ is a new lidar constant, and the subscript g represents the gaseous species to be measured. $\Delta C_{a,a}(R)$ may further be separated into an aerosol component and a component for other known gases absorbing at those wavelengths.

Solving Eq. (10.24) for N yields

$$N(R)=\frac{1}{\Delta C_{a,g}}\left\{\frac{1}{2}\frac{d}{dR}\ln\left[\frac{P_{on}(R)\sigma_{b,off}(R)}{P_{off}(R)\sigma_{b,on}(R)}\right]-\Delta C_{a,m}(R)-\Delta C_{a,a}(R)-\Delta C_{a,og}(R)\right\}\quad,\quad(10.25)$$

where the subscript "*og*" denotes absorption due to other gaseous species. As before, molecular effects may be modeled or otherwise determined through measurements. The particulate aerosol contribution can be determined in the same measurement if using a multiple wavelength lidar with a harmonic providing the DIAL wavelengths and methods presented in this chapter. The final term, due to other gases, is dependent on the wavelengths used and the gas of interest. The relative error introduced is dependent upon the strengths of the measured gas and interfering gases' absorptions. While potentially negligible, the error may be significant (Gimmestad, 2005).

As might be expected, specific laser wavelengths are more useful for different species. Ignoring common, non-aerosol uses for DIAL such as O_3, ultraviolet systems have shown capable of measuring atmospheric SO_2, Cl_2, NO_2, and even Hg (Egeback et al., 1984; Edner et al., 1987, Edner et al., 1989). Visible systems have also been employed to measure SO_2, NO, and NO_2 (Fritzsche and Schubert, 1997; Kolsch et al., 1989), but also industrial and vehicular emissions (Swart and Bergwerff, 1990; Toriumi, Tai, and Takeuchi, 1996). More unique industrial emissions, including hydrocarbons, complex molecules, and high-temperature combustion products, are detectable in the infrared spectrum (Murray and van der Laan, 1978; Rothe, 1980; Killinger and Menyuk, 1981). However, the atmosphere is active in the infrared region. As a result, the potential for species such as water vapor, which are highly variable with strong absorption features to interfere, is large, and wavelengths must be chosen carefully. A number of studies have shown the feasibility and analyzed potential errors in infrared DIAL measurements (Menyuk and Killinger, 1983; Ambrico et al., 2000). The potential uncertainty has led to slower adoption of infrared DIAL techniques in favor of UV and visible measurements (Gimmestad, 2005).

10.4 Raman

Raman scattering presents another multiwavelength approach to lidar retrievals of aerosol parameters. Unlike DIAL or High Spectral Resolution Lidar (HSRL) approaches using two laser-generated wavelengths, Raman scattering utilizes shifts in a single wavelength due to inelastic scattering with particles. Raman scattering processes produce shifts of constant wavenumber as a photon excites the scatterer into (typically) higher rotational or vibrational states. Though wavelength shifts to shorter wavelengths are possible, under typical atmospheric conditions, shifts to longer wavelengths, or Stokes scattering, corresponding to a decrease in photon energy are favored. Contrast this with Mie scattering theory, which assumes elastic collisions with spherical particles where no energy is lost in the collision. Raman scattering is on the order of 500–1000 times weaker than molecular scattering for the vibrational case, and is most effectively used in regimes where signal-to-noise is usually high, such as in the lower atmosphere where scattering is more likely or during the night when solar background is at a minimum. Nitrogen (N_2) is the most commonly used Raman scatterer, though oxygen (O_2) scattering is not uncommon (Whiteman et al., 1992). One of the first atmospheric applications of Raman scattering, in this case water vapor, was accomplished by Melfi et al., (1969).

Numerous ground-based Raman lidar systems are active throughout the world, such as the Department of Energy Atmospheric Radiation Measurement (ARM) Climate Research Facility (CRF) Raman lidar (CARL) (Goldsmith et al., 1998). CARL has provided a climatological database of aerosol and water vapor profiles since 1996 over the ARM Southern Great Plains (Turner et al., 2001). CARL demonstrates the viability of Raman lidar systems when compared with traditional airborne Sun photometer measurements, producing aerosol extinction profiles with systematic uncertainties of about 6% (355 nm). This is within the typical range of 15–20%, or 0.025 km^{-1}, whichever is larger, for state-of-the-art instruments (Schmid et al., 2006). Multiwavelength Raman lidar systems have been used to retrieve vertical profiles of aerosol optical and microphysical parameters such as single-scattering albedo, aerosol absorption, refractive index, and effective radius. Both simulations and field measurements have confirmed the accuracy of these retrievals by Raman lidar (Müller et al., 2001a, 2001b, 2002; Veselovskii et al., 2002; Wandinger et al., 2002).

As noted earlier, the shift in wavelength is constant in the wavenumber domain and is independent of whether the scattering is Stokes or anti-Stokes. Considering the case of nitrogen, the vibrational transition is 2330.7 cm^{-1}. A tripled Nd:YAG laser output with fundamental frequency 1064 nm would shift from 354.7 nm to 386.7 nm, while a doubled Nd:YAG laser will shift from 532 nm to 607.3 nm, assuming Stokes scattering. Since such shifts are unique for atmospheric molecular constituents, they are useful in separating aerosol from molecular scattering effects.

Consider that transmission for photons traveling from the lidar is at the wavelength transmitted from the laser, while the return signal is at the Raman-shifted wavelength. Thus, Eq. (10.1) can be written for Raman scattering as

$$P_R(R) = P_0 \frac{c\Delta t}{2} \frac{A\,\eta_R}{R^2} O_R(R)\sigma_{b,R}(R) \exp\left\{-\int_0^R [\sigma_e(r,\lambda_0) + \sigma_e(r,\lambda_R)]\mathrm{d}r\right\}, \qquad (10.26)$$

where the subscript "R" is used to denote Raman components, and the subscript "0" denotes the laser wavelength components.

The backscatter coefficient for the Raman scattering may be expressed as

$$\sigma_{b,R} = N_R(R)\frac{dC_R}{d\Omega}(\pi, \lambda_0), \qquad (10.27)$$

where $\dfrac{dC_R}{d\Omega}(\pi,\lambda_0)$ is the Raman backscatter cross section.

Substituting Eq. (10.27) into Eq. (10.26) and solving for extinction yields

$$\sigma_e(R,\lambda_0)+\sigma_e(R,\lambda_R)=\frac{d}{dR}\ln\left[\frac{N_R(R)}{X_R(R)}\right]+\frac{d}{dR}\ln[O(R,\lambda_R)]\ .\qquad(10.28)$$

Separating molecular and aerosol components further provides

$$\sigma_{e,a}(R,\lambda_0)+\sigma_{e,a}(R,\lambda_R)=\frac{d}{dR}\ln\left[\frac{N_R(R)}{X_R(R)}\right]-\sigma_{e,m}(R,\lambda_0)-\sigma_{e,m}(R,\lambda_R)\,,\qquad(10.29)$$

assuming total laser/receiver overlap. In regions close to the receiver, the overlap function can be a major source of uncertainty due to the logarithm of a rapidly decreasing function as one approaches the receiver, as well as a mismatch of the etendue of the receiver and transmitter.

To simplify future expressions, overlap will be assumed complete ($O(R) = 1$). Also note that only Raman scatterer number density is dependent upon range in the backscatter coefficient expression. For nitrogen or oxygen, number densities are typically modeled using a standard atmosphere or directly measured through means such as balloon soundings of temperature and pressure.

To further consolidate Eq. (10.29), the Ångström exponent is introduced. The Ångström exponent relates the extinction at two different wavelengths as

$$\frac{\sigma_e(R,\lambda_1)}{\sigma_e(R,\lambda_2)}=\left(\frac{\lambda_2}{\lambda_1}\right)^{\alpha(R)},\qquad(10.30)$$

where $\alpha(R)$ denotes the Ångström exponent at distance R from the lidar.

Substituting Eq. (10.30) into Eq. (10.29) for the transmitted aerosol extinction yields

$$\sigma_{e,a}(R,\lambda_0)=\frac{\dfrac{d}{dR}\ln\left[\dfrac{N_R(R)}{X_R(R)}\right]-\sigma_{e,m}(R,\lambda_0)-\sigma_{e,m}(R,\lambda_R)}{1+\left(\dfrac{\lambda_0}{\lambda_R}\right)^{\alpha(R)}}.\qquad(10.31)$$

In cases where the difference between wavelengths is small, such as in rotational Raman scattering, the denominator can be approximated as 2.

The corresponding backscatter coefficient at the emitted wavelength can also be derived. The signals from both elastic backscatter (P) and Raman scattering (P_R) of the desired wavelength are retrieved at the desired altitude (R) and at a reference altitude (R_0)

where no aerosol is assumed to exist. This is typically a region of the upper troposphere, though care should be taken in the event of aerosol loading common from fires or volcanic events. Begin by forming the ratio (Ansmann et al., 1992),

$$
\frac{P_R(R_0)P(R)}{P_R(R)P(R_0)} = \left[\frac{O_R(R_0)O(R)}{O_R(R)O(R_0)}\right]\left[\frac{\sigma_b(R_0,\lambda_R)\sigma_b(R,\lambda_0)}{\sigma_b(R,\lambda_R)\sigma_b(R_0,\lambda_0)}\right]\frac{\exp\left[-\int_{R_0}^{R}\sigma_e(r,\lambda_0)dr\right]}{\exp\left[-\int_{R_0}^{R}\sigma_e(r,\lambda_R)dr\right]}, \quad (10.32)
$$

using Eq. (10.1) for the elastic signal and Eq. (10.26) for the Raman signal. Note that in the case that the overlap for both wavelengths is the same ($O=O_R$) the expression is insensitive to overlap conditions. Using this assumption, substituting Eq. (10.27) for the Raman back-scatter coefficients, separating molecular and aerosol components from the elastic back-scatter coefficients, and solving for backscatter at range R yields

$$
\sigma_{b,a}(R,\lambda_0)+\sigma_{b,m}(R,\lambda_0)=\left[\frac{P_R(R_0)P(R)}{P_R(R)P(R_0)}\right]\left[\frac{N_R(R)}{N_R(R_0)}\right]\left[\sigma_{b,a}(R_0,\lambda_0)+\sigma_{b,m}(R_0,\lambda_0)\right]
$$
$$
\times \frac{\exp\left[-\int_{R_0}^{R}\sigma_e(r,\lambda_R)dr\right]}{\exp\left[-\int_{R_0}^{R}\sigma_e(r,\lambda_0)dr\right]}. \quad (10.33)
$$

Equation (10.33) can be further simplified by choosing a particle-free region of the atmosphere as the reference altitude in order to eliminate the aerosol contribution to backscatter at that altitude. The molecular contributions to backscatter may then be modeled as before. The extinction profiles at the two wavelengths can be estimated from Eq. (10.31) and the spectral dependence assumed in Eq. (10.30). Finally, the lidar ratio could be determined once extinction and backscatter are known.

The total molecular contribution to backscatter around the laser wavelength consists of an elastic component and a distribution of wavelength-shifted components that result from quantum excitations in the rotational energy of the scattering molecules (primarily O_2 and N_2). The laser-induced excitations can either enhance or reduce molecular rotational energy, producing discrete spectral lines of scattered light with higher (Stokes) or lower (anti-Stokes) wavelengths, respectively, relative to the wavelength of the laser. The distribution of scattered radiation around the central laser line depends on wavelength and temperature. The Raman cross section in Eq. (10.27) characterizing the molecular back-scatter for a given rotational Raman line (λ_R) relative to the central wavelength (λ_0) is a function of not only wavelength but also temperature and thus altitude (Whiteman, 2003; Di Girolamo et al., 2004), so the computation of the backscatter coefficient in Eq. (10.33) using portions of the rotational Raman spectrum requires knowledge of both the number density and temperature profiles. However, since the intensity distribution for the pure rotational Raman spectrum near the laser excitation wavelength depends on temperature, it is possible to extract temperature information from measurements of multiple rotational Raman bands. The power ratio for two Raman signals with different wavelengths (rotational quantum numbers J_1 and J_2) in the pure rotational Raman spectrum is given as a function of temperature (T) by the expression (Penney et al., 1976)

$$\frac{P_{J_2}(R)}{P_{J_1}(R)} = \exp\left[\frac{a}{T'(R)} + b\right] \quad , \tag{10.34}$$

where a and b are calibration constants that can be determined from local radiosonde data. The exact functional form of Eq. (10.34) is strictly valid for single Raman lines but can also be used to approximate the relationship between signals measured through physical passband filters with nonzero spectral widths (Di Girolamo et al., 2004). Figure 10.4 shows measurements of two rotational Raman signals and one elastic scattering signal in the UV taken at night on June 24, 2009 in Hampton, Virginia. Visible in the elastic profile, but not in the molecular returns, is a cloud from about 2.4-2.9 km altitude. Figure 10.5 illustrates the temperature profile retrieved using Eq. (10.34) in comparison with a radiosonde profile.

Alternatively, given a temperature profile and suitable calibration constants, the total Raman response expected in one passband can be determined from the measured Raman signal transmitted through a filter with a different passband. Since the effective Raman backscatter crosssection corresponding to the signal measured through a practical filter must be integrated over the spectral range of the filter passband, it can be difficult to determine precisely for a real lidar system. By exploiting the relationship between portions of the rotational Raman spectrum (Eq. 10.34), the measured Raman signal can be converted to a corresponding signal from another passband for which the precise Raman cross section is known. The ratio of the elastic to Raman lidar equations, evaluated at range R, gives

$$\frac{P(R)}{P_R(R)} = \frac{C_1}{C_2}\left[\frac{O(R)}{O_R(R)}\right]\left[\frac{\sigma_b(R,\lambda_0)}{\sigma_b(R,\lambda_R)}\right]\frac{\exp\left[-\int_0^R \sigma_e(r,\lambda_0)dr\right]}{\exp\left[-\int_0^R \sigma_e(r,\lambda_R)dr\right]}, \tag{10.35}$$

where C_1 and C_2 are the elastic and Raman lidar constants, respectively. Assuming that the lidar system is aligned such that the overlap function is insensitive to wavelength ($O=O_R$) and the atmospheric transmissions at the laser and Raman-shifted wavelengths are approximately the same (so that the ratio of exponentials goes to unity), Eq. (10.35) can be rearranged to yield

$$\sigma_b(R,\lambda_0) = \frac{C_2}{C_1} \times \frac{P(R)\sigma_b(R,\lambda_R)}{P_R(R)} \quad , \text{ or}$$

$$\sigma_{b,a}(R,\lambda_0) + \sigma_{b,m}(R,\lambda_0) =$$

$$\frac{C_2}{C_1} \times \frac{P(R)\sigma_b(R,\lambda_R)}{P_{J_1}(R)\exp\left[\dfrac{a}{T'(R)} + b\right]} = \frac{P(R)\sigma_b(R,\lambda_R)}{P_{J_1}(R)\exp\left[\dfrac{C_3}{T'(R)} + C_4\right]}, \qquad (10.36)$$

where the last expression comes from substituting P_{J_2} in Eq. (10.34) for P_R in Eq. (10.35), which is considered to be the Raman response over a passband for which a good cross section is known. The calibration constants, C_3 and C_4, can be obtained experimentally from clear-sky (aerosol-free) measurements with small elastic (aerosol) returns such that the aerosol backscatter coefficient in Eq. (10.36) can be neglected. With two precalibrated Raman passbands for retrieving temperature according to Eq. (10.34), an elastic scattering measurement (P), number density profile, and appropriate values for the cross section and calibration constants, Eq. (10.36) can be used to obtain the backscatter coefficient. Figure 10.6 shows aerosol backscatter coefficient retrievals for the same night as in Figure 10.4 using the rotational Raman method and the Fernald elastic inversion method with assumed

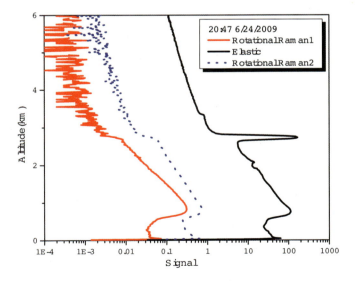

Figure 10.4 Two rotational Raman scattering profiles at 354.2 and 353.35 nm and one elastic scattering profile at 355 nm measured over Hampton University on June 24, 2009.

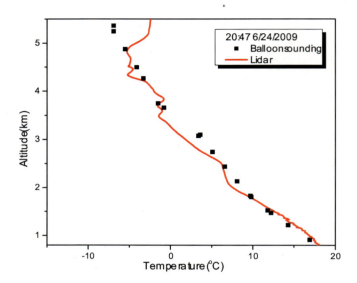

Figure 10.5 Temperature profile retrieved from Raman lidar measurements compared with a radiosonde at 20:47 (local time) over Hampton University on June 24, 2009.

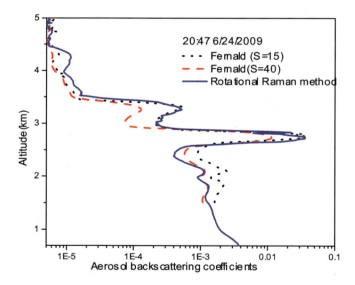

Figure 10.6 Atmospheric aerosol backscatter coefficients obtained with the rotational Raman method and the Fernald/Klett technique with assumed lidar ratios of 15 and 40 sr over Hampton University on June 24, 2009.

lidar ratios of 15 and 40 sr. The rotational Raman profile indicates that the lidar ratios above and below the cloud seen at 2.5 km altitude are different. The cloud has a lidar ratio of 15 sr, while aerosol loading increases the extinction coefficient and the lidar ratio to about 40 sr below the cloud. Additionally, reliable retrievals from the rotational Raman method extend to lower altitudes than those obtained using the Fernald method because the ratio in Eq. (10.35) greatly diminishes the effects of the overlap function at close range if the lidar detector system is aligned consistently for all wavelengths.

10.5 HSRL

As with Raman scattering presented in the previous section, High Spectral Resolution Lidar (HSRL) presents a multiwavelength approach to separating molecular and aerosol contributions to lidar retrievals. Thermal motion in particles causes Doppler broadening of the range of wavelengths accepted by the particles for scattering and absorption. Since this effect is dependent upon the particle's mass, larger aerosol particles will exhibit lower degrees of broadening than atmospheric molecular constituents. The high spectral resolution technique makes use of the differences in linewidth broadening to separate molecular Rayleigh components of scattering from Mie-regime scattering of aerosols. The derivation is precisely the same as presented in Section 10.4, where the subscript "R" would instead denote "Rayleigh" as opposed to "Raman". As a result, no derivation will be presented for the lidar equation, though the underlying physics and techniques for measuring at high spectral resolution will be discussed. The strongest difference, however, is High Spectral Resolution Lidar (HSRL) provides a means to directly measure the aerosol contribution to the signal separate from the molecular contribution.

The process by which molecules scatter light in a range outside their fundamental wavelength is known as "broadening". The three forms of broadening exhibited in the atmosphere are natural, pressure, and Doppler broadening. Natural broadening is a side-effect of Heisenberg uncertainty, where the knowledge of stored energy versus the lifetime of the excited state is constrained via the uncertainty principle. However, the effect is negligible outside of extreme atmospheric reaches where individual molecules can be considered free.

Pressure broadening is a form of dampening whereby collisions lead to de-excitations of molecular states. For pressure broadening to be significant, the mean free path between molecules must be short enough that the time between collisions is shorter than the lifetime of the excited state. Doppler broadening, by contrast, is a Doppler shift of absorbed and emitted radiation due to random, thermal motion. Though the effect is less significant than pressure broadening at low, near- Earth altitudes, the effect becomes exponentially more important with altitude.

The Lorentz profile for pressure broadening is given in terms of the spectral shape factor (Φ) by

$$\Phi_L(f) = \frac{\alpha_L}{\pi\left[(f - f_0)^2 + \alpha_L^2\right]} , \tag{10.37}$$

where f is the broadened frequency of the photon, f_0 is the initial frequency, and α_L is the Lorentzian Full-Width Half-Max (FWHM). The magnitude of the broadening takes the form of a Cauchy or Lorentz distribution centered about the initial photon wavelength.

Comparatively, the Doppler profile for the absorption cross section is given by

$$C_a(f) = S\Phi_D(f) = \frac{S_L}{\alpha_D\sqrt{\pi}} \exp\left[-\frac{(f-f_0)}{\alpha_D^2}\right], \tag{10.38}$$

where S_L represents the line strength of the profile and α_D is the Doppler width defined as

$$\alpha_D = \frac{f_0 v_0}{c},$$

where v_0 is the particle's speed and c is the speed of light. Note that in this case, the shape of the profile is Gaussian with line width $\alpha_D\sqrt{\ln 2}$.

Finally, combining the two probabilistic distributions in Eqs (10.37) and (10.38) into a single profile known as the Voigt profile produces (Thomas and Stamnes, 1999)

$$C_a(f) = S\Phi_V(f) = \frac{S_a}{\pi^{3/2}\alpha_D} \int_{-\infty}^{\infty} \frac{e^{-y^2}\,dy}{(f-y)^2 + a^2}, \tag{10.39}$$

where a is known as the damping ratio, α_L/α_D.

Since aerosols are far more massive than molecules, the degree of broadening will tend to be far lessened. Thus, an atmospheric scattering or absorption profile with aerosols present will see a sharp peak near the fundamental wavelength if the aerosol is sensitive to that wavelength, and then a broader range of accepted wavelengths due to the Rayleigh scattering molecular component. Due to the relatively small nature of the broadening effects, a high degree of measurement precision is necessary to achieve the spectral resolution to measure on- and off-peak. The aerosol contribution to broadening is also often confined within the linewidth of the laser (Eloranta, 2005). To achieve this precision, Fabry-Perot interferometers or absorption filters are used to greatly limit the accepted wavelengths and produce a retrieved spectrum.

Fiocco et al. (1971) first showed the possibility of using a scanning Fabry-Perot interferometer for HSRL measurements. Improvements to this technique using a stationary Fabry-Perot interferometer have also been demonstrated (Shipley et al., 1983; Grund and Eloranta, 1991). The molecular component is commonly predicted from model estimations, and perturbations from this curve are taken as aerosol contributions. The contributions from molecular and aerosol scattering are ratioed as in the Raman technique presented in Section 10.4. The scanning technique requires longer measurement times to produce a usable spectrum. Furthermore, narrowband filters required for the desired spectral resolution can reject too much of the molecular contribution. This requirement lowers lidar efficiency and decreases signal-to-noise.

The stationary technique alleviates much of this problem by focusing solely on the region central to the laser's frequency. Aerosol and molecular contributions are measured simultaneously and independently, thereby lowering the amount of time necessary for accurate measurements. The tuning is also non-specific and may be used at any wavelength, though any initial setup of the etalons requires a great deal of precision to maintain the method's benefits.

Alternatively, methods using atomic and molecular absorption filters to produce the narrow wavelength bands have also been used. Initial techniques used heated vapor filters to measure atmospheric temperature, backscatter, and optical depth (Shimizu et al., 1983). Improvements simplifying this technique and reducing power requirements occurred when the atomic barium filters were replaced by molecular iodine filters (Piironen and Eloranta, 1994). It should be noted that this change, while more efficient, also corresponded to the

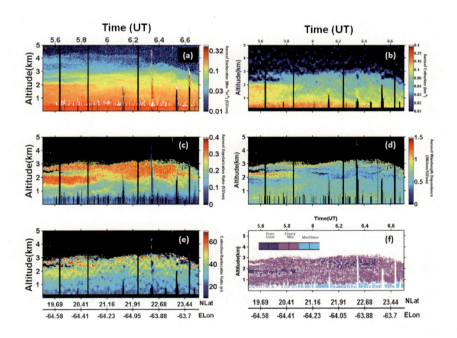

Figure 10.7 NASA Langley airborne HSRL measurements of aerosol backscatter (a), extinction (b), depolarization (c), backscatter Ångström exponent (d), and lidar ratio (e) acquired on August 24, 2010 during a flight over the Atlantic Ocean north of St. Croix, USVI. This segment covers a distance of about 430 km. The white areas below 1 km in (a) represent clouds within the marine boundary layer; these clouds attenuate the laser beams producing the small dark vertical bands below the clouds shown in these panels. The HSRL measurements of aerosol intensive parameters shown in (c–e) were used in a qualitative classification scheme to infer the aerosol type shown in (f). In this case, maritime (i.e. sea salt) aerosols are below about 600 m and Saharan dust is present between 600 m and 3 km. (Courtesy of Richard A. Ferrare).

growing availability of Nd:YAG lasers, capable of producing a 532 nm beam, to which the filters could be thermally-tuned.

Like Raman systems, ground-based and airborne HSRL systems have been used to study aerosols and clouds. Since 2005, the Arctic High Spectral Resolution Lidar (AHRSL) is one such ground-based HSRL system and is part of NOAA's SEARCH (Study of Environmental Arctic Change) contribution to the Canadian Network for the Detection of Atmospheric Change (CANDAC) facility (Eloranta, 2005, 2006). The AHSRL has measured calibrated backscatter, optical thickness, and depolarization profiles (532 nm), which are valuable for investigating aerosol properties over the Arctic (O'Neill et al., 2008; Saha et al., 2010).

NASA Langley Research Center (LaRC) (Hair et al., 2008; Obland et al., 2008) and the German Aerospace Center (DLR) (Esselborn et al., 2008) have also developed airborne HSRL systems. The LaRC HSRL has acquired over 1000 hours of data to date during more than 300 research flights since 2006, and measures backscatter color ratio ($\sigma_{b,1064}/\sigma_{b,532}$) with an additional elastic channel at 1064 nm, as well as depolarization at both 532 nm and 1064 nm. Systematic errors for derived aerosol extinction profiles are 15–20% for visible when compared with airborne sun photometer and *in situ* measurements (Rogers et al, 2009). Figure 10.7 shows an example of the suite of measurements acquired by the LaRC HSRL on August 24, 2010 during a flight segment over the Atlantic Ocean north of St. Croix, USVI. Figures 10.7a and 10.7b show aerosol extensive parameters (backscatter and extinction (532 nm), respectively) and Figures 10.7c, 10.7d, and 10.7e show aerosol intensive parameters (linear particulate depolarization (532 nm), backscatter Ångström exponent (1064/532 nm), and aerosol lidar ratio (532 nm), respectively). The white areas at altitudes between 0.5 and 1 km in Figure 10.7a correspond to small cumulus clouds in the marine boundary layer. Intensive parameters are insensitive to aerosol amount and variations in these parameters correspond to changes in the aerosol optical and microphysical properties associated with different aerosol types. In this example, a layer of Saharan dust extends from about 0.6 to 3 km above the marine boundary layer. The HSRL measurements show that the dust has higher particulate depolarization and lidar ratio than the sea salt aerosols within the marine boundary layer below 0.5 km. LaRC HSRL measurements of aerosol intensive parameters were used in a qualitative classification procedure to group the data into major significant categories (Burton et al., 2011). Eight distinct types with different aerosol intensive properties were identified. Figure 10.7f shows how this classification scheme identifes the dust and maritime aerosols in this example. The HSRL measurements have also been used to derive aerosol optical depth (AOD) and apportion AOD to these different aerosol types (Burton et al., 2011). DLR airborne HSRL measurements of aerosols including dust (Esselborn et al., 2009) have proven valuable for evaluating satellite AOD measurements (Kahn et al., 2009).

10.6 Depolarization

Previous sections have dealt with retrieval of optical properties relating to number density and composition or type of scattering or absorbing aerosols. Depolarization techniques provide a means by which the non-sphericity of scattering aerosols may be determined.

Basic Mie theory assumes spherical scatterers which do not change the state of linear polarization of light relative to the incident beam when reradiating in the backscatter direction (Bohren and Huffman, 1983). However, atmospheric aerosols commonly take shapes that may be fractal, as in the case of freshly formed soot; irregular, as in the case of desert dust; or a variety of other shapes such as needles, discs, ellipsoids, and otherwise seemingly random. These deviations from spherical shapes can serve to alter the polarization plane of backscattered light relative to the incident plane of linearly polarized light.

The depolarization ratio at range R, $d(R)$, is defined as

$$d(R) = \frac{\sigma_{b,\perp}(R)}{\sigma_{b,\parallel}(R)} \exp\left[\tau_\parallel - \tau_\perp\right], \tag{10.40}$$

where $\sigma_{b,\perp}$ is the orthogonal component of the returned backscatter coefficient, $\sigma_{b,\parallel}$ is the parallel component, and $\tau_\perp$ and $\tau_\parallel$ are the respective optical depths. Known typical values of depolarization combined with other information such as lidar ratio, extinction coefficient, and backscatter coefficient, can help to identify unknown or ambiguous aerosol layers. Time series measurements of depolarization in persistent layers can also illustrate aerosol evolution, such as the change in volcanic aerosol from initial tephra and SO_2 to more spherical liquid sulfate aerosols.

Depolarization measurements are commonly used in cloud studies to differentiate water vapor from ice. However, aerosol depolarization measurements are known to show a strong dependence upon the aerosol size parameter, $x = 2\pi r/\lambda$, where r is the particle's radius, and λ the wavelength of the incident photon (Mischchenko and Sassen, 1998). Given the wide range of sizes aerosols may take, ranging from Rayleigh scattering regime through the Mie regime and into geometric optics, the extra information on size that can be provided through depolarization measurements is invaluable. This is especially true if combined with multiple wavelength ratio techniques.

10.7 Lidar networks

Whereas the preceding sections have described lidar theory and the various methods by which one may retrieve parameters related to aerosols, such as size distribution, number density, and detection of specific molecular species, the following sections will cover technical applications regarding lidar implementation. Lidar stations tend to be limited to a single location for each measurement. While mobile lidar stations have been constructed in the beds of trucks and vans, they tend to be kept stationary, except for short periods where they are used for a local/regional measurement campaign. Airborne and satellite-carried systems, to be discussed in the following section, expand coverage to larger regional and even global scales, but each profile is still fixed to a single location at a set time. Networks of lidars, however, while comprised of individual fixed lidar stations, provide a means to cover a large geographical area with simultaneous measurements performed with similar equipment at each station. Some of the current leading lidar networks include the European Aerosol Research Lidar Network (EARLINET), the Global Atmosphere Watch (GAW) Aerosol Lidar Observation Network (GALION), and Goddard Space Flight Center's Micropulse Lidar Network (MPLNET).

EARLINET began in 2000 with the objective of establishing "a quantitative comprehensive statistical database of the horizontal, vertical, and temporal distribution of aerosols on a continental scale, provide aerosol data with unbiased sampling, for important selected processes and air-mass history, together with a comprehensive analyses of these data" (Schneider, 2000). While the original project period only lasted until 2003, the network has continued with volunteered support from member stations. The current implementation, EARLINET-ASOS (Advanced Sustainable Observation System), continues the EARLINET project and advances its original goals to provide multi-year continental scale datasets using state-of-the-art lidar techniques (Pappalardo, 2006). An additional project, EARLINET-CALIPSO, is also ongoing since 2008 and is aimed at providing a long-term record of groundbased lidar measurements to provide a validation record for satellite missions in response to the 28 April 2006 launch of the spaceborne lidar aboard the CALIPSO satellite, which will be discussed in detail in the next section. EARLINET is currently comprised of 15 member stations spanning Europe, a map of which can be seen in Figure 10.8. While validation efforts have been on going since CALIPSO's launch, EARLINET-CALIPSO formalized the network validation program (Pappalardo et al., 2010).

Members of EARLINET follow a set of quality standards listed in the EARLINET constitution, which can be found on their website listed at the end of this chapter. Members are expected to compare their systems and data to other member stations, pursue joint research projects, and facilitate the exchange and availability of data with other member stations. All lidars operate at any of 1064 nm, 532 nm, and/or 355 nm wavelengths for elastic backscatter retrievals. Some stations also perform Raman retrievals of aerosol backscatter and extinction, and additional retrievals of depolarization and Raman water vapor and temperature. Further system capabilities may include scanning and portability. These capabilities are defined in the NetCDF files provided by each station. These data are taken at scheduled intervals as determined by the EARLINET members in order to facilitate the goal of a four-dimensional lidar data set in space and time.

A recent example of EARLINET's effectiveness was its coordination of measurements in response to the eruptions of Eyjafjallajökull beginning 14 April 2010. The eruption itself injected over 250 million cubic meters of tephra that was lofted to a height of 9 km, which the network was able to aid in quantifying and tracking. For the two weeks following the eruption, until 30 April 2010, lidar stations in EARLINET coordinated rapid-fire communication between all members in order to watch for and track the volcanic plume and its evolution (Ansmann et al., 2010; Guerrero-Rascado et al., 2010; Wiegner et al., 2012; Groß et al., 2012). The system, by providing quicklook images via website, further enabled the flow of data between centers in near real time. Updates were also provided as they became available to Volcanic Ash Advisory Centers (VAACs). These data aided in validation of models, and helped quantify the extent of the plume's travel, rate of fallout, divergence, and other physical and dynamic characteristics exceedingly important to air traffic over Europe.

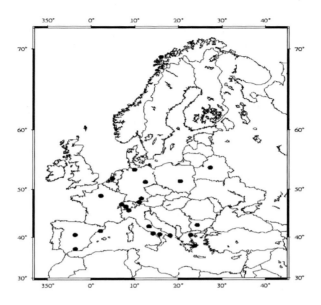

Figure 10.8 The locations of all EARLINET stations (represented by dots) as of this writing. Image produced by Pappalardo et al. (2010).

The World Meteorological Organization's (WMO) GALION network seeks to repeat the success of EARLINET on the global scale. GALION's objectives are in line with the GAW aerosol program as its goal is to determine the spatiotemporal distribution of aerosol properties related to climate forcing and air quality up to multidecadal time scales (Bösenberg et al., 2008). Data collected by GALION is intended for use in aiding the understanding of aerosol effects on climate and human health by way of air quality. These data feed climatologies, transport models and model assessments, aerosol effects on radiation, air quality forecasts, and satellite validation efforts. Though still under development, GALION members have held two workshop meetings so far, discussing measurement framework and results to date (Hoff and Pappalardo, 2010). Currently, approximately 100 lidar stations are listed to be part of GALION, with a bias toward the Northern Hemisphere in makeup due to station availability. Further information on GALION, GAW, and the WMO can be found at the WMO website and University of Maryland, Baltimore County's (UMBC) website at the end of this chapter.

NASA's MPLNET, also listed as a GALION partner, represents a specialized lidar network comprised of micropulse lidars. Micropulse lidars are made to be eye-safe and operate at extremely high repetition rates using a solid state laser emitting energies on the order of microjoules at up to 10 kHz (Spinhirne, 1993, 1995). Proposed in 2000, MPLNET utilizes the advantages of smaller, more portable MPL systems in order to measure aerosol and cloud optical properties such as extinction, backscatter, and layer heights (Welton et al., 2001), as well as exploring the retrieval of multiple scattering effects in cloud layers (Berkoff and Welton, 2008). MPLNET currently consists of approximately 20 MPL stations operating on a defined analysis schedule and processing scheme (Campbell et al., 2002; Welton et al., 2010). The network has thus far been successful at measuring aerosol and cloud layers both in the PBL and lofted, polar stratospheric clouds, and their associated optical properties (Campbell et al., 2003; Campbell et al., 2008; Campbell and Shiobara, 2008; Campbell and Sassen, 2008; Campbell et al., 2008). Data are provided through the MPLNET website, provided at the end of the chapter, and are held to format standards comparable to those for satellite-based systems. Data products include raw data, backscatter signals, layer heights, profiles of extinction, backscatter, and optical depth, lidar ratio, and feature masks for clouds, aerosols, PBL height, and other relevant retrieval information.

10.8 Spacebased lidar

Lidar carries the advantages of high vertical resolution backscatter profiles, the ability to probe between clouds, and the potential to penetrate through optically thin clouds in the troposphere. While ground-stations provide useful stationary data, and aircraft provide regional data, spacecraft lidar can produce near-global data. These benefits make highly detailed spaceborne lidar retrievals down to the surface not only possible but likely.

The first airborne lidar measurements were taken from a T-33 aircraft from NASA Langley Research Center (LaRC). Simultaneous operation of a ground-based lidar provided validation measurements of the horizontal-looking system (Lawrence et al., 1968). While the original intention of this mission was detection of clear air turbulence, the development of lidar brought aerosol applications to airborne measurements as well. A year later, downward-looking lidar measurements of aerosols were made from aircraft as part of the Barbados Oceanographic and Meteorological Experiment (Uthe and Johnson, 1971).

LaRC later produced the first upward looking aircraft lidar as well as part of the validation effort for the Stratospheric Aerosol Measurement-II (SAM-II) in 1978 (McCormick, 1979). As a precursor to orbital measurements, high-altitude aircraft were flown with lidars onboard. Carried aboard the WB-57 and developed by NASA Goddard Space Flight Center (GSFC), the flight allowed for autonomous control of the aircraft and provided information about high-altitude, automated measurements necessary for spaceborne lidar retrievals (Spinhirne et al., 1982). The combination of airborne measurements provided the basis for global retrievals of aerosol constituents such as desert dust from the Sahara; Polar Stratospheric Clouds (PSCs); and stratospheric aerosols, such as from volcanic eruptions.

All of this previous work set the stage for spaceborne lidars: the LIDAR In-space Technology Experiment (LITE), L'atmosphere par Lidar Sur Saliout (The Atmosphere by Lidar on Saliout) (ALISSA), Geoscience Laser Altimeter System (GLAS), and Cloud-Aerosol

LIDAR and Pathfinding Satellite Observation (CALIPSO) instruments. LITE acted as the proof of concept mission, flying aboard Space Shuttle Discovery flight STS-64. Launched September 9, 1994, LITE employed three wavelengths using a Nd-YAG laser operating at 1064, 532, and 355 nm. Also, a boresighted camera was attached to the instrument in order to colocate actual imagery of the scenes to provide context to the lidar measurements. Additional research objectives included measurements of clouds and aerosols in the stratosphere and troposphere, PBL height, and temperature and density profiling of the stratosphere (McCormick et al., 1993). Operating for ten days, LITE provided information on the size variability, distribution, and range extent of cloud and aerosol particles in the atmosphere, and its data were used extensively in developing long-duration spaceborne lidar. Some of the novel applications include its use for determining surface wind speeds from sea surface directional reflectance measurements (Menzies et al., 1998), and its use in characterizing a super typhoon (Kovacs and McCormick, 2003).

Two years following LITE, the French-Russian endeavor ALISSA was launched on the PRIRODA module destined for MIR. Launched April 26, 1996, ALISSA utilized four Nd-YAG lasers operating at 532 nm, with objectives of determining vertical structure of clouds with absolute measurement of cloud top altitude (Chanin, 1999). A key feature is the use of four lidars operating one at a time in order to guarantee data continuity. Difficulties with MIR eventually caused interruptions in the data and the eventual end of the mission when the space station was brought back to Earth in 1999. GLAS was the first long-duration spaceborne lidar. Launched January 13, 2003, GLAS operated at 532 and 1064 nm with the main objective of using laser altimetry to measure ice-sheet topography, changes thereof, and other associated cloud and atmospheric properties (Spinhirne and Palm, 1996). Aerosol measurements were taken predominantly using the 532 nm channel. The number of observations it was able to make each year was greatly reduced (measurements were taken for approximately one to two months per year) due to a laser diode pump problem. GLAS ended its science mission in February 2010 with the failure of the last of its three lasers.

Most recently, CALIPSO was launched April 28, 2006 as a joint French-US endeavor between LaRC, France's CNES, and Hampton University. CALIPSO was placed in the Afternoon Train (A- Train) satellite constellation along with CloudSat, in order to perform simultaneous measurements of aerosol and cloud properties such as both total and layer optical depth, chemical species concentrations, polarization, and cloud properties (Goddard Space Flight Center, 2003). This information is crucial to understanding aerosol direct and indirect effects on Earth's albedo for the effects of climate change, and for determining the types, states, and development of aerosol and cloud layers. CALIPSO utilizes two Nd-YAG lasers for redundancy as part of the the Cloud-Aerosol Lidar with Orthogonal Polarization (CALIOP) instrument. Also on board are the the Infrared Imaging Radiometer (IIR) and a Wide Field Camera (WFC). All instruments operate continuously, with the exception of WFC which requires daylight to image. In addition to continuous lidar measurements, CALIOP also offers polarization measurements at 532 nm, providing both perpendicular and parallel detector channels. This distinction allows for differentiation between ice cloud and spherical aerosol particles.

As with LITE, IIR and WFC provide context to measurements taken by CALIOP. Figure 10.9 shows a typical CALIPSO curtain profile along its orbital track. The most striking feature is the distinction in the plume of Saharan dust on the right-hand side of the image.

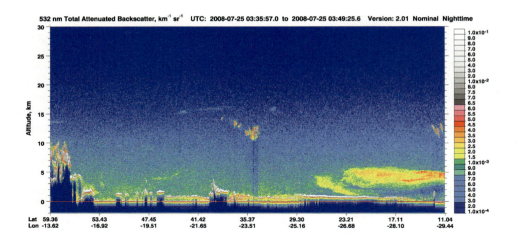

Figure 10.9 CALIPSO curtain plot taken July 25, 2008 showing an aerosol plume traveling from the Sahara. Cloud and stratospheric features are also visible.

Comparatively with the clouds shown both at high and low altitudes, particularly in the 35° latitude region, and the stratospheric features above, the aerosols are clearly visible and information about the extent of transport at profile time is provided by the plume's edge. As of this writing, CALIPSO has fired billions of shots, and only after nearly three years of near-continuous operation switched to the second laser in March 2009. CALIPSO still operates well within expected performance criteria and is expected to provide a long record of cloud and aerosol lidar measurements from space.

The CALIOP lidar on CALIPSO measures total attenuated elastic backscatter at 532- and 1064-nm wavelengths in addition to depolarization in the 532-nm channel. Since CALIOP is not a Raman lidar, an assumed value for the extinction-to-backscatter (lidar) ratio is required to derive the aerosol extinction coefficients (Liu et al., 2005). In order to obtain values of the lidar ratio for the various surface and atmospheric conditions encountered by the CALIPSO spacecraft as it orbits the Earth, the retrieval algorithm attempts to classify observed atmospheric regions with enhanced backscatter according to general aerosol type. It is assumed that the aerosol category assigned to each region consists of a mixture of particles with certain properties typically found in mesoscale air masses rather than pure aerosol systems that mix externally. The content of an aerosol mixture is determined by regional sources, wind patterns, hydration state, and chemical processes. A decision tree (Liu et al., 2005, page 48) is used to select the appropriate aerosol category assigned to a feature in a given measurement based on the mean attenuated backscatter coefficient, depolarization ratio, altitude, location, and surface type below the feature. A tabulated set of parameters corresponding to the aerosol properties of the mixture is used to compute the lidar ratio for the observed feature from Mie theory.

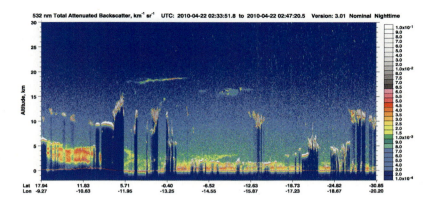

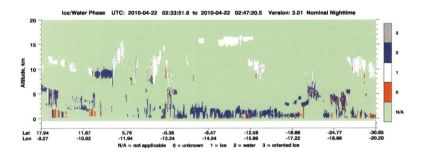

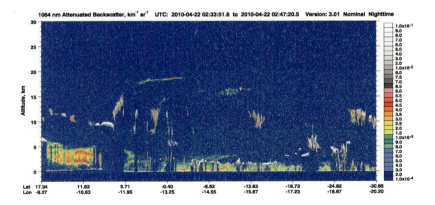

Figure 10.10 The basic measurements of CALIOP, including the total (top panel) and perpendicular (second panel) attenuated backscatter coefficients (km⁻¹sr⁻¹) at 532 nm as well as the total attenuated backscatter coefficient (km⁻¹sr⁻¹) at 1064 nm (bottom panel), are shown for a nighttime section of CALIPSO's orbit over parts of western Africa and the southern Atlantic Ocean on April 22, 2010.

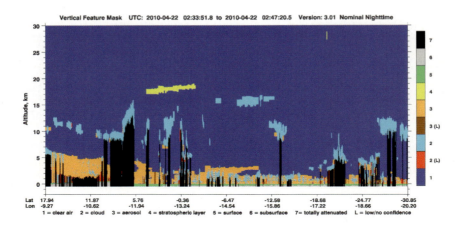

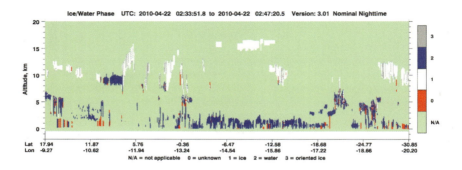

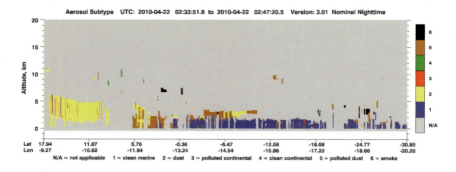

Figure 10.11 The classification of high-backscatter features into aerosol and cloud types according to CALIOP's scene classification algorithm for the same section of CALIPSO's orbit as shown in Figure 10.8. The vertical feature mask (top panel) discriminates clouds and aerosols, which are then subclassified according to cloud ice/water phase (second panel) and aerosol type (bottom panel).

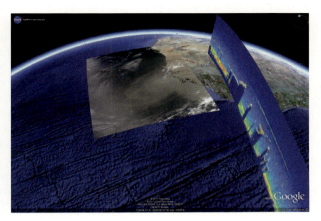

Figure 10.12 An image (left) taken by the Moderate Resolution Imaging Spectroradiometer (MODIS) instrument on NASA's Aqua satellite near the west coast of Africa on April 22, 2010, showing Saharan dust plumes being transported westward along the trade winds beneath high convective tropical clouds, and the composite (right) of this image with the CALIOP 532-nm backscatter curtain image (vertical scale exaggerated) using Google Earth.

The aerosol types and their associated optical and physical properties, which are used in the CALIOP retrieval algorithm, were identified by a clustering analysis of measured parameters obtained from AERONET, a global network of multi-wavelength sun photometers, and other sources (Liu et al., 2005). The six aerosol categories most commonly observed by AERONET are labeled as background or clean continental (containing small amounts of sulfates [SO_4], nitrates [NO_3], ammonium [NH_4], and organic carbon), marine or sea salt [$NaCl$], smoke (containing soot and organic carbon from biomass burning), desert dust or mineral soil, polluted dust (mixture of desert dust and smoke), and polluted continental (containing urban pollution).

Figures 10.10 and 10.11 show the basic data measured by CALIOP and the results of cloud particle and aerosol classification for a section of CALIPSO's orbit over western Africa and the Atlantic Ocean on April 22, 2010. Total attenuated backscatter ($km^{-1}sr^{-1}$) at 532 nm is shown in the first panel of Figure 10.10. The second panel shows the component of the total backscatter at 532 nm that is oriented perpendicular to the polarization of the transmitted laser pulse. Information from these two measurements yields the depolarization ratio, a quantity that indicates aerosol and cloud particle shape and thermodynamic phase. The last panel shows total attenuated backscatter profiles from CALIOP's 1064-nm channel. The color ratio between the 1064- and 532-nm channels provides information on particle size distribution.

Figure 10.11 illustrates the results of the cloud and aerosol classification algorithm for high backscatter features observed in this scene. The vertical feature mask (first panel) discriminates clouds and aerosols and identifies basic features such as surface returns and total attenuation of the lidar signal below optically thick clouds. The second panel illustrates the phase (water/ice) of the water content in the clouds identified in the vertical feature mask. The third panel shows the subdivision of aerosols identified in the vertical feature mask. Evident in the figure extending southward from Africa into the Atlantic Ocean (to about 10°S latitude) near the surface to about 5 km altitude is a thick aerosol layer whose composition was identified primarily as (partially polluted) dust. Saharan dust plumes form over western Africa by convection of the hot, dry desert air and transport mineral dust westward across the Atlantic Ocean along the trade winds towards the Caribbean. It is thought that sea surface cooling from Saharan dust interferes with hurricane development in the Atlantic Ocean. The polluted content appears mostly over the Atlantic Ocean west of southern Africa. Below the dust layer south of the equator is a layer identified as clean marine (sea salt) near the surface. Also evident in the figure are high tropical convective clouds (mostly ice) and a stratospheric feature near the equator at about 18 km altitude (probably a thin cirrus cloud).

The image shown on the left in Figure 10.12 was taken on April 22, 2010 off the western coast of Africa by the Moderate Resolution Imaging Spectroradiometer (MODIS) on NASA's Aqua satellite, which orbits ahead of CALIPSO in the A-Train constellation. Low-level Saharan dust (brown) can be seen traveling westward from Africa into the Atlantic Ocean below high-altitude convective clouds (white). The image on the right is a composite of the image obtained by MODIS and the total attenuated backscatter measurements obtained from CALIOP's 532-nm channel overlaid on a picture of the Earth. The composite image was assembled using Google Earth.

10.9 Summary

Lidar instruments are important tools for studying atmospheric aerosols and cloud systems and their role in global climate, and offer several advantages over passive techniques for atmospheric remote sensing. Because they are active remote-sensing devices, employing their own light sources, lidars can make range-resolved observations along their line of sight by measuring the transit time of their laser pulses. For zenith- and nadir-looking systems this yields information on the vertical structure of the atmosphere, which is lacking in pictures and column-integrating passive sensors. Lidars are capable of making both daytime and nighttime observations of multi-layer aerosol and cloud systems with high vertical resolution. Raman and HSRL multi-wavelength observations can be used to retrieve both backscatter and extinction coefficients and to derive concentrations of atmospheric aerosols and gaseous chemical species such as carbon dioxide and water vapor, all of which are important components for understanding radiative forcing effects and global climate. Polarization-sensitive lidars can be exploited to distinguish aerosol types such as ash plumes from volcanic eruptions, biomass burning, pollution, sea salt, and dust. Depolarization from clouds gives an indication of their microphysical composition and thermodynamic phase based on cloud particle shape, allowing for the distinction of water and ice clouds as well as nitric acid polar stratospheric clouds, which is important for understanding cloud impact on both the atmospheric radiative budget and high-latitude ozone depletion. With the advent of space-based platforms, lidars are now capable of examining the vertical structure of the atmosphere on a global scale and, unlike many passive satellite instruments including occultation devices, can penetrate down into the lower troposphere to probe the planetary boundary layer, characterizing the particulate composition of the lower atmosphere and providing a detailed picture of the atmospheric components that affect the Earth's weather and climate.

Data Website References

http://earthobservatory.nasa.gov/NaturalHazards/view.php?id=43852 (MODIS image)

http://rapidfire.sci.gsfc.nasa.gov/ (NASA/GSFC, MODIS Rapid Response, MODIS image)

http://modis.gsfc.nasa.gov/ (MODIS site)

http://www-calipso.larc.nasa.gov (CALIPSO site)

http://www.arm.gov/ (Department of Energy Atmospheric Radiation Measurement (ARM) Climate Research Facility (CRF) site)

http://mplnet.gsfc.nasa.gov/ (Micropulse Lidar Network (MPLNET) site)

http://www.earlinet.org/ (European Aerosol Research Lidar Network (EARLINET) site)

http://www.meteo.physik.uni-muenchen.de/~stlidar/quicklooks/European-quicklooks. html (EARLINET Quicklook Images site)

http://www.wmo.int (World Meteorological Organization site)

http://alg.umbc.edu/galion/ (GAW Aerosol Lidar Observation Network (GALION) site)

Acknowledgments

The authors wish to thank Michael T. Hill and Jia Su from Hampton University, for their contributions of data graphics for this chapter. In addition, the authors wish to thank Richard A. Ferrare of the NASA Langley Research Center for his contributions on Raman and HSRL field campaigns and HSRL data graphics.

11 Conclusion: Results and suggestions for future research

Jacqueline Lenoble, Lorraine A. Remer and Didier Tanré

11.1 Summary

Aerosols are a complex and important component of the atmosphere. These particles consist of various chemical compositions (homogeneous or inhomogeneous), shapes and sizes, and they affect human health, the environment, visibility, and atmospheric chemistry. A major concern is their influence on climate, directly by modifying the Earth's radiation budget, and indirectly by modifying cloudiness, cloud properties, precipitation and atmospheric circulations. The aerosol influence does not depend only on their total amount, as it is the case for the gaseous components of the atmosphere, but on their chemical composition, shapes, and sizes.

Because of aerosol importance, and spatial and temporal variability, aerosols require constant characterization and monitoring, and a global perspective. In situ measurements are critical to provide details of understanding, but these observations are relatively sparse and infrequent. A global perspective can be achieved only by remote sensing, performed either from the Earth's surface or by satellite-borne instruments. Ground-based and satellite remote sensing provide complementary information, with the ground-based instruments providing validation for satellite retrievals and sometimes a broader suite of retrieved parameters than can be achieved from space, while satellite remote sensing provides the true global perspective. Remote sensing measurements are based on the complex impact of aerosol on radiation. This complexity makes retrieval of aerosol from remote sensing data a difficult task.

The simplest remote sensing instrument is the ground-based sunphotometer, which provides the aerosol total optical depth, i.e. an information on the total amount of aerosol weighted by their extinction coefficient, at one or several wavelengths. Even in this case,

315

the required quality of measurements is dependent on the design, fabrication, calibration and operation of the instrument (Chapter 4, Section 4.2). The optical depth is obtained directly, but when one tries to extract further information, as the aerosol size distribution, from the spectral distribution of optical depth, one is faced with a difficult inversion problem (Chapter 5, Section 5.2). Figure 5.1.b shows the overlap of the extinction kernels for four wavelengths between 0.44 and 1.02 μm, meaning that they all carry similar information; therefore increasing the number of wavelengths in this interval cannot improve the retrieval. Extending the spectral interval to around 2.0 μm brings some more information, but does not permit retrievals of parameters such as single scattering albedo or refractive index. However, more information can be sought by adding independent observations, such as the sky radiance distribution, and the sky polarization to the direct sun observation. The analysis of this large set of data requires complicated retrieval algorithms, but enables retrieval of many more aerosol characteristics (Chapter 5, Section 5.8). Table 11.1 lists some prominent sources of publicly-available aerosol information derived from suborbital (ground-based or airborne) instruments.

Instrument	Network or platform	Period	Wavelengths or bands (μm)*	Parameter	Reference
AATS	Various airborne platforms	1985-present	14(0.354–2.139)	τ, α	Matsumoto et al. (1987) Livingston et al. (2005)
Cimel	AERONET	1993-present	7(0.34–1.02)	τ, α, ϖ %spherical dV/dlnr, $m_r + im_i, p(\Theta)$	Holben et al. (1998) Dubovik and King (2000)
Prede POM	SKYNET	1998-present	7(0.315–1.02)	τ, α, ϖ %spherical dV/dlnr, $m_r + im_i, p(\Theta)$	http://atmos. cr.chiba-u.ac. jp/
Microtops II	Marine Aerosol Network	2004-present	5(0.34–1.02)	τ, α	Smirnov et al. (2009)

* Primary bands used in aerosol retrieval. If the sensor contains additional bands aerosol retrievals often also make use of a wider spectral range than what is listed here for cloud masking, etc.

Table 11.1 Suborbital (ground-based or airborne) passive instruments, platforms and networks, providing total column characterization of aerosol

The occultation instruments are dedicated to characterizing the high atmospheric layers. They measure directly a slant aerosol optical thickness, and are based on the same principles as ground based sunphotometry (Chapter 4, Section 4.3). The aerosol data record produced by the occultation instruments provides nearly a 30 year characterization of stratospheric aerosol, covering multiple important volcanic eruptions (Chapter 7, Sections 7.2 and 7.3). Table 11.2 lists the important occultation instruments.

Earth-viewing nadir observations began with instruments using one or two wavelengths, either in the visible and near-IR or in the near-UV. These heritage instruments were used to retrieve the total aerosol optical depth and, in the case of the near-UV instruments, the Absorbing Aerosol Index. The retrieval algorithms were generally based on look-up tables (LUT), required several assumptions in the retrievals and were limited to specific land surface types (Chapter 7). Nowadays, several sophisticated spaceborne instruments observe aerosols; most of them use the backscattered solar radiation between near ultraviolet to near infrared and make use of multiangle views of the same scene and/or polarization (Chapters 7 and 8). Table 11.3 provides a long list of earth-viewing passive sensors using

Sensor	Platform	Period	Wavelengths or bands (μm)*	Parameter	Reference
SAM	Apollo-Soyuz	1975-1975	0.83	σ_e	McCormick et al. (1979)
SAM-2	Nimbus-7	1978-1993	1.00	σ_e	McCormick et al. (1979)
SAGE	AEM-B	1979-1981	2(0.45,1.00)	σ_e	Chu and McCormick (1979)
SAGE-2	ERBS	1984-2005	4(0.386–1.02)	σ_e, r_{eff}, SAD	Chu et al. (1989)
POAM-2	SPOT-3	1993-1996	5(0.353–1.060)	σ_e	Lumpe et al. (1997)
HALOE	UARS	1991-2005	4(2.45,3.40, 3.46, 5.26)	σ_e, r_{eff}	Hervig et al. (1998)
POAM-3	SPOT-4	1998-2005	5(0.354–1.020)	σ_e	Lumpe et al. (2002)
SAGE-3	METEOR-3M	2001-2005	9(0.385–1.545)	σ_e	Thomason et al. (2007)

* Primary bands used in aerosol retrieval. If the sensor contains additional bands aerosol retrievals often also make use of a wider spectral range than what is listed here for cloud masking, etc.

Table 11.2 Occultation sensors and their platforms used in aerosol measurements from space, providing profiles through the stratosphere and upper troposphere

Sensor	Platform	Period	Wavelengths or bands (μm)*	Parameter	Reference
MSS	Landsat ERTS-1	1972-1978	4(0.5–1.1)	τ	Griggs (1975)
VISSR	GOES-1-12	1975-present	1(0.65)	τ	Knapp et al. (2002) Prados et al. (2007)
VISSR	GMS-1-5	1977-2005	1(0.67)	τ	Masuda et al. (2002) Wang et al. (2003)
CZCS	NIMBUS-7	1978-1986	4(0.443--0.67)	Atmos corr	Fraser et al. (1997)
TOMS	NIMBUS-7	1978-1993	6 (0.312–0.380)	τ_{UV}, AAI	Torres et al. (2002) J. Herman et al. (1997)
AVHRR	NOAA-6-16	1979-present	2(0.65,0.85)	τ	Stowe et al. (1997) Mishchenko et al. (1999b)
TM	Landsat-5	1982-present	7(0.452–2.347)	τ	Tanré et al. (1988)
VIRS	TRMM	1997-present	2(0.63, 1.61)	τ, α	Ignatov and Stowe (2000)
ATSR-2	ERS-2	1995-2011	4(0.55–1.6)	τ	Veefkind et al. (1999)
GOME	ERS-2	1995-2003	spectrometer (0.24–0.79)	τ, AAI	Torricella et al. (1999), de Graaf et al. (2005)
TOMS	Earth Probe	1996-2005	6(0.309–0.360)	τ_{UV}, ϖ, AAI	Torres et al. (2002)
POLDER-1	ADEOS	1996-1997	7(0.443–0.97)	τ, α, η, %spherical	Herman et al. (1997)
SeaWiFs	OrbView-2	1997-present	3(0.510–0.865)	τ	Gordon and Wang (1994) Sayer et al. (2011)

Table 11.3 Passive shortwave Earth-viewing sensors and their platforms used in aerosol measurements from space, providing total column measurements

MODIS	TERRA	2000-present	8 (0.41–2.13)	τ, η	Remer et al. (2005) Levy et al. (2010)
MISR	TERRA	2000-present	4 (0.45–0.87)	$\tau, \alpha,$ SML,ϖ, %spherical, plume ht	Martonchik et al. (2009) Kahn et al. (2010)
MODIS	AQUA	2002-present	8 (0.41–2.13)	τ, η	Remer et al. (2005)
AATSR	ENVISAT	2002-present	4(0.55-1.6)	τ	Grey et al. (2006)
MERIS	ENVISAT	2002-present	4(0.412–0.865)	τ	Vidot et al. (2008)
SCIAMA-CHY	ENVISAT	2002-present	spectrometer (0.24–2.4)	AAI	De Graaf et al. (2005)
POLDER-2	ADEOS-2	2002-2003	7(0.443–0.97)	$\tau, \alpha, \eta,$ %spherical	Herman et al. (1997)
GLI	ADEOS-2	2002-2003	10(0.38–0.865)	τ, α	Murakami et al. (2006)
SEVIRI	MSG-1	2002-present	3(0.635–1.640)	τ	Popp et al. (2007)
OMI	AURA	2004-present	3(0.27–0.5)	$\tau, \varpi,$ AAI	Torres et al. (2007)
POLDER-3	PARASOL	2004-present	7(0.443–1.02)	$\tau, \alpha, \eta,$ %spherical	Herman et al. (1997) Tanré et al. (2011)
GOME-2	Metop-A	2006-present	spectrometer 0.24–0.79	$\tau,$ AAI	De Graaf et al. (2005)
CAI	GOSAT	2009-present	4(0.380–1.60)	τ	Sano et al. (2009)
VIIRS	NPP	2011-present	9(0.412–2.25)	τ, α	Northrup Grumman Space Technology, ATBD RevF (2010)

* Primary bands used in aerosol retrieval. If the sensor contains additional bands aerosol retrievals often also make use of a wider spectral range than what is listed here for cloud masking, etc.

observations in the shortwave spectrum that have been used for aerosol retrieval. Most of these sensors have either a long history of aerosol retrievals or were designed for aerosol retrieval, and many, but not all, make their aerosol products available to the public. A few sensors in Table 11-3 are included for historical interest.

Research on characterizing aerosol based on the observation of longwave (terrestrial) radiation shows very promising results (Chapter 9). However, to date the only operational

Sensor	Platform	Period	Wavelengths or bands (μm)*	Parameter	Reference
CLAES	UARS	1991–1993	8(5.3–12.8)	σ_e	Roche et al. (1993) Massie et al. (1996)
ISAMS	UARS	1991–1992	6.21,12.1	σ_e	Taylor et al. (1993)
HIRDLS	Aura	2004-present	5(7.1–17.4)	σ_e	Khosravi et al. (2009)

* Primary bands used in aerosol retrieval. If the sensor contains additional bands aerosol retrievals often also make use of a wider spectral range than what is listed here for cloud masking, etc.

Table 11.4 Infrared sensors and their platforms used in aerosol measurements from space, providing characterization of aerosol in stratosphere and upper troposphere

Sensor	Platform	Period	Wavelengths or bands (μm)*	Parameter	Reference
MVIRI	Meteosat	1982–2006	10.5–12.5	IDDI	Legrand et al. (2001)
AIRS	Aqua	2002-present	3.74–4.61, 6.20–8.22, 8.8–15.4	$\tau(10\mu m)$, altitude, r_{eff}	Pierangelo et al. (2004b; 2005a), Peyridieu et al. (2010a)
IASI	Metop	2007-present	3.62–15.5	$\tau(10\mu m)$, altitude, r_{eff}	Peyridieu (2010c)

* Primary bands used in aerosol retrieval. If the sensor contains additional bands aerosol retrievals often also make use of a wider spectral range than what is listed here for cloud masking, etc.

Table 11.5 Infrared sensors and their platforms used in aerosol measurements from space, providing total column retrievals of aerosol

products retrieved from the longwave part of the spectrum are retrievals of aerosol extinction in the upper atmosphere. These are listed in Table 11.4. Some interesting aerosol retrievals through the total atmospheric column using longwave radiation, still experimental, are listed in Table 11.5.

Lidar observations using a laser as the source of radiation have the main advantage of providing information on the aerosol vertical profile (Chapter 10). While suborbital lidar has been producing important insight on aerosol vertical distribution for decades, the organization of lidar instruments into networks or the data into easily accessible archives began more recently. Table 11.6 lists a few representative suborbital lidar systems with archived data. There have been only three space-based lidars designed for aerosol characterization and these are listed in Table 11.7.

Lidar type	Network or platform	Period	Wavelengths or bands (μm)	Parameter	Reference
Backscatter micropulse	MPLNet	2000-present	0.523 or 0.527	σ_b	Welton et al. (2001)
Raman	EARLINET	2000-present	0.351/0.355	σ_b	Matthias et al. (2004)
HSRL	Airborne LaRC B200	2006-present	0.532, 1.064	σ_b, σ_e, depol	Hair et al. (2008)

Table 11.6 Suborbital lidar instruments and networks, providing profiles through at least the lower atmosphere

Sensor	Platform	Period	Wavelengths or bands (μm)	Parameter	Reference
LITE	Space Shuttle Discovery	1994-1994	3(0.355,0.532, 1.064)	σ_b	McCormick et al. (1993) Gu et al, (1997)
GLAS	ICEsat	2003-2003	2(0.532, 1.064)	σ_b	Spinhirne and Palm (1996) Spinhirne et al. (2005)
CALIOP	CALIPSO	2006-present	2(0.532,1.064)	σ_b, depol	Z. Liu et al. (2005) Winker et al. (2009)

Table 11.7 Satellites and lidar instruments used in aerosol measurements from space, providing profiles through the entire column

Symbols defined below are applicable for all tables, 11.1 through 11.7.

(Acronyms are defined in the list of symbols and acronyms)

τ	Aerosol optical thickness that includes a value in the midvisible but may include values across the indicated range of wavelengths
τ_{UV}	Aerosol optical thickness available only for ultraviolet wavelengths
α	Ångström exponent
ϖ	The single scattering albedo
% spherical	The percentage of the coarse mode τ due to spherical particles
η	The fraction of the total τ at 550nm due to fine particles (Section 8.3)
m_r	The real part of the refractive index
m_i	The imaginary part of the refractive index
$p(\Theta)$	Phase function
σ_e	Extinction coefficient
σ_b	Backscattering coefficient
r_{eff}	Particle effective radius
SAD	Surface area density
AAI	Absorbing aerosol index (Section 7.5)
SML	The ability to distinguish Small, Medium and Large sizes
Atmos corr	Instrument derived aerosol only as a by product and that atmospheric correction of surface reflectance was the primary product (Section 6.5)
depol	Measurements of depolarization (Section 10.6)
IDDI	Infrared Difference Dust Index (Section 9.5)
plume ht.	Plume height (Section 8.8)

11.2 Results

Remote sensing observations have provided new insight and better understanding of the global aerosol system. For example, discovery of arctic haze by ground based sunphotometers (chapter 6), observation of volcanic particles dispersion around the globe, and slow decrease after El Chichon and Pinatubo eruptions by SAGE instruments (chapter 7, section 7.3), cross-oceanic transport of desert dust and other particles (figures in Chapter 8). Furthermore, the day-to-day work of these sensors acquiring data and the application of inversions and retrieval algorithms to the data create an ever-growing climatology of aerosol properties. We see individual events (Figures 7.6, 7.7, 7.8, 8.4), the long-term average conditions (Figures 6.7, 8.5, 8.7) and the anomalies from those conditions (Figure 8.5), and the gradual trends of aerosol characteristics over time (Figures 7.4, 7.9, 8.2).

Global aerosol climatologies from various instruments are now available. Availablity does not necessarily indicate useability for a particular application such as estimating the aerosol effect on climate forcing. What confidence do we have in these results ? How accurate are the aerosol optical depth retrievals ? What of the other particle properties such as single scattering albedo and particle size ? How much confidence do we have in reported trends such as shown in Figure 7.4 ? Before blindly using remote sensing aerosol products in climate studies or other applications we need to quantify the accuracy or statistical confidence of the product or parameter. This raises the issue of validation of remote sensing products, an on-going effort addressed by all groups involved in providing data and measurement analysis. The references listed in the tables of this chapter provide a starting point to obtain information on the aerosol retrieval, data archive, uncertainties and limitations of that instrument and products. Some limitations are also discussed in Chapter 8.

11.3 Algorithms vs. Products; Validation vs. intercomparison

A distinction must be made between validating a product and validating an algorithm. Algorithms are mathematical constructs that turns an idealized set of measurements of the radiation field into information of aerosol particles theoretically embedded in that field. In a perfect world, validating an algorithm would be the same as validating the products resulting from that algorithm, but the world is not perfect. Input radiances are subject to instrumental defects : calibration drift, crosstalk from nearby channels, missing and bad detectors, and point-spread functions that smear light reflected from one pixel into nearby pixels. Furthermore, real-world algorithms must contend with identifying the scenes that are appropriate for retrieval for a certain algorithm and masking inappropriate scenes. Situations such as clouds, sun glint, snow have traditionally been masked by various retrieval algorithms, and decisions as to where to draw the threshold between appropriate and inappropriate scenes are highly subjective. Finally, data aggregation from instrument pixel size to standard product size, and then from standard product size to climate appropriate spatial and temporal means introduces a new set of subjective decisions that can create differences in global mean aerosol optical depth of 40% (Levy et al., 2009).

Algorithms can be validated as part of the process of validating products, but the distinction must be understood. A highly capable instrument measuring multispectral multiangular and polarization with the right algorithm should return the most information of the aerosol field with the highest accuracy. However, that highly capable algorithm may not produce a better aerosol product than a single wavelength, single angle radiometer if the calibration on the polarimeter is poor, the pixels are unregistered, etc. Therefore both algorithms and products should be validated, but we note that it is much easier to validate an algorithm than a product.

In order to validate a product measured by a specific instrument, it is necessary to have another independent and reliable measure of the same parameter, given by another instrument, at the same time and in the same place. This situation almost never occurs.

Most generally space observations are «validated» using ground based observations, from networks like AERONET, in coincidence as close as possible, both in time and in position. How close in time and space is close enough? A first difficulty is the spatio-temporal collocations of the two types of measurements (Ichoku et al., 2002b). A second issue is the different data screening by the two instruments. For example, ground-based sunphotometers and satellite instruments screen for clouds differently. The collocation of the two instruments will only occur when both instruments are reporting data. This tells you nothing about the aerosol retrieval from the satellite when the satellite is reporting data but the ground-based instrument is not. Such cases suggest but do not prove cloud contamination in the aerosol retrieval. This is an example of validating the retrieval, but not the product.

Most importantly, the only quantity directly obtained by ground based instruments, is the aerosol total optical depth, as measured by well calibrated sunphotometers. All other quantities (size distribution, information on sphericity, single scattering albedo, etc.) are actually retrieved from measurements, by algorithms, which are neither simpler nor more trustable, than the algorithms used for space borne instruments. It is therefore more sensible to speak of «intercomparison» between instruments, than of real validation.

Intercomparison can be performed between different space borne instruments, as well. Here there is again the problem of coincidence of the observations and differences in data screening. For example, Kahn et al. (2009) compare MISR and MODIS products. They find that MISR and MODIS each make successful aerosol retrievals about 15% of the time, discarding a majority of retrieval opportunities because of the presence of clouds, inappropriate surface conditions, etc. However, each sensor chooses a different 15%, collocating for only 6-7% of the total overlapping possibilities.

A third type of intercomparison is between remote sensing data and corresponding in situ measurements. Here the challenge is the comparison between ambient and often total column measurements with samples of particles disturbed from ambient conditions, taken from the partial column. All the issues with spatio-temporal collocations and different screening procedures remain. Still, in situ measurements can get to the heart of the particle properties and provide both a constraint when these properties are retrieved and on the assumptions inherent in the retrievals when the particle properties are assumed.

Another promising approach to validating algorithms has been proposed by Kokhanovsky et al. (2010), and relies on using synthetic data. Starting with an aerosol model, a forward calculation is made to produce the entire suite of reflectances and polarized radiances (if applicable) that would be measured by each sensor. This output is computed for each wavelength and view angle of the sensor in question. Then these properties are inverted using the instrument's operational inversion algorithm. The retrieved quantities are compared with the original aerosol model parameters. Such a comparison provides important insight into algorithms but not products. The comparison is highly theoretical, not accounting for retrievals tuned for the real world through empirical assumptions. Also such an exercise does not deal with such real-world retrieval issues as clouds, calibration drift etc.

Validation is an on-going process that requires multiple approaches. It is unrealistic that an aerosol data product can be declared 'validated' and used indiscriminantly for the remainder of the satellite mission. Even if the retrieval algorithms were frozen in time, the sensors are not. Satellite sensor calibration can and does drift in time, which makes long-term trend studies such as shown in Figure 7-4 very difficult to interpret (Zhang and Reid, 2010).

11.4 Recommendations for future work

The priority of any future work is to continue to maintain the capability that we have today. We need continuation of the satellite product records for as long the sensors are operating and on orbit, and we need the continuation of the complementary suite of suborbital networks and airborne sensors to serve as a source of validation and intercomparison. These data records require continual evaluation and examination for changing calibration and refinements of the algorithms.

As sensors age and the data stream ends, new sensors need to be launched. Rather than being carbon copies of existing sensors we should push technology forward and enhance the capability of future aerosol space missions. Increasing spectral range, decreasing pixel size, enhancing angular views and adding polarization provides new information for more comprehensive retrievals that provide a more complete characterization of the aerosol system. Expanding space-based lidar from current backscattering measurements to directly measured extinction profiles offers opportunity for a much clearer 3-dimensional characterization of the global aerosol, including vertical profiles of aerosol absorption properties. A multi-beam or scanning lidar can expand the limited lidar coverage currently available. Adding high temporal resolution from geostationary platforms, expanding oceanic suborbital observations to include sky radiance and inversions, and implementing some of the more advanced and experimental retrieval ideas described in Section 8.8 and Chapter 9, are just some of the innovations that should be possible in the near future.

As new technology is being developed to provide this additional capability, new algorithms must also be generated. The old LUT approach is entirely appropriate when a sensor ingests a small set of inputs that must be constrained with *a priori* assumptions. However, if a sensor is providing multispectral, multiangle, polarization information, there is sufficient information to apply optimal estimation methods without need to overly constrain the retrieval with severe assumptions (Sections 8.5.3 and 8.8). These new algorithms must be in development now in order to meet the challenges of the next generation of sensors.

Even as sensors and algorithms advance, there will always be need to consider associating different sensors and platforms together to make use of complementary information and to be able to intercompare. Section 8.9.5 touches on some of the benefits of combining information from different instruments, and these benefits will continue even as aerosol remote sensing from each particular instrument becomes more sophisticated.

Validation and intercomparison will be a constant need in the future, as it is now. Sensors require intercomparison. Algorithms require intercomparison. Data products require intercomparison. There needs to be more concern for sensor calibration issues, more algorithm intercomparison using synthetic data, more opportunities to compare final products and not just aerosol optical thickness. In situ data should not be ignored because it provides valuable information on the aerosol particles themselves, and as retrievals become more sophisticated and provide more aerosol parameters, in situ characterization of particle properties are the only hope of constraining parameters such as phase function and complex refractive index. Technology development in the realm of in situ sampling must keep up with the technology development of remote sensing instrumentation.

Aerosol observations from remote sensing are playing an increasingly important role in numerical modeling at all scales. Aerosol remote sensing data are used to constrain global climate and regional air quality models, and aerosol products are being operationally assimilated into global-scale forecast models (Section 8.9.5). These hybrid systems that depend on a constant stream of global aerosol information to provide a better representation of the global aerosol system will continue to require daily operational satellite-derived aerosol products.

Finally as the community moves forward, continuity with present data records must be maintained. There should be opportunity for overlapping data sets between old and new sensors. Therefore, there is no time to waste. Examination of the tables in this chapter points out that there is no current occultation instrument to continue the long-term upper atmosphere record. NPP-VIIRS has just been launched to continue the data record from MODIS, but what of the additional capabilities associated with MISR, POLDER, etc.? What happens when CALIOP expires? To maintain a continuous data record, replacement sensors need to be launched before aging sensors die.

The field of aerosol remote sensing has been growing over centuries, but has accelerated greatly over the past 30 years. There is no end in sight. We have not fully exploited the information content available to be measured and interpreted from the interaction between radiation and a dispersion of suspended particles. There is much work left for the future.

Combined Bibliography

Abbot, C. G., *Solar Radiation and Weather Studies*, Smithsonian Miscellaneous Collection, **94**, No. 10, 1935.

Abbot, C. G., and L. B. Aldrich, *Annals of the Astrophysical Observatory of the Smithsonian Institute* 3, Government Printing Office, Washington, D.C., 1913.

Acarreta, J. R., J. F. De Haan and P. Stamnes, "Cloud pressure retrieval using the O-2-O-2 absorption band at 477 nm", *J. Geophys. Res.*, **109** (2004), D05204, doi: 10.1029/2003JD003915.

Ackerman, S. A., "Using the radiative temperature difference at 3.7 and 11 μm to track dust outbreaks", *Remote Sens. Environ.*, **27** (1989), 129–133.

Ackerman, S. A., and K. Strabala, "Satellite remote sensing of H2SO4 aerosol using the 8- to 12-μm window region: Application to Mount Pinatubo", *J. Geophys. Res.*, **99** (1994), D9, doi: 10.1029/94JD01331.

Ackerman, S. A., H-L. Huang, P. Antonelli, R. Holz, H. Revercomb, D. Tobin, K. Baggett, and J. Davies, " Detection of clouds and aerosols using infrared hyperspectral observations ", International Conference on Interactive Information and Processing Systems (IIPS) for Meteorology, Oceanography, and Hydrology, (2004), Seattle, WA, 11–15.

Adames, A. F., M. Reynolds, A. Smirnov, D. S. Covert, and T. P. Ackerman, "Comparison of moderate resolution imaging spectroradiometer ocean aerosol retrievals with ship-based sun photometer measurements from the Around the Americas expedition", *J. Geophys. Res.*, **116** (2011), D16303, doi: 10.1029/2010JD015440.

Ahn, C., O. Torres, and P. K. Bhartia, "Comparison of OMI UV aerosol products with Aqua-MODIS and MISR observations in 2006", *J. Geophys. Res.*, **113** (2008), D16S27, doi: 10.1029/2007JD008832.

Ambrico P. F., A. Amodeo, P. Di Girolamo, and N. Spinelli, "Sensitivity analysis of differential absorption lidar measurements in the mid-infrared region", *Appl. Opt.*, **39:36** (2000), 6847–6865.

327

Amuroso, A., M. Cacciani, A. DiSarra, and G. Fiocco, "Absorption cross sections of ozone in the 590–610 nm region at $T = 230$K and $T = 299$K", *J. Geophys. Res.*, **95** (1990), 20,565–20,568.

Anderson, J., C. Brogniez, L. Cazier, V. Saxena, J. Lenoble, and M. P. McCormick, "The characterization of aerosols from simulated SAGE III measurements applying two retrieval techniques", *J. Geophys. Res.*, **105**, D2 (2000), 2013–2027.

Ångström, A. K., "On the atmospheric transmission of sun radiation and on dust in the air", *Geografis Annal.*, **2** (1929), 156–166.

Ångström, A. K., "The quantitative determining of radiant light by the method of electrical compensation," *Phys. Rev.* **1** (1932): 365–371.

Ångström, A. K., "Apparent solar constant variations and their relation to the variability of atmospheric transmission," *Tellus*, **22** (1970a), 205–218.

Ångström, A. K., "On determinations of the atmospheric turbidity and their relation to pyrheliometric measurements", *Adv. in Geophys.*, **14** (1970b), 269.

Ansmann A., U. Wandinger, M. Riebesell, C. Weitkamp, and W. Michaelis, "Independent measurement of extinction and backscatter profiles in cirrus clouds by using a combined Raman elastic-backscatter lidar", *Appl. Opt.*, **31** (1992), 7113–7131.

Ansmann, A., M. Tesche, S. Groß, V. Freudenthaler, P. Seifert, A. Hiebsch, J. Schmidt, U. Wandinger, I. Mattis, D. Müller, and M. Wiegner, "The 16 April 2010 major volcanic ash plume over central Europe: EARLINET lidar and AERONET photometer observations at Leipzig and Munich, Germany", *Geophys. Res. Lett.*, **37** (2010), L13810, doi:10.1029/2010GL043809.

Aumann, H. H., M. T. Chahine, C. Gautier, M. Goldberg, E. Kalnay, L. McMillin, H. Revercomb, P. W. Rosenkranz, W. L. Smith, D. H. Staelin, L. Strow and J. Susskind, "AIRS/AMSU/HSB on the Aqua Mission: Design, science objectives, data products and processing systems", *IEEE Trans. Geosci. Rem. Sens.*, **41** (2003), 253–264.

Babenko, V. A., L. G. Astafyeva, and V. N. Kuzmin, *Electromagnetic Scattering in Disperse Media: Inhomogeneous and Anisotropic Particles*. Praxis Publishing, Chichester, UK, 2003.

Balkanski, Y., M. Schulz, T. Claquin, and S. Guibert, "Reevaluation of mineral aerosol radiative forcings suggests a better agreement with satellite and AERONET data", *Atmos. Chem. Phys.*, **7** (2007), 81–95, doi: 10.5194/acp-7-81-2007.

Baran, A. J., J. S. Foot, and P. C. Dibben, "Satellite detection of volcanic sulphuric acid aerosol", *Geophys. Res. Lett.*, **20(17)** (1993), 1799–1801, doi: 10.1029/93GL01965.

Barnsley, M. J., J. J. Settle, M. A. Cutter, D. R. Lobb, and F. Teston, "The PROBA/CHRIS mission: A low-cost smallsat for hyperspectral multiangle observations of the earth surface and atmosphere" IEEE Trans. Geosci. Rem. Sens., 42(7), (2004), 1512–1520, doi: 10.1109/TGRS.2004.827260.

Bauman, J. J., P. B. Russell, M. A. Geller, and P. Hamill, "A stratospheric aerosol climatology from SAGE II and CLAES measurements: 1. Methodology", *J. Geophys. Res.* **108**, (2003a), D13, 4382, doi: 10.1029/2002JD002992.

Bauman, J. J., P. B. Russell, M. A. Geller, and P. Hamill, "A stratospheric aerosol climatology from SAGE II and CLAES measurements: 2. Results and comparisons, 1984–1999", *J. Geophys. Res.*, **108,** D13 (2003b), 4383, doi: 10.1029/2002JD002993.

Benedetti, A., J. J. Morcrette, O. Boucher, A. Dethof, R. J. Engelen, M. Fisher, H. Flentje, N. Huneeus, L. Jones, J. W. Kaiser, S. Kinne, A. Mangold, M. Razinger, A. J. Simmons, and M. Suttie, "Aerosol analysis and forecast in the European Centre for Medium-Range Weather Forecasts Integrated Forecast System: 2. Data assimilation", *J. Geophys. Res.*, **114** (2009) D13205, doi:10.1029/2008JD011115.

Berk, A., L. S. Bernstein, and D. C. Robertson, "MODTRAN: A moderate resolution model for LOWTRAN 7", Air Force Geophysics Laboratory Technical Report GL-TR-89-0122, Geophysics Directorate, Phillips Laboratory, Hanscom Air Force Base, Massachusetts, 38 pp. (1989).

Berkoff, T. A., and E. J. Welton, "A method to obtain multiple scattering measurements using a micro-pulse lidar", in: *Proceedings of the 24th International Laser Radar Conference*, Boulder, Colorado, 173–176, 2008.

Bernath, P. F., *Spectra of Atoms and Molecules,* 2nd edition, Oxford University Press, 2005.

Berthet, G., J.-B. Renard, V. Catoire, M. Chartier, C. Robert, N. Huret, F. Coquelet, Q. Bourgeois, E. D. Rivière, B. Barret, F. Lefèvre, and A. Hauchecorne, "Remote sensing measurements in the polar vortex: Comparison to in situ observations and implications for the simultaneous retrievals and analysis of the NO_2 and OClO species", *J. Geophys. Res.*, **112** (2007) D21310, doi: 10.1029/2007JD008699.

Bevan, S. L., P. R. J. North, S. O. Los, and W. M. F. Grey, "A global dataset of atmospheric aerosol optical depth and surface reflectance from AATSR", *Rem. Sens. Environ.* (2011), doi: 10.1016/j.rse.2011.05.024.

Bevington, P. R., *Data Reduction and Error Analysis for the Physical Sciences*, McGraw-Hill, 1969.

Bhartia, P., J. Herman, R. McPeters, and O. Torres, "Effect of Mount Pinatubo aerosols on total ozone measurements from backscatter ultraviolet (BUV) experiments", *J. Geophys. Res.*, **98**, D10 (1993), 18,547–18,554.

Bi, L., P. Yang, G. W. Kattawar, and R. Kahn, "Single-scattering properties of triaxial ellipsoidal particles for a size parameter range from the Rayleigh to geometric-optics regimes", *Appl. Opt.*, **48** (2009), 114–126.

Bigelow, D. S., J. R. Slusser, A. F. Beaubien and J. H. Gibson, "The USDA ultraviolet radiation monitoring program", *Bull. Amer. Meteorol. Soc.*, **79** (1998), 601–615.

Bingen, C., D. Fussen, and F. Vanhellemont, "A global climatology of stratospheric aerosol size distribution parameters derived from SAGE~II data over the period 1984–2000: 1. Methodology and climatological observations", *J. Geophys. Res.*, **109** (2004), D06201, doi: 10.1029/2003JD003518.

Bodhaine, B. A., N. B. Wood, E. G. Dutton, and J. R. Slusser, "On Rayleigh optical depth calculations", *J. Atmos. Ocean. Technol.*, **16** (1999), 1854–1861.

Bohren, C. F., and D. R. Huffman, *Absorption and Scattering of Light by Small Particles*, John Wiley & Sons Inc., 1983.

Bösenberg, J., R. Hoff, A. Ansmann, D. Müller, J. C. Antuña, D. Whiteman, N. Sugimoto, A. Apituley, M. Hardesty, J. Welton, E. Eloranta, Y. Arshinov, S. Kinne, and V. Freudenthale, *Plan for the Implementation of the GAW Aerosol Lidar Observation Network GALION* (Hamburg, Germany, 27–29 March 2007) (WMO TD No. 1443), (2008).

Boussara A. E., D. A. Degenstein, R. L. Gattinger R., and E.J. Llewellyn, "Stratospheric aerosol retrieval with optical spectrograph and infrared imaging system limb scatter measurements", *J. Geophys. Res.*, **112**, (2007) D10, D10217, doi: 10.1029/2006JD008079.

Bovensmann, H., J. P. Burrows, M. Buchwitz, J. Frerick, S. Noël, V. V. Rozanov, K. V. Chance, and A. P. H. Goede, "SCIAMACHY: Mission objectives and measurement modes", *J. Atmos. Sci.*, **56** (1999), 127–150, doi: 10.1175/1520-0469(1999)056<0127: SMOAMM>2.0.CO;2.

Bréon, F. M., and N. Henriot, "Spaceborne observations of ocean glint reflectance and modeling of wave slope distributions", *J. Geophys. Res.*, **111** (2006), C06005, doi: 10.1029/2005JC003343.

Bréon, F. M., D. Tanré, P. Lecomte, and M. Herman, "Polarized reflectance of bare soils and vegetation – measurements and models", *IEEE Trans. Geosci. Rem. Sens.*, **33** (1995), 487–499.

Bréon, F. M., F. Maignan, M. Leroy, and I. Grant, "Analysis of hot spot directional signatures measured from space", *J. Geophys. Res.*, **107** (2002), 4282–4296.

Brindley, H. E., and J. E. Russell, "Improving GERB scene identification using SEVIRI: Infrared dust detection strategy", *Rem. Sens. Environ.*, **104:4** (2006), 426–446.

Brogniez, C., C. Piétras, M. Legrand, P. Dubuisson, and M. Haeffelin, "A high-accuracy multiwavelength radiometer for in situ measurements in the thermal infrared. Part II : Qualification in field experiments", *J. Atmos. Oceanic Technol.*, **20** (2000), 1023–1033.

Brogniez, C., A. Bazureau, J. Lenoble, and W. P. Chu, "SAGE III measurements: A study on the retrieval of ozone, nitrogen dioxide and aerosol extinction coefficients", *J. Geophys. Res.*, **107**, (2002) D24, doi: 10.1029/2001JD001576.

Brogniez, C., C. Pietras, M. Legrand, P. Dubuisson, and M. Haeffelin, "A high-accuracy multiwavelength radiometer for in situ measurements of the thermal infrared. Part II:

Behavior in field experiments", *J. Atmos. Ocean. Tech.*, **20** (2003), 1023–1033, doi: 10.1175/1520-042620<1023:AHMRFI>2.0.CO;2.

Brogniez, C., V. Buchard, and F. Auriol, "Validation of UV-visible aerosol optical thickness retrieved from spectroradiometer measurements", *Atmos. Chem. Phys.*, **8** (2008), 4655–4663.

Brooks, N., and M. Legrand, "Dust variability over Northern Africa and rainfall in the Sahel", in: S. J. McLaren and D. R. Kniveton (Eds), *Linking Climate Change to Land Surface Change*, 1–25, Kluwer Academic Publishers, 2000.

Bruegge, C. J., R. N. Halthore, B. Markham, M. Spanner, and R. Wringley, "Aerosol optical depth retrievals over the Konza Prairie", *J. Geophys. Res.*, **97**, D17 (1992), 18,743–18,758.

Bulgin, C. E., P. I. Palmer, C. J. Merchant, R. Siddans, S. Gonzi, C. A. Poulsen, G. E. Thomas, A. M. Sayer, E. Carboni, R. G. Grainger, E. J. Highwood, and C. L. Ryder, "Quantifying the response of the ORAC aerosol optical depth retrieval for MSG SEVIRI to aerosol model assumptions," *J. Geophys. Res.*, **116**, (2011) D05208, doi:10.1029/2010JD014483.

Burrows, J. P., M. Weber, M. Buchwitz, V. Rozanov, A. Ladstätter-Weissenmayer, A. Richter, R. DeBeek, R. Hoogen, K. Bramstedt, K.-U. Eichman, M. Eisinger, D. Perner, "The Global Ozone Monitoring Experiment (GOME): Mission concept and first scientific results", *J. Atmos. Sci.*, **56** (1999a), 151–175, doi: 10.1175/1520-0469(1999)056<0151:TGOMEG>2.0.CO;2.

Burrows, J. P., A. Richter, A. Dehn, B. Deters, S. Himmelmann, and J. Orphal, "Atmospheric remote-sensing reference data from GOME-2: Temperature-dependent absorption cross sections of O_3 in the 231–794 nm range", *J. Quant. Spectrosc. Radiat. Transfer*, **61** (1999b), 509–517.

Burton, S. P., R. A. Ferrare, C. A. Hostetler, J. W. Hair, R. R. Rogers, M. D. Obland, A. L. Cook, D. B. Harper, and K. D. Froyd, "Aerosol classification of airborne high spectral resolution lidar measurements – 1. Methodology and examples", *Atmospheric Measurement Techniques Discussions*, 4 (2011), 5631–5688.

Cabannes, J., *La Diffusion Moléculaire de la Lumière*, Les Presses Universitaires de France, Paris, 1929.

Cahalan, R. F., "The Kuwait oil fires as seen by Landsat", *J. Geophys. Res.*, **97** (D13) (1992), 14,565–14,571.

Cahalan, R. F., L. Oreopoulos, A. Marshak, K. F. Evans, A. B. Davis, R. Pincus, K. H. Yetzer, B. Mayer, R. Davies, T. P. Ackerman, H. W. Barker, E. E. Clothiaux, R. G. Ellingson, M. J. Garay, E. Kassianov, S. Kinne, A. Macke, W. O'Hirok, A. N. Partain, G. L. Stephens, F. Szczap, E. E. Takara, T. Varnai, G. Y. Wen, and T. B. Zhuraleva, "The I3RC. Bringing together the most advanced radiative transfer tools for cloudy atmospheres", *BAMS*, **86** (2005), 1275–1293.

Cahoon, D. R., Jr., B. J. Stocks, J. S. Levine, W. R. Cofer III, and J. M. Pierson, "Satellite analysis of the severe 1987 forest fires in northern China and southeastern Siberia", *J. Geophys. Res.*, **99** (1994) (D9), 18,627–18,638, doi: 10.1029/94JD01024.

Cairns, B., B. E. Carlson, R. Ying, A. A. Lacis, and V. Oinas, "Atmospheric correction and its application to an analysis of Hyperion data", *IEEE Trans. Geosci. Rem. Sens.*, **41** (2003), 1232–1245.

Cairns, B., F. Waquet, K. Knobelspiesse, J. Chowdhary, and J.-L. Deuzé, "Polarimetric remote sensing of aerosols over land surfaces", in: Alexander A. Kokhanovsky, Gerrit de Leeuw (Eds) *Satellite Aerosol Remote Sensing Over Land*, Springer-Praxis, 2009.

Callies, J., E. Corpaccioli, M. Eisinger, A. Hahne, and A. Lefebvre, "GOME-2 – Metop's second-generation sensor for operational ozone monitoring", *ESA Bulletin-European Space Agency*, **102** (2000), 28–36.

Campbell, J. R., and K. Sassen, "Polar stratospheric clouds at the South Pole from five years of continuous lidar data: Macrophysical, optical and thermodynamic properties", *J. Geophys. Res.*, **113** (2008), D20204, doi: 10.1029/2007JD009680.

Campbell, J. R., and M. Shiobara, "Glaciation of a mixed-phase boundary layer cloud at a coastal Arctic site as depicted in continuous lidar measurements", *Polar Sci.*, **2** (2008), 121–127.

Campbell, J. R., K. Sassen, and E. J. Welton, "Elevated cloud and aerosol layer retrievals from micropulse lidar signal profiles", *J. Atmos. Oceanic Technol.*, **25** (2008a), 685–700.

Campbell, J. R., D. L. Hlavka, E. J. Welton, C. J. Flynn, D. D. Turner, J. D. Spinhirne, V. S. Scott III, and I. H. Hwang, "Full-time, Eye-Safe Cloud and Aerosol Lidar Observation at Atmospheric Radiation Measurement Program Sites: Instrument and Data Processing:, *J. Atmos. Oceanic Technol.*, **19** (2002), 431–442.

Campbell, J. R., E. J. Welton, J. D. Spinhirne, Q. Ji, S.-C. Tsay, S. J. Piketh, M. Barenbrug, and B. N. Holben, "Micropulse Lidar observations of tropospheric aerosols over northeastern South Africa during the ARREX and SAFARI-2000 Dry Season experiments", *J. Geophys. Res.*, (2003), **108**, 8497, doi:10.1029/2002JD002563.

Campbell, J. R., E. J. Welton, and J. D. Spinhirne, "Continous lidar monitoring of polar stratospheric clouds at the South Pole", *Bull. Amer. Meteorol. Soc.,* **90** (2008b), 613–617, doi: 10.1175/2008BAMS2754.1.

Caquineau, S., A. Gaudichet, L. Gomes, and M. Legrand, "Mineralogy of Saharan dust transported over northwestern tropical Atlantic Ocean in relation with sources regions", *J. Geophys. Res.*, **107** (2002), doi: 10.1029/ 2000JD000247.

Carlson, T. N., and S. G. Benjamin, "Radiative heating rates for Saharan dust", *J. Atmos. Sci.*, **37** (1980), 193–213.

Carn, S. A., L. L. Strow, S. de Souza-Machado, Y. Edmonds, and S. Hannon, "Quantifying tropospheric volcanic emissions with AIRS: The 2002 eruption of Mt. Etna (Italy)", *Geophys. Res. Lett.*, **32** (2005), L02301, doi: 10.1029/2004GL021034.

Chahine, M. T., "Determination of temperature profile in an atmosphere from its outgoing radiance", *J. Opt. Soc. Am.*, **12** (1968), 1634–1637.

Chalon G., F. R. Cayla, and D. Diebel, "IASI: An advance sounder for operational meteorology", in: *IAF 2001 Conference Proceedings*, 2001.

Chamaillard, K., S. G. Jennings, C. Kleefeld, D. Ceburnis, and Y. J. Yoon, "Light backscattering and scattering by nonspherical sea-salt aerosols", *J. Quant. Spectrosc. Radiat. Transfer* **79–80** (2003), 577–597.

Chami, M., R. Santer and E. Dilligeard, "Radiative transfer model for the computation of radiance and polarization in an atmosphere-ocean system: Polarization properties of suspended matter for remote sensing", *Appl. Opt.*, **40** (2001), 2398–2416.

Chand, D., R. Wood, T. L. Anderson, S. K. Satheesh, and R. J. Charlson, "Satellite-derived direct radiative effect of aerosols dependent on cloud cover", *Nature Geosci.*, **2** (2009), 181–184.

Chandrasekhar, S, 1950, *Radiative Transfer*, Oxford University Press; republished by Dover Publications, 1960.

Chanin, M. L. C., A. Hauchecorne, C. Malique, D. Nedeljkovic, J. E. Blamont, M. Desbois, G. Tulinov, and V. Melnikov, "First results of the ALISSA lidar on board the MIR platform", *Earth and Planetary Science*, **328:06** (1999), 359–366.

Charlson, R. J., Ackerman, A. S., Bender, F. A.-M., Anderson, T. L., and Liu, Z., "On the climate forcing consequences of the albedo continuum between cloudy and clear air", *Tellus B.*, **59** (2007), 715–727, doi: 10.1111/j.1600-0889.2007.00297.x.

Chaumat, L., C. Standfuss, B. Tournier, R. Armante, and N. A. Scott, "*4A/OP Reference Documentation,* NOV-3049-NT-1178-v4.0, NOVELTIS, LMD/CNRS", *CNES*, 2009.

Chédin, A., N. Scott, C. Wahiche, and P. Moulinier, "The improved initialization inversion method – a high resolution physical method for temperature retrievals from satellites of the TIROS-N series", *J. Clim. Appl. Meteorol.*, **24** (1985), 128–143.

Chédin, A., S. Serrar, A. Hollingsworth, R. Armante, and N. A. Scott, "Detecting annual and seasonal variations of CO2, CO and N2O from a multi-year collocated satellite-radiosonde data-set using the new Rapid Radiance Reconstruction Network (3R-N) model", *J. Quant. Spectrosc. Radiat. Transfer*, **77** (2003), 285–299.

Chédin, A., E. Péquignot, S. Serrar, and N. A. Scott, "Simultaneous determination of continental surface emissivity and temperature from NOAA 10/HIRS observations: Analysis of their seasonal variations", *J. Geophys. Res.*, **109** (2004), D20110, doi: 10.1029/2004JD004886.

Chen, W.-T., R. Kahn, D. Nelson, K. Yau, and J. Seinfeld, "Sensitivity of multi-angle imaging to optical and microphysical properties of biomass burning aerosols", *J. Geophys. Res.*, **113** (2008), D10203, doi: 10.1029/2007JD009414.

Chevallier, F., F. Chéruy, N. A. Scott, and A. Chédin, "A neural network approach for a fast and accurate computation of a longwave radiative budget", *J. Appl. Meteor.*, **37** (1998), 1385–1397.

Chiapello, I., G. Bergametti, L. Gomes, B. Chatenet, F. Dulac, J. Pimenta, and E. S. Soares, "An additional low layer transport of Sahelian and Saharan dust over the North-Eastern Tropical Atlantic", *Geophys. Res. Lett.*, **22** (1995), 3191–3194.

Chin, M., R. Kahn, and S. Schwartz, "Atmospheric aerosol properties and climate impacts", US. Climate Change Science Program. Synthesis and Assessment Product 2.3, 2009.

Chomette, O., M. Legrand, and B. Marticorena, "Determination of the windspeed threshold for the emission of desert dust using satellite remote sensing in the thermal infrared", *J. Geophys. Res.*, **104** (1999), 31,207–31,215.

Chou, C., P. Formenti, M. Maille, P. Ausset, G. Helas, M. Harrison, and S. Osborne, "Size distribution, shape, and composition of mineral dust aerosols collected during the African Monsoon Multidisciplinary Analysis Special Observation Period 0: Dust and Biomass-Burning Experiment field campaign in Niger, January 2006", *J. Geophys. Res.*, **113** (2008), D00C10, doi: 10.1029/2008JD009897.

Chowdhary J., B. Cairns, M. Mishchenko, and L. D. Travis, "Retrieval of aerosol properties over the ocean using multispectral and multiangle photopolarimetric measurements of the research Scanning Polarimeter", *Geophys. Res. Lett.*, **28** (2001), 243–246.

Chowdhary, J., B. Cairns, M. I. Mishchenko, P. V. Hobbs, G. Cota, J. Redemann, K. Rutledge, B. N. Holben, and E. Russell, "Retrieval of aerosol scattering and absorption properties from photo-polarimetric observations over the ocean during the CLAMS experiment", *J. Atmos. Sci.*, **62** (2005), 1093–1117, doi: 10.1175/JAS3389.1.

Chowdhary J., B. Cairns, and L. D. Travis, "Contribution of water-leaving radiance to multiangle, multispectral polarization observations over the open ocean: Bio-optical model results for case 1 waters", *Appl. Opt.*, **45** (2006), 5542–5567.

Chowdhary, J., B. Cairns, F. Waquet, K. Knobelspiesse, M. Ottaviani, J. Redemann, L. Travis, and M. Mishchenko, "Sensitivity of multiangle, multispectral polarimetric remote sensing over open ocean to water-leaving radiance: Analyses of RSP data acquired during the MILAGRO campaign", *Remote Sens. Environ.* (2011 submitted).

Chu, W. P., and M. P. McCormick, "Inversion of stratospheric aerosol and gaseous constituents from spacecraft solar extinction data in the 0.38–1.0 μm wavelength region", *Appl. Opt.*, **18** (1979), 1404–1413.

Chu, W. P., M. P. McCormick, J. Lenoble, C. Brogniez, and P. Pruvost, "SAGE II inversion algorithm", *J. Geophys. Res.*, **94** (1989), 8339–8351.

Chylek, P., V. Ramaswamy, and R. J. Cheng, "Effect of graphitic carbon on the albedo of clouds", *J. Atmos. Sci.*, **41** (1984), 3076–3084.

Claquin, T., M. Schulz, and Y. J. Balkanski, "Modeling the mineralogy of atmospheric dust sources", *J. Geophys. Res.*, **104**, (1999) D18, 22,243–22,256, doi: 10.1029/1999JD900416.

Clarisse, L., D. Hurtmans, A. J. Prata, F. Karagulian, C. Clerbaux, M. de Mazière, and P.-F. Coheur, "Retrieving radius, concentration, optical depth, and mass of different types of aerosols from high-resolution infrared nadir spectra", *Appl. Opt.*, **49** (2010), 3713–3722.

Coddington, O., P. Pilewskie, J. Redemann, S. Platnick, P. B. Russell, S. Schmidt, W. Gore, J. Livingston, G. Wind, and T. Vukicevic, "Examining the impact of aerosols on the retrieval of cloud optical properties from passive remote sensors", *J. Geophys. Res.*, **115** (2010), D10211, doi: 10.1029/2009JD012829.

Cox, C., and W. Munk, "Measurement of the roughness of the sea surface from photographs of the sun's glitter", *J. Opt. Soc. Amer.*, **44** (1954), 838–850.

Crevoisier, C., S. Heilliette, A. Chedin, S. Serrar, R. Armante, and N. A. Scott, "Midtropospheric CO2 concentration retrieval from AIRS observations in the tropics", *Geophys. Res. Lett.*, **31** (2004), L17106.

Dave, J. V., : "Effect of aerosols on the estimation of total ozone in an atmospheric column from the measurements of its ultraviolet radiance", *J. Atmos. Sci.*, **35** (1978) 899–911.

Dave, J. V, and P. Furukawa, "Intensity and polarization of the radiation emerging from an optically thick Rayleigh atmosphere", *J. Opt. Soc. Amer.*, **56** (1966), 394–400.

Dave, J. V., and C. L. Mateer, "A preliminary study on the possibility of estimating total atmospheric ozone from satellite measurements", *J. Atm. Sci.*, **24** (1967), 414–427.

D'Almeida, G. A., P. Koepke, and E. P. Shettle, *Atmospheric Aerosols. Global Climatology and Radiative Characteristics*. A. Deepak Publ., Hampton, Va., USA., 1991.

Deepshikha, S., S. K. Satheesh, and J. Srinivasan, "Dust aerosols over India and adjacent continents retrieved using METEOSAT infrared radiance. Part I: Sources and regional distribution", *Ann. Geophys.*, **24** (2006a), 37–61.

Deepshikha, S., S. K. Satheesh, and J. Srinivasan, "Dust aerosols over India and adjacent continents retrieved using METEOSAT infrared radiance. Part II: Quantification of wind dependence and estimation of radiative forcing", *Ann. Geophys.*, **24** (2006b), 63–79.

de Graaf, M., and P. Stammes, "SCIAMACHY Absorbing Aerosol Index – calibration issues and global results from 2002–2004", *Atmos. Chem. Phys.*, **5** (2005), 2385–2394.

de Graaf, M., P. Stammes, O. Torres, and R. B. A. Koelemeijer, "Absorbing Aerosol Index: Sensitivity analysis, application to GOME and comparison with TOMS", *J. Geophys. Res.*, **110,** (2005), D01201, doi: 10.1029/2004JD005178.

de Graaf, M., L. G. Tilstra, P. Wang, and P. Stammes, "Retrieval of the aerosol direct radiative effect over clouds from spaceborne spectrometry", *J. Geophys. Res.*, **117** (2012), D07207, doi:10.1029/2011JD017160.

de Graaf, M., P. Stammes, and E. A. A. Aben, "Analysis of reflectance spectra of UV-absorbing aerosol scenes measured by SCIAMACHY", *J. Geophys. Res.*, **112** (2007), D02206, doi: 10.1029/2006JD007249.

de Haan, P. Bosma, and J. Hovenier, "The adding method for multiple scattering computations of polarized light", *Astron. Astrophys.*, **183** (1987), 371–391.

de Paepe, B., and S. Dewitte, "Dust aerosol optical depth retrieval over desert surface, using the SEVIRI window channels", *Journal of Atmospheric and Oceanic Technology*, **26** (2009), 704–718.

de Souza-Machado, S. G., L. L. Strow, S. E. Hannon, and H. E. Motteler, "Infrared dust spectral signatures from AIRS", *Geophys. Res. Lett.*, **33** (2006), L03801, doi: 10.1029/2005GL024364.

Deschamps. P. Y., F. M. Bréon, M. Leroy, A. Podaire, A. Bricaud, J. C. Buriez, and G. Seze, "The POLDER mission: Instrument characteristics and scientific objectives", *IEEE Trans. Geosci. Remote Sens.*, **32** (1994), 598–615.

Deuzé, J. L., C. Devaux, M. Herman, R. Santer, and D. Tanré, "Saharan aerosols over the south of France: Characterization derived from satellite data and ground based measurements", *J. Appl. Meteorol.*, **27** (1988), 680–686.

Deuzé, J. L., M. Herman, and R. Santer, "Fourier series expansion of the transfer equation in the ocean-atmosphere system", *J. Quant. Spectrosc. Radiat. Transfer*, **41** (1989), 483–494.

Deuzé, J. L., M. Herman, P. Goloub, D. Tanré, and A. Marchand, "Characterization of aerosols over ocean from POLDER/ADEOS-1", *Geophys. Res. Lett.*, **26:10** (1999), 1421–1424.

Deuzé, J. L., P. Goloub, M. Herman, A. Marchand, G. Perry, S. Susana, and D. Tanré, "Estimate of the aerosol properties over the ocean with POLDER", *J. Geophys. Res.*, **105** (2000), 15,329–15,346.

Deuzé, J. L., F. M. Breon, C. Devaux, P. Goloub, M. Herman, B. Lafrance, F. Maignan, A. Marchand, F. Nadal, G. Perry, and D. Tanré, "Remote sensing of aerosols over land surfaces from POLDER-ADEOS-1 polarized measurements", *J. Geophys. Res.* **106**, D5, (2001), 4913–4926.

Devaux, C., A. Vermeulen, J. L. Deuzé, P. Dubuisson, M. Herman, R. Santer, and M. Verbrugghe, "Retrievalof aerosol single-scattering albedo from ground-based measurements: Application to observational data", *J. Geophys. Res.* **103** (1998), 8753–8761.

Di Girolamo, P., R. Marchese, D. N. Whiteman, and B. B. Demoz, "Rotational Raman lidar measurements of atmospheric temperature in the UV", *Geophys. Res. Lett.*, **31** (2004), L01106, doi: 10.1029/2003GL018342.

Diner, D. J., J. C. Beckert, T. H. Reilly, C. J. Bruegge, J. E. Conel, R. Kahn, J. V. Martonchik, T. P. Ackerman, R. Davies, S. A. W. Gerstl, H. R. Gordon, J.-P. Muller, R. Myneni, R. J. Sellers, B. Pinty, and M. M. Verstraete, "Multiangle Imaging SpectroRadiometer (MISR) description and experiment overview", *IEEE Trans. Geosci. Rem. Sens.*, **36** (1998), 1072–1087.

Diner, D. J., W. Abdou, T. Ackerman, K. Crean, H. R. Gordon, R. A. Kahn, J. V. Martonchik, S. McMuldroch, S. R. Paradise, B. Pinty, M. M. Verstraete, M. Wang, and R. A. West, "MISR level 2 aerosol retrieval algorithm theoretical basis", JPL D-11400, Rev. E, 2001.

Diner, D. J., J. V. Martonchik, R. A. Kahn, B. Pinty, N. Gobron, D. L. Nelson, and B. N. Holben, "Using angular and spectral shape similarity constraints to improve MISR aerosol and surface retrievals over land", *Rem. Sens. Environ.*, **94** (2005), 155–171.

Diner, D. J., W. A. Abdou, T. P. Ackerman, K. Crean, H. R. Gordon, R. A. Kahn, J. V. Martonchik, S. R. Paradise, B. Pinty, M. M. Verstraete, M. Wang, and R. A. West, "Multiangle Imaging SpectroRadiometer Level 2 Aerosol Retrieval Algorithm Theoretical Basis", Revision F. Jet Propulsion Laboratory, California Institute of Technology JPL D-11400, 2006.

Dirksen, R. J., K. F. Boersma, J. de Laat, P. Stammes, G. R. van der Werf, M. V. Martin, and H. M. Kelder, "An aerosol boomerang: Rapid around-the-world transport of smoke from the December 2006 Australian forest fires observed from space", *J. Geophys. Res.*, **114** (2009), D21201, doi: 10.1029/2009JD012360.

Dubovik, O., "Optimization of numerical inversion in photopolarimetric remote sensing", in: G. Videen, Y. Yatskiv, and M. Mishchenko (Eds), *Photopolarimetry in Remote Sensing*, Kluwer Academic Publishers, Dordrecht, The Netherlands, 2004, 65–106.

Dubovik, O., and M. D. King, "A flexible inversion algorithm for retrieval of aerosol optical properties from sun and sky radiance measurements", *J. Geophys. Res.*, **105** (2000), 20673–20696.

Dubovik O., A. Smirnov, B. N. Holben, M. D. King, Y. J. Kaufman, T. F. Eck, and I. Slutsker, "Accuracy assessment of aerosol optical properties retrieval from AERONET sun and sky radiance measurements", *J. Geophys. Res.*, **105** (2000), 9791–9806.

Dubovik, O., B. N. Holben, T. F. Eck, A. Smirnov, Y. J. Kaufman, M. D. King, D. Tanré, and I. Slutsker, "Variability of absorption and optical properties of key aerosol types observed in worldwide locations", *J. Atmos. Sci.*, **59** (2002), 590–608.

Dubovik, O., A. Sinyuk, T. Lapyonok, B. N. Holben, M. Mishchenko, P. Yang, T. F. Eck, H. Volten, O. Munoz, B. Veihelmann, W. J. van der Zende, J. F. Léon, M. Sorokin, and I. Slutsker, "Application of spheroid models to account for aerosol particle nonsphericity in remote sensing of desert dust", *J. Geophys. Res.*, **111** (2006), D11208, doi: 10.1029/2005JD006619.

Dubovik, O., T. Lapyonok, Y. J. Kaufman, M. Chin, P. Ginoux, R. A. Kahn, and A. Sinyuk, "Retrieving global aerosol sources from satellites using inverse modeling", *Atmos. Chem. Phys.*, **8** (2008), 209–250.

Dubovik, O., M. Herman, A. Holdak, T. Lapyonok, D. Tanré, J. L. Deuzé, F. Ducos, A. Sinyuk, and A. Lopatin, "Statistically optimized inversion algorithm for enhanced retrieval of aerosol properties from spectral multi-angle polarimetric satellite observations", *Atmos. Meas. Tech.*, **4** (2011), 975–1018, doi: 10.5194/amt-4-975-2011.

Dubuisson, P., J. C. Buriez, and Y. Fouquart, "High spectral resolution solar radiative transfer in absorbing and scattering media: Application to the satellite simulation", *J. Quant. Spectrosc. Radiat. Transfer*, **55** (1996), 103–126, doi: 10.1016/0022-4073(95)00134-4.

Duncan, B. N., R. V. Martin, A. C. Staudt, R. Yevich, and J. A. Logan, "Interannual and seasonal variability of biomass burning emissions constrained by satellite observations", *J. Geophys. Res.* **108,** (2003) D2,, 4100, doi: 10.1029/2002JD002378.

Dunkelman, L., and R. Scolnik, "Solar spectral irradiance and vertical atmospheric attenuation in the visible and ultraviolet", *J. Opt. Soc. Amer.*, **49:4** (1959).

Echle, G., T. von Clarmann, and H. Oelhaf, "Optical and microphysical parameters of the Mt. Pinatubo aerosol as determined from MIPAS-B mid-IR limb emission spectra", *J. Geophys. Res.*, **103,** (1998) D15, 19,193–19,211, doi: 10.1029/98JD01363.

Eck, T. F., B. N. Holben, J. S. Reid, O. Dubovik, A. Smirnov, N. T. O'Neill, I. Slutsker, and S. Kinne, "Wavelength dependence of the optical depth of biomass burning, urban and desert dust aerosols", *J. Geophys. Res.*, **104** (1999), 31,333–31,350.

Eck, T. F., B. N. Holben, A. Sinyuk, R. T. Pinker, P. Goloub, H. Chen, B. Chatenet, Z. Li, R. P. Singh, S. N. Tripathi, J. S. Reid, D. M. Giles, O. Dubovik, N. T. O'Neill, A. Smirnov, P. Wang and X. Xia, "Climatological aspects of the optical properties of fine/coarse mode aerosol mixtures", *J. Geophys. Res.*, **115** (2010), D19205, doi:10.1029/2010JD014002.

Edie, W. T., D. Dryard, F. E. James, M. Roos, and B. Sadoulet, *Statistical Methods in Experimental Physics*, North-Holland Publishing Company, Amsterdam, 1971.

Edner, H., G. Faris, A. Sunesson, and S. Svanberg, "Atmopheric atomic mercury monitoring using differential absorption lidar techniques", *Appl. Opt.*, **28:5** (1989), 921–930.

Edner, H., K. Fredriksson, A. Sunesson, and W. Wendt, "Monitoring Cl2 using a differential absorption lidar system", *Appl. Opt.*, **26:16** (1987), 3183–3185.

Egeback, A. L., K. A. Fredriksson, and H. M. Hertz, "Dial techniques for the control of sulfur dioxide emissions", *Appl Opt.*, **23:5** (1984), 722–729.

Eloranta, E. E., "Lidar: Range-resolved optical remote sensing of the atmosphere", ch. 5 in *High Spectral Resolution Lidar*, Springer, 2005.

Eloranta, E. W., I. A. Razenkov, J. P. Garcia, and J. Hedrick,"Observations with the University of Wisconsin Arctic high spectral resolution lidar", in: *22nd International Laser Radar Conference*, Matera, Italy, 2006.

Emili, E., A. Lyapustin, Y. Wang, C. Popp, S. Korkin, M. Zebisch, S. Wunderle, and M. Petitta, "High spatial resolution aerosol retrieval with MAIAC: Application to mountain regions", *J. Geophys. Res.,* **116** (2011), D23211, doi:10.1029/2011JD016297.

Esselborn, M., M. Wirth, A. Fix, M. Tesche, and G. Ehret, "Airborne high spectral resolution lidar for measuring aerosol extinction and backscatter coefficients", *Appl. Opt.*, **47**(3) (2008), 346–358.

Esselborn, M., M. Wirth, A. Fix., B. Weinzierl, K. Rasp, M. Tesche, and A. Petzold, "Spatial distribution and optical properties of Saharan dust observed by airborne high spectral resolution lidar during SAMUM 2006". *Tellus Series B-Chemical and Physical Meteorology* (2009), **61**(1), 131–143.

EUMETSAT, 2009, IASI Level 2 Product Guide, EUM/OPS-EPS/MAN/04/0033, available at http://oiswww.eumetsat.org/WEBOPS/eps-pg/IASI-L2/IASIL2-PG-0TOC.htm

Fernald, F. G., "Analysis of atmospheric lidar observations: Some comments", *Appl. Opt.*, **23**:5 (1984), 652–653.

Fernald, F. G., B. M. Herman, and J. A. Reagan, "Determination of aerosol height distributions by lidar", *J. Appl. Meteor.*, **11** (1972), 482–489.

Fiocco, G., and G. Grams, "Observations of the aerosol layer at 20 km by optical radar", *J. Atmos. Sci.*, **21** (1964), 323–324.

Fiocco G., and L. D. Smullin, "Detection of scattering layers in the upper atmosphere (60-140 km) by optical radar", *Nature*, **199** (1963), 1275–1276.

Fiocco, G., G. Benedetti-Michelangeli, K. Maischberger, and E. Madonna, "Measurement of temperature and aerosol-to-molecule ratio in the troposphere by optical radar", *Nature*, **229** (1971), 78–79.

Fischer, J., and H. Grassl, "Radiative transfer in an atmosphere-ocean system: An azimuthal dependent matrix operator approach", *Appl. Optics*, **23** (1984), 1023–1039.

Fischer, J., and H. Grassl, "Detection of cloud-top height from backscattered radiances within the oxygen A band. Part 1: Theoretical study", *J. Appl. Meteorol.*, **30** (1991), 1245–1259.

Fischer, K., "The optical constants of atmospheric aerosol particles in the 7.5–10 μm spectral region", *Tellus*, **28** (1976), 266–274.

Flowerdew, R. J., and J. D. Haigh, "Retrieval of aerosol optical thickness over land using the ATSR-2 dual-look satellite radiometer", *Geophys. Res. Lett.*, **23** (1996), 351–354.

Flowers, E. C., and H. J. Viebrock, "Solar radation: An anomalous decrease of direct solar radiation", *Science*, April 23, **148** (1965).

Flowers, E. C., R. A. McCormick, and K. R. Kurfis, "Atmospheric turbidity over the United States, 1961–1966", *J. Appl. Meteor.*, **8** (1969), 955–962.

Fouquart, Y., B. Bonnel, G. Brogniez, J. Buriez, L. Smith, J.-J. Morcrette, and A. Cerf, "Observations of Saharan aerosols: Results of ECLATS field experiment. Part II: Broadband radiative characteristics of the aerosols and vertical radiative flux divergence", *J.Climate Appl. Meteor.*, **26** (1987), 38–52.

Frangi, J.-P., A. Druilhet, P. Durand, H. Ide, J.-P. Pages, and A. Tinga, "Energy budget of the Sahelian surface layer", *Ann. Geophys.*, **10** (1992), 25–33.

Fraser, R. S., "Satellite measurement of mass of Sahara dust in the atmosphere", *Appl. Opt.*, **15** (1976), 2471–2479.

Fraser, R. S., and Y. J. Kaufman, "The relative importance of aerosol scattering and absorption in remote sensing", *IEEE Trans. Geosci. Rem. Sens.*, **23** (1985), 625–633, doi: 10.1109/TGRS.1985.289380.

Fraser, R. S., Y. J. Kaufman, and R. L. Mahoney, "Satellite measurements of aerosol mass transport", *Atmos. Environ.*, **18** (1984), 2577–2584.

Fraser, R. S., S. Mattoo, E.-N. Yeh, and C. R. McClain, "Algorithm for atmospheric and glint corrections of satellite measurements of ocean pigment", *J. Geophys. Res.*, **102**, D14 (1997), 17, 107–117, 118.

Fredriksson, K. A., and H. M. Hertz, "Evaluation of the DIAL technique for studies on no2 using a mobile lidar system", *Appl. Opt.*, **23: 9** (1984), 1403–1411.

Fritzsche, K., and G. Schubert, "Simultaneous measurement of different pollutants in leipzig using LIDAR", *Laser und Optoelektronik*, **29:5** (1997), 56–61.

Fröhlich, C., *WMO/PMOD sunphotometer: Instructions for manufacture*, WMO Report, Geneva, Switzerland, 1977.

Gangale, G., A. Prata, and L. Clarisse, "On the infrared spectral signature of volcanic ash", *Rem. Sens. Environ.*, **114(2)** (2010), 414–425, doi: 101016/j.rse.2009.09.007, 2010.

Gao, B.-C., and A. F. H. Goetz, "Column atmospheric water vapor and vegetation liquid water retrievals from airborne imaging spectrometer data", *J. Geophys. Res.*, **95** (1990), 3549–3564.

Gao, B.-C., and Y. J. Kaufman, "Selection of the 1.375-μm MODIS channel for remote sensing of cirrus clouds and stratospheric aerosols from space", *J. Atmos. Sci.*, **52** (1992), 4231–237.

Gatebe, C. K., O. Dubovik, M. D. King, and A. Sinyuk, "Simultaneous retrieval of aerosol and surface optical properties from combined airborne- and ground-based direct and diffuse radiometric measurements", *Atmos. Chem. Phys.*, **10** (2010), 2777–2794.

Gel'fand, I. M., and Z. Y. Shapiro, "Representation of the group of rotations of 3-dimensional space and their applications", *Amer. Math. Transl.*, **2** (1956), 207–316.

Geogdzhayev, I. V., M. I. Mishchenko, W. B. Rossow, B. Cairns, and A. A. Lacis, "Global two-channel AVHRR retrievals of aerosol properties over the oceans for the period of NOAA-9 observations and preliminary retrievals using NOAA-7 and NOAA-11 data", *J. Atmos. Sci.*, **59** (2002), 262–278.

Geogdzhayev, I. V., M. I. Mishchenko, L. Liu, and L. Remer, "Global two-channel AVHRR aerosol climatology: Effects of stratospheric aerosols and preliminary comparisons with

MODIS and MISR retrievals", *J. Quant. Spectrosc. Radiat. Transfer*, **88** (2004), 47–59, doi: 10.1016/j.jqsrt.2004.03.024.

Gimmestad, G. G., "Lidar: Range-resolved optical remote sensing of the atmosphere", in: *Differential-Absorption Lidar for Ozone and Industrial Emissions*, Springer, 2005.

Ginoux, Paul, M. Chin, I. Tegen, J. M. Prospero, B. Holben, O. Dubovik, and Shian-Jiann Lin, "Sources and distributions of dust aerosols simulated with the GOCART model", *J. Geophys. Res.*, **106,** D17 (2001), 20,255–20,273.

Glaccum, W., R. L. Lucke, R. M. Bevilacqua, E. P. Shettle, J. S. Hornstein, D. T. Chen, J. D. Lumpe, S. S. Krigman, D. Debrestian, M. D. Fromm, F. Dalaudier, E. Chassefiere, C. Deniel, C. E. Randall, D. W. Rusch, J. J. Olivero, C. Brogniez, J. Lenoble, and R. Kremer, "The Polar Ozone and Aerosol Measurement instrument", *J. Geophys. Res.*, **101**, D9 (1996), 14479–14487.

Gleason, J., N. C. Hsu, and O. Torres, "Biomass burning smoke measured using backscattered ultraviolet radiation: SCAR-B and Brazilian smoke interannual variability", *J. Geophys Res.*, **103,** D24 (1998), 31,969–31,978, doi: 10:1029/98JD00160.

Goddard Space Flight Center, Formation Flying, "The afternoon 'A-Train' satellite constellation, FS-2003-1-053-GFSC", February 2003.

Goldsmith, J. E. M., F. H. Blair, S. E. Bisson, and D. D. Turner, "Turn-key Raman lidar for profiling atmospheric water vapor, clouds, and aerosols", *Appl. Opt.*, **37**(21) (1998), 4979–4990.

Goloub, P., J.-L. Deuzé, M. Herman, and Y. Fouquart, "Analysis of the POLDER polarization measurements performed over cloud covers", *IEEE Trans. Geosci. Rem. Sens.*, **32** (1994), 78–88.

Gonzalez, C. R., J. P. Veefkind, and G. de Leeuw, "Aerosol optical depth over Europe in August 1997 derived from ATSR-2 data", *Geophys. Res. Lett.*, **27:**7 (2000), 955–958.

Goody, R. M., and Y. L. Young, *Atmospheric Radiation (Theoretical Basis)*, 2nd ed., Oxford University Press, 1989.

Gordon, H. R., and A. Y. Morel, *Remote Assessment of Ocean Color for Interpretation of Satellite Visible Imagery: A Review,* Springer-Verlag, New York, 1983.

Gordon, H. R., and M. Wang, "Retrieval of water-leaving radiance and aerosol optical thickness over the oceans with Sea-WiFS: A preliminary algorithm", *Appl. Opt.*, **33** (1994), 443–452.

Govaerts, Y. M., S. Wagner, A. Lattanzio, and P. Watts, "Joint retrieval of surface reflectance and aerosol optical depth from MSG/SEVIRI observations with an optimal estimation approach: 1. Theory", *J. Geophys. Res.*, **115**, (2010) D02203, doi:10.1029/2009JD011779.

Greenblatt, G. D., J. J. Orlando, J. B. Burkholder, and A. R. Ravishankara, "Absorption measurements of oxygen between 330 nm and 1140 nm", *J. Geophys. Res.*, **95** (1990), 18,577–18,582.

Grey, W. M. F., P. R. J. North, S. O. Los, and R. M. Michell, "Aerosol optical depth and land surface reflectance from Multiangle AATSR measurements: Global validation and intersensor comparisons", *IEEE Trans. Geosci. Rem. Sens.*, **44:8** (2006), 2184–2197.

Griggs, M., "Measurements of atmospheric aerosol optical thickness over water using ERTS-1 data", *J. Air Pollut. Control. Assoc.*, **25** (1975), 622–626.

Groß, S., V. Freudenthaler, M. Wiegner, J. Gasteiger, A. Geiß, and F. Schnell, "Dual-wavelength linear depolarization ratio of volcanic aerosols: Lidar measurements of the Eyjafjallajökull plume over Maisach, Germany", *Atmos. Environ.* **48** (2012), 85–96.

Grund, C. J., and E. E. Eloranta, "University of Wisconsin High Spectral Resolution Lidar", *Opt. Eng.*, **30:1** (1991), 6–12.

Gu, Y. Y. Y., C. S. Gardner, P. A. Castleberg, G. C. Papen and M. C. Kelley., "Validation of the lidar in-space technology experiment: Stratospheric temperature and aerosol measurements", *Appl. Opt.*, **36:21** (1997), 5148–5157, doi: 10.1364/AO.36.005148.

Guerrero-Rascado, J. L., M. Sicard, F. Molero, F. Navas-Guzmán, J. Preißler, D. Kumar, J. A. Bravo-Aranda, S. Tomás, M. N. Reba, L. Alados-Arboledas, A. Comerón, M. Pujadas, F. Rocadenbosch, F. Wagner, and A. M. Silva, "Monitoring of the Eyjafjallajökull ash plume at four lidar stations over the Iberian Peninsula: 6 to 8 May 2010", *Cuarta Reunión Española de Ciencia y Tecnología de Aerosoles* (2010), C15 1–7.

Gustafson, B.Å.S., "Scaled analogue experiments in electromagnetic scattering", *Light Scattering Rev.*, **4** (2009), 3–30.

Hair, J. W., C. A. Hostetler, A. L. Cook, D. B. Harper, R. A. Ferrare, T. L. Mack, W. Welch, L. R. Izquierdo, and F. E. Hovis, "Airborne high spectral resolution lidar for profiling aerosol optical properties", *Appl. Opt.*, **47** (2008), 6734–6752.

Hall, F. G., P. J. Sellers, D. E. Strebel, E. T. Kanemasu, R. D. Kelly, B. L. Blad, B. J. Markham, J. R. Wang, and F. Huemmrich, "Satellite remote-sensing of surface-energy and mass balance results from FIFE", *Rem. Sens. Environ.*, **35** (1991), 187–199, doi: 10.1016/0034-4257.

Hamill, P., and C. Brogniez, "Stratospheric aerosol record and climatology", in: L. Thomason and Th. Peter (Eds), *SPARC Assessment of Stratospheric Aerosol Properties, Stratospheric Aerosol Record and Climatology*, WCRP 124, WMO/TD No. 1295, SPARC Report No. 4, 109–175, 2006.

Hamill, P., C. Brogniez, L. Thomason and T. Deshler, "Instrument descriptions", in: L. Thomason and Th. Peter (Eds), *SPARC Assessment of Stratospheric Aerosol Properties*, WCRP 124, WMO/TD No. 1295, SPARC Report No. 4, 109–175, 2006.

Hammad, A., and S. Chapman, "The primary and secondary scattering of sunlight in a plane stratified atmosphere of uniform composition", *Phil. Mag.*, **28** (1939), 99.

Hanel, R. A., B. J. Conrath, D. E. Jennings, and R. E. Samuelson, *Exploration of the Solar System by Infrared Remote Sensing*, 2nd edition, Cambridge University Press, 2003.

Hansell, R. A., S. C. Ou, K. N. Liou, J. K. Roskovensky, S. C. Tsay, C. Hsu, and Q. Ji, "Simultaneous detection/separation of mineral dust and cirrus clouds using MODIS thermal infrared window data", *Geophys. Res. Lett.*, **34** (2007), L11808, doi: 10.1029/2007GL029388.

Hansell, R. A., K. N. Liou, S. C. Ou, S. C. Tsay, Q. Ji, and J. S. Reid, "Remote sensing of mineral dust aerosol using AERI during the UAE: A modeling and sensitivity study", *J. Geophys. Res.*, **113** (2008), D18202, doi: 10.1029/2008JD010246.

Hansen, J. E., and L. D. Travis, "Light scattering in planetary atmospheres", *Space Sci. Rev.*, **16** (1974), 527–610, doi: 10.1007/BF00168069.

Hansen, J. E., and J. W. Hovenier, "Interpretation of the polarization of Venus", *J. Atmos. Sci.*, **31** (1974), 1137–1160.

Hasekamp, O. P., and J. Landgraf, "Retrieval of aerosol properties over the ocean from multispectral single-viewing-angle measurements of intensity and polarization: Retrieval approach, information content, and sensitivity study", *J. Geophys. Res.*, **110** (2005a), D20207, doi: 10.1029/2005JD006212.

Hasekamp, O. P., and J. Landgraf, "Linearization of vector radiative transfer with respect to aerosol properties and its use in satellite remote sensing", *J. Geophys. Res.*, **110** (2005b), D04203.

Hasekamp, O., P. Litvinov, and A. Butz, "Aerosol properties over the ocean from PARASOL multi-angle photopolarimetric measurements", *J. Geophys. Res.*, **116** (2011), doi: 10.1029/ 2010JD015469.

Hauglustaine, D., F. Hourdin, L. Jourdain, M. Filiberti, S. Walters, J. Lamarque, and E. Holland, "Interactive chemistry in the Laboratoire de Météorologie Dynamique general circulation model: Description and background tropospheric chemistry evaluation", *J. Geophys. Res.*, **109** (2004), D04314.

Heilliette, S., C. Pierangelo, C. Crévoisier, and A. Chédin, "A fast solver to compute atmospheric radiative transfer in a scattering atmosphere using the successive order of scattering method", *Technical Note from Laboratoire de Météorologie Dynamique*, 2004.

Heintzenberg, J., "Particle-size distribution and optical-properties of Arctic haze", *Tellus*, **32** (1980), 251–260.

Henyey, L. C., and J. L. Greenstein, "Diffuse radiation in the galaxy", *Astrophys. J.*, **93** (1941), 70–83.

Herber, A., L. W. Thomason, V. F. Radionov, and U. Leiterer, "Comparison of trends in the tropospheric and stratospheric aerosol optical depths in the Antarctic", *J. Geophys Res.*, **98** (1993), 18,441–18,447.

Herman, J. R., and E. A. Celarier, "Earth surface reflectivity climatology at 340 nm to 380 nm from TOMS data", *J. Geophys. Res.*, **102** (1997), 28,003–28,011.

Herman, J. R., P. K. Bhartia, O. Torres, C. Hsu, C. Seftor, and E. Celarier, "Global distribution of UV-absorbing aerosols from Nimbus 7/TOMS data", *J. Geophys. Res.*, **102** (1997), 16,911–16,922.

Herman, M., J. L. Deuzé, C. Devaux, P. H. Goloub, F. M. Bréon, and D. Tanré, "Remote sensing of aerosols over land surfaces, including polarization measurements: Application to some airborne POLDER measurements", *J. Geophys. Res.*, **102** (1997), 17,039–17,049.

Herman, M., J. L. Deuzé, A. Marchand, B. Roger, and P. Lallart, "Aerosol remote sensing from POLDER/ADEOS over the ocean: Improved retrieval using a nonspherical particle model", *J. Geophys. Res.*, **110** (2005), D10S02, doi: 10.1029/2004JD004798.

Hervig, M. E., J. M. Russell III, L. L. Gordley, J. Daniels, S. R. Drayson, and J. H. Park, "Aerosol effects and corrections in the Halogen Occultation Experiment", *J. Geophys. Res.*, **100** (1995), 1067–1079.

Hervig, M. E., T. Deshler, and J. M. Russell, "Aerosol size distributions obtained from HALOE spectral extinction measurements", *J. Geophys. Res.*, **103**, (1998) D21, 1573–1583, doi: 10.1029/97JD03081.

Hess, M., P. Koepke, and I. Shult, "Optical properties of aerosols and clouds: The software package OPAC", *Bull. Amer. Meteor. Soc.*, **79** (1998), 831–844.

Highwood, E., J. M. Haywood, M. D. Silverstone, S. M. Newman, and J. P. Taylor, "Radiative properties and direct effect of Saharan dust measured by the C-130 aircraft during Saharan Dust Experiment (SHADE), 2. Terrestrial spectrum", *J. Geophys. Res.*, **108** (2003), 8578, doi: 10.1029/2002JD002 552.

Higurashi, A., and T. Nakajima, "Development of a two channel aerosol retrieval algorithm on global scale using NOAA AVHRR", *J. Atmos. Sci.*, **56** (1999), 924–941.

Hoff, R. M., and G. Pappalardo, "The GAW Aerosol Lidar Observation Network (GALION) as a source of near-real time aerosol profile data for model evaluation and assimilation", AGU Fall Meeting, 2010, G8+.

Holben, Brent N., "Characteristics of maximum-value composite images from temporal AVHRR data", *Int. J. Rem. Sens.*, **7** (1986). 1417–1434.

Holben, B. N. and Y. E. Shimabukuro, "Linear mixing model applied to coarse spatial-resolution multispectral satellite sensors", *Int. J. Rem. Sens.*, 14 (1993), 2231–2240.

Holben, B. N., Y. J. Kaufman, and J. D. Kendall, "NOAA-11 AVHRR visible and Near-IR bands inflight calibration", *Int. J. Rem. Sens.*, **11** (1990), 1511–1519.

Holben, B. N., T. F. Eck, and R.S. Fraser, "Temporal and spatial variability of aerosol optical depth in the Sahel region in relation to vegetation remote-sensing", *Int. J. Rem. Sens.*, **12** (1991), 1147–1163.

Holben, B. N., A. Setzer, T. F. Eck, A. Pereira, and I. Slutsker, "Effect of dry-season bio-mass burning on Amazon basin aerosol concentrations and optical properties, 1992–1994", *J. Geophys. Res.*, **101**, D14 (1996), 19,465–19, 481.

Holben, B. N., T. F. Eck, I. Slutsker, D. Tanré, J. P. Buis, A. Setzer, E. Vermote, J. A. Reagan, Y. J. Kaufman, T. Nakajima, F. Lavenu, I. Jankowiak, and A. Smirnov, "AERONET – A federated instrument network and data archive for aerosol characterization", *Remote Sens. Environ.*, **66** (1998), 1–16.

Holben, B. N., D. Tanré, A. Smirnov, T. F. Eck, I. Slutsker, B. Chatenet, F. Lavenu, Y. J. Kaufman, J. Van de Castle, A. Setzer, B. Markham, D. Clark, R. Frouin, N. A. Karneli. N. O'Neill, C. Pietras, R. Pinker, K. Voss, and G. Zibordi, "An emerging ground-based aerosol climatology: Aerosol optical depth from AERONET ", *J. Geophys. Res.*, **106** (2001), 12,067–12,097.

Holben, B. N., E. Vermote, Y. J. Kaufman, D. Tame, and V. Kalb, "Aerosol retrieval over land from AVHRR data-application for atmospheric correction." *IEEE Trans. Geosci. Rem. Sens.*, **30:2** (2002), 212–222.

Hollweg, H.-D., S. Bakan, and J. P. Taylor, "Is the aerosol emission detectable in the thermal infrared?", *J. Geophys. Res.*, **111** (2006), D15202, doi: 10.1029/2005JD006432.

Holzer-Popp, T., M. Schroedter, and G. Gesell, "Retrieving aerosol optical depth and type in the boundary layer over land and ocean from simultaneous GOME spectrometer and ATSR-2 radiometer measurements. 1: Method description", *J. Geophys. Res.* **107**, (2002) D21, 4578, doi: 10.1029/2001JD002013.

Hong, G., P. Yang, H.-L. Huang, S. A. Ackerman, and I. N. Sokolik, "Simulation of high-spectral-resolution infrared signature of overlapping cirrus clouds and mineral dust", *Geophys. Res. Lett.*, 33 (2006), L04805, doi: 10.1029/2005GL024381.

Horvath, H. (Editor), "Light scattering: Mie and more – commemorating 100 years of Mie's 1908 publication", *J. Quant. Spectrosc. Radiat. Transfer*, **110** (2009), 783–949.

Hough, J. (Editor), "XI conference on electromagnetic and light scattering by non-spherical particles: 2008", *J. Quant. Spectrosc. Radiat. Transfer*, **110** (2009), 1207–1779.

Hovenier, J.W. (Editor), "Light scattering by non-spherical particles", *J. Quant. Spectrosc. Radiat. Transfer*, **55** (1996), 535–694.

Hovenier, J.W., "Measuring scattering matrices of small particles at optical wavelengths", in: M. I. Mishchenko, J. W. Hovenier, and L. D. Travis (Eds) *Light Scattering by Nonspherical Particles: Theory, Measurements, and Applications*: pp. 355–365, Academic Press, San Diego, 2000.

Hovenier, J. W., and C. V. M. van der Mee, "Basic relationships for matrices describing scattering by small particles", in: M. I. Mishchenko, J. W. Hovenier, and L. D. Travis (Eds) *Light Scattering by Nonspherical Particles: Theory, Measurements, and Applications*, pp. 61–85, Academic Press, San Diego, 2000.

Hovenier, J. W., C. van der Mee, and H. Domke, *Transfer of Polarized Light in Planetary Atmospheres – Basic Concepts and Practical Methods*, Kluwer Academic Publishers, Dordrecht, The Netherlands, 2004.

Hsu, N. C., J. R. Herman, P. K. Bhartia, C. J. Seftor, O. Torres, A. M. Thompson, J. F. Gleason, T. F. Eck, and B. N. Holben, "Detection of biomass burning smoke from TOMS measurements", *Geophys. Res. Lett.*, **23:7** (1996), 745–748.

Hsu, N. C., J. R. Herman, J. Gleason, O. Torres, and C. J. Seftor, "Satellite detection of smoke aerosols over a snow/ice surface by TOMS", *Geophys. Res. Lett.*, **26** (1999), 1165–1168.

Hsu, N. C., S. C. Tsay, M. D. King, and J. R. Herman, "Aerosol properties over bright-reflecting source regions", *IEEE Trans. Geosci. Rem. Sens.*, **42** (2004), 557–569.

Hsu, N. C., S. C. Tsay, M. D. King, and J. R. Herman, "Deep blue retrievals of Asian aerosol properties during ACE-Asia", *IEEE Trans. Geosci. Rem. Sens.*, **44** (2006), 3180–3195.

Hu, X. Q., N. M. Lu, T. Niu, and P. Zhang, "Operational retrieval of Asian sand and dust storm from FY-2C geostationary meteorological satellite and its application to real time forecast in Asia", *Atmos. Chem. Phys.*, **8** (2008), 1649–1659.

Hudson, P. K., E. R. Gibson, M. A. Young, P. D. Kleiber, and V. H. Grassian, "A newly designed and constructed instrument for coupled infrared extinction and size distribution measurements of aerosols", *Aerosol Science and Technology*, **41:7** (2007), 701–710.

Hudson, P. K., E. R. Gibson, M. A. Young, P. D. Kleiber, and V. H. Grassian, "Coupled infrared extinction and size distribution measurements for several clay components of mineral dust aerosol", *J. Geophys. Res.*, **113** (2008), D01201, doi:10.1029/2007JD008791.

Husar R. B., L. L. Stowe, and J. M. Prospero, "Satellite sensing of tropospheric aerosols over the oceans with NOAA AVHRR", *J. Geophys. Res.*, **102** (1997), 16,889–16,909.

Husar, R. B., D. M Tratt, B. A. Schichtel, S. R. Falke, F. Li, D. Jaffe, S. Gasso, T. Gill, N. S. Laulainen, F. Lu, M. C. Reheis, Y. Chun, D. Westphal, B. N. Holben, C. Gueymard, I. McKendry, N. Kuring, G. C. Feldman, C. McClain, R.J. Frouin, J. Merrill, D. DuBois, F. Vignola, T. Murayama, S. Nickovic, W. E. Wilson, K. Sassen, N. Sugimoto, and W. C. Malm, "Asian dust events of April 1998", *J. Geophys. Res.*, **106** (2001), 18,317–18,330, doi:10.1029/2000JD900788

Ichoku, C., R. Levy, Y. J. Kaufman, L. A. Remer, R.-R. Li, V. J. Martins, B. N. Holben, N. Abuhasssan, I. Slutsker, T. F. Eck, and C. Pietras, "Analysis of the performance characteristics of the five-channel Microtops II Sun photometer for measuring aerosol optical thickness and precipitable water vapor", *J. Geophys. Res.*, **107**, D13 (2002a), 4179, doi:10.1029/2001JD001302.

Ichoku, C., D. A. Chu, S. Mattoo, Y. J. Kaufman, L. A. Remer, D. Tanré, I. Slutsker and B. N. Holben, "A spatio-temporal approach for global validation and analysis of MODIS aerosol products", *Geophys. Res. Lett.*, **29** (2002b), 8006, 10.1029/2001GL013206.

Ignatov, A., and L. Stowe, "Physical basis, premises, and self-consistency checks of aerosol retrievals from TRMM VIRS", *J. Appl. Meteor.*, **39** (2000), 2259–2277.

Ignatov, A., L. Stowe, S. M. Sakerin, and G. K. Korotaev, "Validation of the NOAA/NESDIS satellite aerosol product over the North Atlantic in 1989", *J. Geophys. Res.*, **100,** D3 (1995a), 5123–5132.

Ignatov, A., L. Stowe, R. Singh, D. Kabanov, and I. Dergileva, "Validation of NOAA AVHRR aerosol retrievals using sun-photometer measurements from RV Akademik Vernadsky in 1991", *Adv. Space Res.*, **16:10** (1995b), 95–98.

Ingold, T., C. Mätzler, N. Kämpfer, and A. Heimo, "Aerosol optical depth measurements by means of a sun photometer network in Switzerland", *J. Geophys. Res.*, **106,** D21 (2001), 27,537–27,554.

IPCC (Intergovernmental Panel on Climate Change), *Changes in Atmospheric Constituents and in Radiative Forcing in Climate Change*, Cambridge University Press, New York, 2007.

Jacobowitz, H., L. Stowe, G. Ohring, A. Heidinger, K. Knapp, and N. R. Nalli, "The Advanced Very High Resolution Radiometer Pathfinder Atmosphere (PATMOS) climate data set: A resource for climate research", *Bull. Amer. Meteor. Soc.*, **84** (2003), 785–793.

Jacquinet-Husson, N., N. A. Scott, A. Chédin, L. Crépeau, R. Armante, V. Capelle, J. Orphal, A. Coustenis, C. Boone, N. Poulet-Crovisier, A. Barbee, M. Birk, L. R. Brown, C. Camy-Peyret, C. Claveau, K. Chance, N. Christidis, C. Clerbaux, P. F. Chœur, V. Dana, L. Daumont, M. R. De Backer-Barilly, G. Di Lonardo, J. M. Flaud, A. Goldman, A. Hamdouni, M. Hess, M. D. Hurley, D. Jacquemart, I. Kleiner, P. Kopke, J. Y. Mandin, S. Massie, S. Mikhailenko, V. Nemtchinov, A. Nikitin, D. Newnham, A. Perrin, V. I. Perevalov, S. Pinnock, L. Regalia-Jarlot, C P. Rinsland, A. Rublev, F. Schreier, L. Schult, K. M. Smith, S. A. Tashkun, J. L. Teffo, R. A. Toth, V. G. Tyuterev, J. V. Auwera, P. Varanasi, and G. Wagner, "The GEISA spectroscopic database: Current and future archive for Earth and planetary atmosphere studies." *J. Quant. Spectrosc. Rad. Transf.*, **109** (2008), 1043–1059.

Jaffe, D. A., T. Anderson, D. Covert, R. Kotchenruther, B. Trost, J. Danielson, W. Simpson, T. Berntsen, S. Karlsdottir, D. Blake, J. Harris, G. Carmichael, and I. Uno, "Transport of Asian air pollution to North America", *Geophys. Res. Lett.*, **26** (1999), 711–714.

James, F. *Statistical Methods in Experimental Physics*, 2nd Edition, World Scientific, New Jersey, 2006.

Jethva, H., and Torres, O., "Satellite-based evidence of wavelength-dependent aerosol absorption in biomass burning smoke inferred from Ozone Monitoring Instrument", *Atmos. Chem. Phys.*, **11** (2011), 10,541–10,551, doi: 10.5194/acp-11-10541-2011.

Jordan, C. F., "Derivation of leaf area index from quality measurements of light on the forest floor", *Ecology*, **50** (1969), 663–666.

Kahn, R., P. Banerjee, and D. McDonald, "The sensitivity of multiangle imaging to natural mixtures of aerosols over ocean", *J. Geophys. Res.*, **106** (2001),18,219–18,238.

Kahn, R., B. Gaitley, J. Martonchik, D. Diner, K. Crean, and B. Holben, "MISR global aerosol optical depth validation based on two years of coincident AERONET observations." *J. Geophys. Res.*, **110** (2005), doi: 10:1029/2004JD004706.

Kahn, R. A., W.-H. Li, C. Moroney, D. J. Diner, J. V. Martonchik, and E. Fishbein, "Aerosol source plume physical characteristics from space-based multiangle imaging", *J. Geophys. Res.*, **112** (2007), D11205, doi: 10.1029/2006JD007647.

Kahn, R., Y. Chen, D. L. Nelson, F.-Y. Leung, Q. Li, D. J. Diner, and J. A. Logan, "Wildfire smoke injection heights – Two perspectives from space", *Geophys. Res. Lett.*, **35** (2008), doi: 10.1029/2007GL032165.

Kahn, R. A., D. L. Nelson, M. Garay, R. C. Levy, M. A. Bull, D. J. Diner, J. V. Martonchik, S. R. Paradise, E. G. Hansen, and L. A. Remer, "MISR Aerosol product attributes, and statistical comparisons with MODIS", *IEEE Trans. Geosci. Rem. Sens.*, **47** (2009a), 4095–4114.

Kahn, R., A. Petzold, M. Wendisch, E. Bierwirth, T. Dinter, M. Esselborn, M. Fiebig, B. Heese, P. Knippertz, D. Muller, A. Schladitz, and W. von Hoyningen-Huene, "Desert dust aerosol air mass mapping in the western Sahara: Using particle properties derived from space-based multi-angle imaging", *Tellus*, **61B** (2009b), 239–251, doi: 10.1111/j.1600-0889.2008.00398.x.

Kahn, R. A., B. J. Gaitley, M. J. Garay, D. J. Diner, T. Eck, A. Smirnov, and B. N. Holben, "Multiangle Imaging SpectroRadiometer global aerosol product assessment by comparison with the Aerosol Robotic Network", *J. Geophys. Res.*, **115** (2010), D23209, doi: 10.1029/2010JD014601.

Kahn, R. A., M. J. Garay, D. L. Nelson, R. C. Levy, M. A. Bull, D. J. Diner, J. V. Martonchik, E. G. Hansen, L. A. Remer, and D. Tanré, "Response to 'Toward unified satellite climatology of aerosol properties'. 3. MODIS versus MISR versus AERONET", *J. Quant. Spectrosc. Radiat. Transfer.*, **112** (2011), 901–909, doi: 10.1016/j.jqsrt.2009.11.003.

Kahnert, F. M., "Numerical methods in electromagnetic scattering theory", *J. Quant. Spectrosc. Radiat. Transfer*, **79–80** (2003), 775–824.

Kahnert, M., "Modelling the optical and radiative properties of freshly emitted light absorbing carbon within an atmospheric chemical transport model", *Atmos. Chem. Phys.*, **10** (2010), 1403–1416.

Kalashnikova, O. V., and R. Kahn, "Ability of multiangle remote sensing observations to identify and distinguish mineral dust types: Part 2. Sensitivity over dark water", *J. Geophys. Res.*, **111** (2006), D11207, doi: 10.1029/2005JD006756.

Kalashnikova, O. V., and R. A. Kahn, "Mineral dust plume evolution over the Atlantic from MISR and MODIS aerosol retrievals", *J. Geophys. Res.*, **113** (2008), D24204 , doi: 10.1029/2008JD010083.

Karagulian, F., L. Clarisse, C. Clerbaux, A. J. Prata, D. Hurtmans, and P. F. Coheur, "Detection of volcanic SO_2, ash and H_2SO_4 using the Infrared Atmospheric Sounding Interferometer (IASI)", *J. Geophys. Res.*, **115** (2010), D00L02, doi:10.1029/2009JD012786.

Kasten, F., "A new table and approximation formula for the relative optical airmass", *Arch. Meteor. Geophys. Bioklim*, **B14** (1965), 206–223.

Kaufman, Y. J., "Satellite sensing of aerosol absorption", *J. Geophys. Res.*, **92** (1987), 4307–4317, doi: 10.1029/JD092iD04p04307.

Kaufman, Y. J., and R. S. Fraser, "Light extinction by aerosols during summer air pollution", *J. Appl. Meteorol.*, **22** (1983a), 1694–1706.

Kaufman, Y. J., and R. S. Fraser, "Different atmospheric effects on remote sensing of uniform and nonuniform surfaces", *Adv. Space Res.*, **2** (1983b), 147–155.

Kaufman, Y. J., and T. Nakajima, "Effect of Amazon smoke on cloud microphysics and albedo – Analysis from satellite imagery", *J. Appl. Meteor. (Squires special issue)*, **32** (1993), 729–744.

Kaufman, Y. J., and C. Sendra, "Algorithm for atmospheric corrections of visible and near IR satellite imagery", *Int. J. Rem. Sens.*, **9** (1988), 1357–1381.

Kaufman, Y. J., and L. A. Remer, "Detection of forests using Mid-IR reflectance: An application for aerosol studies", *IEEE Trans. Geosci. Remote Sens.*, **32** (1994), 672–683.

Kaufman, Y. J., and D. Tanré, "Strategy for direct and indirect methods for correcting the aerosol effect on remote sensing: From AVHRR to EOS-MODIS", *Remote Sens. Environ.*, **55** (1996), 65–79.

Kaufman, Y. J., C. J. Tucker, and I. Fung, "Satellite measurements of large-scale air pollution: Methods", *J. Geophys. Res.*, **95** (1990), 9927–9939, doi: 10.1029/JD095iD07p09927.

Kaufman Y. J., A. Gitelson, A. Karniely, E. Ganor, R. S. Fraser, T. Nakajima, S. Mattoo, and B. N. Holben, "Size distribution and scattering phase function of aerosol particles derived from sky brightness measurements", *J. Geophys. Res.*, **99** (1994), 10,341–10,356.

Kaufman, Y. J., A. Wald, L. A. Remer, B.-C. Gao, R.-R. Li, and L. Flynn, "The MODIS 2.1 μm channel-correlation with visible reflectance for use in remote sensing of aerosols", *IEEE Trans. Geosci. Remote Sens.*, **35** (1997a), 1286–1298.

Kaufman, Y. J., D. Tanré, L. A. Remer, E. F. Vermote, A. Chu, and B. N. Holben, "Operational remote sensing of tropospheric aerosol over land from EOS moderate resolution imaging spectroradiometer", *J. Geophys. Res.*, **102,** D14 (1997b), 17,051–17,067.

Kaufman, Y. J., D. Tanré, O. Dubovik, A. Karnieli, and L. A. Remer, "Absorption of sunlight by dust as inferred from satellite and ground–based remote sensing", *Geophys. Res. Lett.*, **28** (2001), 1479–1482, doi: 10.1029/2000GL012647.

Kaufman Y. J., O. Boucher, D. Tanré, M. Chin, L. A. Remer, and T. Takemura, "Aerosol anthropogenic component estimated from satellite data", *Geophys. Res. Lett.* (2005), L17804, doi:10.1029/2005GL023125.

Kent, G. S., and M. P. McCormick, "SAGE and SAM II measurements of global strato-spheric aerosol optical depth and mass loading", *J. Geophys. Res.*, **89**, D4 (1984), 5303–5314, doi: 10.1029/JD089iD04p05303.

Kerr, R. A., "Global pollution: Is the Arctic Haze actually industrial smog?" *Science*, **205** (1979), 290–293, doi: 10.1126/science.205.4403.290.

Key, J., and A. J. Schweiger, "Tools for atmospheric radiative transfer: Streamer and FluxNet", *Computers and Geosciences*, **24:5** (1998), 443–451.

Khosravi, R., A. Lambert, H. Lee, J. Gille, J. Barnett, G. Francis, D. Edwards, C. Halvorson, S. Massie, C. Craig, C. Krinsky, J. McInerney, K. Stone, T. Eden, B. Nardi, C. Hepplewhite, W. Mankin, and M. Coffey, "Overview and characterization of retrievals of temperature, pressure, and atmospheric constituents from the High Resolution Dynamics Limb Sounder (HIRDLS) measurements", *J. Geophys. Res.*, **114** (2009), D20304, , doi:10.1029/2009JD011937.

Killinger, D. K., and N. Menyuk, "Remote probing of the atmosphere using a CO_2 DIAL system", *Quantum Electronics*, **17:9** (1981), 1917–1929.

Kim, D.-H., B.-J. Sohn, T. Nakajima, T. Takamura, T. Takemura, B.-C. Choi, and S.-C. Yoon, "Aerosol optical properties over east Asia determined from ground-based sky radiation measurements", *J. Geophys. Res.*, **109** (2004), D02209, doi: 10.1029/2003JD003387.

Kim, S.-W., S.-C. Yoon, E. G. Dutton, J. Kim, C. Wehrli, and B. N. Holben, "Global surface-based sun photometer network for long-term observations of column aerosol optical properties: Intercomparison of aerosol optical depth", *Aerosol Sci. Tech.*, **42** (2008), 1–9, doi: 10.1080/02786820701699743.

King, M. D. "Sensitivity of constrained linear inversions to the selection of the Lagrange multiplier", *J. Atmos. Sci.*, **39** (1982), 1356–1369.

King, M. D., D. M. Byrne, B. M. Herman, and J. A. Reagan, "Aerosol size distributions obtained by inversion of spectral optical depth measurements", *J. Atmos. Sci.*, **35** (1978), 2153–2167.

King, M. D., Tsay, S.-C., Ackerman, S., and Larsen, N., "Discriminating heavy aerosol, clouds, and fires during SCAR-B: Application of airborne multispectral data," *J. Geophys. Res.*, **103** (1998), 31, 989–31, 999.

King, M. D., Y. J. Kaufman, D. Tanré, and T. Nakajima, "Remote sensing of tropospheric aerosols from space: Past, present and future", *Bull. Amer. Meteor. Soc.*, **80** (1999), 2229–2259.

Klett, James D, "Stable analytical inversion solution for processing lidar returns", *Appl. Opt.*, **20:2** (1981), 211–220.

Kleiber, P. D., V. H. Grassian, M. A. Young, and P. K. Hudson, "T-matrix studies of aerosol particle shape effects on IR resonance spectral line profiles and comparison with experiment", *J. Geophys. Res.*, **114** (2009), D21209, doi: 10.1029/2009JD012710.

Kleidman, R. G., A. Smirnov, R. C. Levy, S. Mattoo, and D. Tanré, "Evaluation and wind-speed dependence of MODIS aerosol retrievals over open ocean", *IEEE Trans. Geosci. Rem. Sens.*, **50** (2011), 1–7, doi: 10.1109/TGRS.2011.2162073.

Kleipool, Q. L., M. R. Dobber, J. F. de Haan and P. F. Levelt, "Earth surface reflectance climatology from 3 years of OMI data", *J. Geophys. Res.*, **113** (2008), D18308, doi: 10.1029/2008JD010290.

Klusek, C., S. Manickavasagam, and M. P. Mengüç, "Compendium of scattering matrix element profiles for soot agglomerates", *J. Quant. Spectrosc. Radiat. Transfer*, **79–80** (2003), 839–859.

Knapp, K. R., and T. H. Vonder Haar, "Calibration of the eighth geostationary observational environmental satellite (GOES-8) Imager visible sensor", *J. Atmos. Oceanic Technol.*, **17** (2000), 1639–1644.

Knapp, K. R., T. H. Vonder Haar, and Y. J. Kaufman, "Aerosol optical depth retrieval from GOES-8: Uncertainty study and retrieval validation over South America", *J. Geophys. Res.*, **107**, D7 (2002), 4055, doi: 10.1029/2001JD000505.

Knobelspiesse, K., B. Cairns, J. Redemann, R. W. Bergstrom, and A. Stohl, "Simultaneous retrieval of aerosol and cloud properties during the MILAGRO field campaign", *Atmos. Chem. Phys.*, **11** (2011), 6245–6263, doi: 10.5194/acp-11-6245-2011.

Koepke, P., "Effective reflectance of oceanic white caps", *Appl. Opt.*, **23** (1984), 1816–1824.

Kokhanovsky, A. A., J. L. Deuzé, D. J. Diner, O. Dubovik, F. Ducos, C. Emde, M. J. Garay, R. G. Grainger, A. Heckel, M. Herman, I. L. Katsev, J. Keller, R. Levy, P. R. J. North, A. S. Prikhach, V. V. Rozanov, A. M. Sayer, Y. Ota, D. Tanré, G. E. Thomas, and E. P. Zege, "The inter-comparison of major satellite aerosol retrieval algorithms using simulated intensity and polarization characteristics of reflected light", *Atmos. Meas. Tech.*, **3** (2010), 909–932, doi: 10.5194/amt-3-909-2010.

Kolokolova, L., B.Å.S. Gustafson, M. Mishchenko, and G. Videen (Eds), "Electromagnetic and light scattering by nonspherical particles 2002", *J. Quant. Spectrosc. Radiat. Transfer*, **79–80** (2003), 491–1198.

Kölsch, H. J., P. Rairoux, J. P. Wolf, and L. Wöste, "Simultaneous NO and NO_2 DIAL measurement using BBO crystals", *Appl. Opt.*, **28:11** (1989), 2052–2056.

Koppers, G. A. A., J. Jansson, and D. P. Murtagh, "Aerosol optical thickness retrieval from GOME data in the oxyen A-Band, in: *Third ERS Symposium on Space at the Service of our Environment, Vols. II and III*, published as an *ESA* special publication, **414** (1997), 693–696.

Koren, I., L. A. Remer, Y. J. Kaufman, Y. Rudich, and J. V. Martins, "On the twilight zone between clouds and aerosols", *Geophys. Res. Lett.*, **34** (2007), L08805, doi: 10.1029/2007GL029253.

Kovacs, T. A., and M. P. McCormick, "Observations of Typhoon Melissa during the Lidar In-Space Technology Experiment (LITE)", *J. Appl. Meteor.*, **42** (2003), 1003–1013.

Krishna Moorthy, K., Prabha R. Nair, and B. V. Krishna Murthy, "Multiwavelength solar radiometer network and features of aerosol depth at Trivandrum", *Ind. J. Radio & Space Phy.*, **18** (1989), 194–201.

Kruglanski, M., M. De Mazière, A. C. Vandaele, and D. Hurtmans, "Boundary layer aerosol retrieval from thermal infrared nadir sounding – Preliminary results", *Advances in Space Research*, **37:12** (2006), 2160–2165.

Kuščer, I., and M. Ribarič, "Matrix formalism in the theory of diffusion of light", *Opt. Acta*, **6** (1959), 42–51.

Kusmierczyk-Michulec, J., and G. de Leeuw, "Aerosol optical thickness retrieval over land and water using Global Ozone Monitoring Experiment (GOME) data", *J. Geophys. Res.*, **110** (2005), D10S05, doi: 10.1029/2004JD004780.

Lacis, A., and V. Oinas, "A description of the correlated k distribution method for modelling nongray gaseous absorption, thermal emission, and multiple scattering in vertical inhomogeneous atmospheres", *J. Geophys. Res.*, **96** (1991), 9027–9063.

Lambert, A., R. G. Grainger, J. J. Remedios, C. D. Rodgers, M. Corney, and F. W. Taylor, "Measurements of the evolution of the Mt. Pinatubo aerosol cloud by ISAMS", *Geophys. Res. Lett.*, **20** (1993), 1287–1290.

Lambert, A., R. G. Grainger, C. D. Rodgers, F. W. Taylor, J. L. Mergenthaler, J. B. Kumer, and S. T. Massi, "Global evolution of the Mt. Pinatubo volcanic aerosols observed by the infrared limb-sounding instruments CLAES and ISAMS on the Upper Atmosphere Research Satellite", *J. Geophys. Res.*, **102**, D1 (1997), 1495–1512.

Langley, S. P., *Researches on solar heat and its absorption by the Earth's atmosphere. A Report of the Mount Whitney Expedition,* Professional Papers of the Signal Service, No. 15, Govt. Printing Office, Washington, 1884.

Lapina, K., C. L. Heald, D. V. Spracklen, S. R. Arnold, J. D. Allan, H. Coe, G. McFiggans, S. R. Zorn, F. Drewnick, T. S. Bates, L. N. Hawkins, L. M. Russell, A. Smirnov, C. D. O'Dowd, and A. J. Hind, "Investigating organic aerosol loading in the remote marine environment", *Atmos. Chem. Phys.*, **11** (2011), 8847–8860, doi: 10.5194/acp-11-8847-2011.

Lawrence J. D. Jr., M. P. McCormick, S. H. Melfi, and D. P. Woodman, "Laser backscatter correlation with turbulent regions of the atmosphere", *Appl. Phys. Lett.*, **12** (1968), 72.

Lazarev, A. I., V. V. Kovanelok, and S. V. Avakyan, *Investigation of Earth from Manned Spacecraft*, Leningrad, Gidrometeoizdat, 1987.

Legrand, M., and O. Pancrati, Thermal infrared radiometry, microphysical properties and geochemical nature of mineral dust, in: H. Fischer and B.-J. Sohn (Eds) *IRS'2004: Current Problems in Atmospheric Radiation*, A. Deepak Publishing, 359–362, 2006.

Legrand, M., G. Cautenet, and J.-C. Buriez, "Thermal impact of Saharan dust over land. Part II: Application to satellite IR remote sensing", *J. Appl. Meteor.*, **31** (1992), 181–193.

Legrand, M., J.-J. Bertrand, and M. Desbois, "Dust over West Africa: A characterization by satellite data", *Ann. Geophys.*, **3** (1985), 777–783.

Legrand, M., C. Piétras, G. Brogniez, M. Haeffelin, N. K. Abuhassan, and M. Sicard, "A high-accuracy multiwavelength radiometer for in situ measurements in the thermal infrared. Part I: Characterization of the instrument", *J. Atmos. Oceanic Technol.*, **17** (2000), 1203–1214.

Legrand, M., A. Plana-Fattori, and C. N'doumé, "Satellite detection of dust index using the IR imagery of Meteosat, 1. Infrared difference dust index", *J. Geophys. Res.*, **106** (2001), 18,251–18,274.

Lenoble, J. (Ed.), *Radiative Transfer in Scattering and Absorbing Atmospheres: Standard Computational Procedures*, Deepak Publishing, 1985.

Lenoble, J., *Atmospheric Radiative Transfer*, Deepak Publishing, 1993.

Lenoble, J., M. Herman, J. L. Deuzé, B. Lafrance, R. Santer, and D. Tanré, "A successive order of scattering code for solving the vector equation of transfer in the earth's atmosphere with aerosols", *J. Quant. Spectrosc. Radiat. Transfer*, **107** (2007), 479–507.

Lenoble, J., C. Brogniez, A. de La Casinière, T. Cabot, V. Buchard, and F. Guirado. "Measurements of UV aerosol optical depth in the French Southern Alps", *Atmos. Chem. Phys.*, **8** (2008), 6597–6602.

Léon, J.-F., and M. Legrand, "Mineral dust sources in the surroundings of the North Indian Ocean", *Geophys. Res. Lett.*, **30** (2003), 1309, doi: 10.1029/ 2002GL016690.

Leroy, M., J. L. Deuzé, F. M. Bréon, O. Hautecoeur, M. Herman, J. C. Buriez, D. Tanré, S. Bouffies, P. Chazette, and J. L. Roujean, "Retrieval of atmospheric properties and surface bidirectional reflectances over land from POLDER/ADEOS", *J. Geophys. Res.*, **102** (1997), 17,023–17,038.

Levelt, P. F., E. Hilsenrath, G. W. Leppelmeier, G. H. J. van den Ooord, P. K. Bhartia, J. Taminnen, J. F. de Haan, and J. P. Veefkind, "Science objectives of the Ozone Monitoring Instrument", *IEEE Trans. Geosci. Rem. Sens.*, Special Issue of the EOS-Aura Mission, **44:5** (2006), 1093–1101.

Levy, R. C., L. A. Remer, D. Tanré, Y. J. Kaufman, C. Ichoku, B. N. Holben, J. M. Livingston, P. B. Russell, and H. Maring, "Evaluation of the Moderate Resolution Imaging Spectroradiometer (MODIS) retrievals of dust aerosol over the ocean during PRIDE", *J. Geophys. Res.*, **108**, D19, (2003), 8594, doi: 10.1029/2002JD002460.

Levy, R. C., L. A. Remer, S. Mattoo, E. F. Vermote, and Y. J. Kaufman, "A second-generation algorithm for retrieving aerosol properties over land from MODIS spectral reflectance", *J. Geophys. Res.*, **112** (2007), D13211.

Levy, R., G. Leptoukh, R. Kahn, V. Zubko, A. Gopalan, and L. Remer, "A critical look at deriving monthly aerosol optical depth from satellite data", *IEEE Trans. Geosci. Remote Sens.*, No. 9, 47 (2009), 2942–2956.

Levy, R. C., L. A. Remer, R. G. Kleidman, S. Mattoo, C. Ichoku, R. Kahn, and T. F. Eck, "Global evaluation of the Collection 5 MODIS dark-target aerosol products over land", *Atmos. Chem. Phys.*, **10** (2010), 10,399–10,420, doi: 10.5194/acp-10-10399-2010.

Li, J., J. R. Anderson, and P. R. Buseck, "TEM study of aerosol particles from clean and polluted marine boundary layers over the North Atlantic" *J. Geophys. Res.*, **108** (2003), 4189.

Li, J., P. Zhang, T. J. Schmit, J. Schmetz, and W. P. Menzel, "Quantitative Saharan dust event with SEVIRI on Meteosat-8", *Int. J. Rem. Sens.*, **28:10** (2007), 2181–2186.

Li, J., B. E. Carlson, and A. A. Lacis, "A study on the temporal and spatial variability of absorbing aerosols using Total Ozone Mapping Spectrometer and Ozone Monitoring Instrument Aerosol Index data", *J. Geophys. Res.*, **114** (2009), D09213, doi: 10.1029/2008JD011278.

Lichtenberg, G., Q. Kleipool, J. M. Krijger, G. van Soest, R. van Hees, L. G. Tilstra, J.R. Acarreta, I. Aben, B. Ahlers, H. Bovensmann, K. Chance, A.M.S. Gloudemans, R.W.M. Hoogeveen, R.T.N. Jongma, S. Noel, A. Pieters, H. Schrijver, C. Schrijvers, C.E. Sioris, J. Skupin, S. Slijkhuis, P. Stammes, and M. Wuttke, "SCIAMACHY Level 1 data: Calibration concept and in-flight calibration", *Atmos. Chem. Phys.*, **6** (2006), 5347–5367, www.atmos-chem-phys.net/6/5347/2006/

Liou, K. N., *An Introduction to Atmospheric Radiation*, Academic Press, New York, 1980.

Liou, K. N., *Radiation and Cloud Processes in the Atmosphere*, Oxford University Press, 1992.

Litvinov, P., O. Hasekamp, B. Cairns, and M. Mishchenko, "Reflection models for soil and vegetation surfaces from multiple-viewing angle photopolarimetric measurements", *J. Quant. Spectrosc. Radiat. Transfer*, **111** (2010), 529–539.

Litvinov, P., O. P. Hasekamp, and B. Cairns, "Models for surface reflection of radiance and polarized radiance: Comparison with airborne multi-angle photopolarimetric measurements and implications for modeling top-of-atmosphere measurements", *Rem. Sens. Environ.*, **115** (2011), 781–792.

Liu, L., and M. I. Mishchenko, "Effects of aggregation on scattering and radiative properties of soot aerosols", *J. Geophys. Res.*, **110** (2005), D11211.

Liu, L., and M. I. Mishchenko, "Scattering and radiative properties of complex soot and soot-containing aggregate particles", *J. Quant. Spectrosc. Radiat. Transfer*, **106** (2007), 262–273.

Liu, L., and M. I. Mishchenko, "Toward unified satellite climatology of aerosol properties: Direct comparisons of advanced level 2 aerosol products", *J. Quant. Spectrosc. Radiat. Transfer*, **109** (2008), 2376–2385, doi:10.1016/j.jqsrt.2008.05.003.

Liu, L., M. I. Mishchenko, and W. P. Arnott, "A study of radiative properties of fractal soot aggregates using the superposition T-matrix method", *J. Quant. Spectrosc. Radiat. Transfer*, **109** (2008), 2656–2663.

Liu, Z., A. H. Omar, Y. Hu, M. A. Vaughan, and D. M. Winker, *CALIOP Algorithm Theoretical Basis Document Part 3: Scene Classification Algorithms, PC-SCI-202* Part 3, NASA Langley Research Center, Hampton VA, 2005.

Livingston, J. M., B. Schmid, P. B. Russell, J. A. Eilers, R. W. Kolyer, J. Redemann, S.A. Ramirez, J.-H. Yee, W. H. Swartz, C. R. Trepte, L. W. Thomason, M. C. Pitts, M. A. Avery, C. E. Randall, J. D. Lumpe, R. M. Bevilacqua, M. Bittner, T. Erbertseder, R. D. McPeters, R. E. Shetter, E. V. Browell, J. B. Kerr, and K. Lamb, "Retrieval of ozone column content from airborne sun photometer measurements during SOLVE II: Comparison with coincident satellite and aircraft measurements" Special Issue on the SOLVE II/VINTERSOL campaign, *Atmos. Chem. Phys.*, **5** (2005), 2035–2054, SRef-ID: 1680-7324/acp/2005-5-2035, 2005.

Livingston, J. M., J. Redemann, P. B. Russell, O. Torres, B. Veihelmann, P. Veefkind, R. Braak, A. Smirnov, L. Remer, R. W. Bergstrom, O. Coddington, K. S. Schmidt, P. Pilewskie, R. Johnson, and Q. Zhang, "Comparison of aerosol optical depths from the Ozone Monitoring Instrument (OMI) on Aura with results from airborne sunphotometry, other space and ground measurements during MILAGRO/INTEX-B", *Atmos. Chem. Phys.*, **9** (2009), 6743–6765.

Long, C. S., and L. L. Stowe, "Using the NOAA/AVHRR to study stratospheric aerosol optical thicknesses following the Mt. Pinatubo Eruption", *Geophys. Res. Lett.*, **21**:20 (1994), 2215–2218, doi: 10.1029/94GL01322.

Lowenthal, D. H., R. D. Borys, J. C. Chow, F. Rogers, and G. E. Shaw, "Evidence for long-range transport of aerosol from the Kuwaiti oil fires to Hawaii", *J. Geophys. Res.*, **97**, D13 (1992), 14,573–14,580.

Lucht, W., C. B. Schaaf, and A. H. Stahler, "An algorithm for the retrieval of albedo from space using semiempirical modes", *IEEE Trans. Geosci. Rem. Sens.*, **38** (2000), 977–998, doi: 10.1109/36.841980.

Lucke, R. L., D. R. Korwan, R. M. Bevilacqua, J. S. Hornstein, E. P. Shettle, D. T. Chen, M. Daehler, J. D. Lumpe, M. D. Fromm, D. Debrestian, B. Neff, M. Squire , G. König-Langlo, and J. Davies, "The Polar Ozone and Aerosol Measurement (POAM) III instrument and early validation results", *J. Geophys. Res.*, **104**, D15 (1999), 18,785–18,799, doi: 10.1029/1999JD900235.

Lumme, K. (Editor), "Light scattering by non-spherical particles", *J. Quant. Spectrosc. Radiat. Transfer*, **60** (1998), 301–500.

Lumpe, J. D., R. M. Bevilacqua, K. W. Hoppel, S. S. Krigman, D. L. Kriebel, C. E. Randall, D. W. Rusch, C. Brogniez, R. Ramananahérisoa, E. P. Shettle, J. J. Olivero, J. Lenoble, and P. Pruvost, "POAM II retrieval algorithm and error analysis", *J. Geophys. Res.*, **102**, D19 (1997), 23,593–23,614.

Lumpe, J. D., R. M. Bevilacqua, K. W. Hoppel, and C. E. Randall, "POAM III retrieval algorithm and error analysis", *J. Geophys. Res.*, **107** (2002), 4575, doi: 10.129/2002JD002137.

Lyapustin, Alexei, D. L. Williams, B. Markham, J. Irons, B. Holben, and Y. Wang, "A method for unbiased high-resolution aerosol retrieval from Landsat", *J. Atmos. Sci.*, **61** (2004), 1233–1244, doi: 10.1175/1520-0469(2004)061.

Lyapustin, A., Y. Wang, I. Laszlo, R. Kahn, S. Korkin, L. Remer, R. Levy, and J. Reid. "Multi-angle implementation of atmospheric correction (MAIAC): Part 2. Aerosol algorithm", *J. Geophys. Res.*, **116** (2011a), D03211, doi: 10.1029/2010JD014986.

Lyapustin, A., J. Martonchik, Y. Wang, I. Laszlo, and S. Korkin, "Multi-angle implementation of atmospheric correction (MAIAC): Part 1. Radiative transfer basis and look-up tables", *J. Geophys. Res.*, **116** (2011b), D03210, doi: 10.1029/2010JD014985.

Mackowski, D. W., "A simplified model to predict the effects of aggregation on the absorption properties of soot particles", *J. Quant. Spectrosc. Radiat. Transfer*, **100** (2006), 237–249.

Mackowski, D. W., and M. I. Mishchenko, "Calculation of the *T* matrix and the scattering matrix for ensembles of spheres", *J. Opt. Soc. Am. A.*, **13** (1996), 2266–2278.

Maignan, F., F.-M. Bréon, and R. Lacaze, "Bidirectional reflectance of Earth surfaces: Evaluation of analytical models using a large set spaceborne measurements with emphasis on the Hot Spot", *Rem. Sens. Environ.*, **90** (2004), 210-220.

Maignan, F., F.-M. Bréon, E. Fédèle, and M. Bouvier, "Polarized reflectances of natural surfaces: Spaceborne measurements and analytical modeling", *Rem. Sens. Environ.*, **113** (2009), 2642–2650.

Maiman, T. H., "Stimulated optical radiation in ruby", *Nature*, **187** (1960), 493.

Marchuk, G. I., N. A. Mikhailov, M. A. Nazaraliev, R. A.Darbinjan, B. A. Kargin, and B. S. Elepov, *The Monte Carlo Methods in Atmospheric Optics*, Springer, New York, 1980.

Maring, H., D. Savoie, M. Izaguirre, L. Custals, and J. S. Reid, "Mineral dust aerosol size distribution change during atmospheric transport", *J. Geophys. Res.*, **108** (2003), 8592, doi: 10.1029/2002JD002 536.

Markham, B. L., J. S. Schafer, B. N. Holben, and R. N. Halthore, "Atmospheric aerosol and water vapor characteristics over north central Canada during BOREAS", *J. Geophys. Res.*, **102**, D24 (1997), 29,737–29,745, doi: 10.1029/97JD00241.

Markowicz, K. M., P. J. Flatau, A. M. Vogelmann, P. K. Quinn, and E. J. Welton, "Clear-sky infrared radiative forcing at the surface and the top of the atmosphere", *Q. J. R. Meteorol. Soc.*, **129** (2003), 2927–2947, doi: 10.1256/qj.02.224.

Marticorena, B., G. Bergametti, B. Aumont, Y. Callot, C. N'doumé, and M. Legrand, "Modeling the atmospheric dust cycle: 2. Simulation of Saharan dust sources", *J. Geophys. Res.*, **102** (1997), 4387–4404.

Marticorena, B., G. Bergametti, and M. Legrand, "Comparison of emission models used for the large-scale simulation of the mineral dust cycle", *Contr. Atmos. Phys.*, **72** (1999), 151–160.

Marticorena, B., P. Chazette, G. Bergametti, F. Dulac, and M. Legrand, "Mapping the aerodynamic roughness length of desert surfaces from the POLDER/ADEOS bi-directional reflectance product", *Int. J. Rem. Sens.*, **25** (2004), 603–626, doi: 10.1080/0143116031000116976.

Martonchik, J. V., and D. J. Diner, "Retrieval of aerosol optical properties from multi-angle satellite imagery", *IEEE Trans. Geosci. Rem. Sensing*, **30** (1992), 223–230.

Martonchik, J. V., D. J. Diner, R. A. Kahn, T. P. Ackerman, M. E. Verstraete, B. Pinty, and H. R. Gordon, "Techniques for the retrieval of aerosol properties over land and ocean using multiangle imaging", *IEEE Trans. Geosci. Rem. Sens.*, **36** (1998), 1212–1227.

Martonchik, J. V., D. J. Diner, K. Crean, and M. Bull, "Regional aerosol retrieval results from MISR", *IEEE Trans. Geosci. Rem. Sens*, **40** (2002), 1520–1531.

Martonchik, J. V., D. J. Diner, R. A. Kahn, B. J. Gaitley, and B. N. Holben, "Comparison of MISR and AERONET aerosol optical depths over desert sites", *Geophys. Res. Lett.*, **31** (2004), L16102, doi: 10.1029/2004GL019807.

Martonchik, J. V., R. A. Kahn, and D. J. Diner, "Retrieval of aerosol properties over land using MISR observations, in: A. A. Kokhanovsky, and G. de Leeuw (Eds), *Satellite Aerosol Remote Sensing Over Land*, Springer, Berlin, 2009.

Massie, S. T., J. C. Gille, D. P. Edwards, P. L. Bailey, L. V. Lyjak, C. A. Craig, C. P. Cavanaugh, J. L. Mergenthaler, A. E. Roche, J. B. Kumer, A. Lambert, R. G. Grainger, C. D. Rodgers, F. W. Taylor, J. M. Russell III, J. H. Park, T. Deshler, M. E. Hervig, E. F. Fishbein, J. W. Waters, and W. A. Lahoz, "Validation studies using multiwavelength cryogenic limb array etalon spectrometer (CLAES) observations of stratospheric aerosol", *J. Geophys. Res.*, **101** (1996), 9757–9773, doi: 10.1029/95JD03225.

Masuda, K., T. Takashima, and Y. Takayama, "Emissivity of pure and sea waters for the model sea surface in the infrared window regions", *Rem. Sens. Environ.*, **24** (1988), 313–329.

Masuda, K., Y. Mano, H. Ishimoto, M. Tokuno, Y. Yoshizaki, D. Tanré, and N. Okawara, "Assessment of the nonsphericity of mineral dust from geostationary satellite measurements", *Rem Sens. Environ.*, **82** (2002), 238–247, doi: 10.1016/S0034-4257(02)00040-8.

Matthias, V., D. Balis, J. Bosenberg, R. Eixmann, M. Iarlori, L. Komguem, I. Mattis, A. Papayannis, G. Pappalardo, M. R. Perrone, and X. Wang, "Vertical aerosol distribution over Europe: Statistical analysis of Raman lidar data from 10 European Aerosol Research Lidar Network (EARLINET) stations," *J. Geophys. Res.*, **109** (2004), D18201, doi:10.1029/2004JD004638.

Matsumoto, T., P. Russell, C. Mina, W. Van Ark, and V. Banta, "Airborne tracking sunphotometer", *J. Atmos. Ocean. Tech.*, **4** (1987), 336–339.

Mauldin, L. E. III, N. H. Zaun, M. P. McCormick, J. H. Guy, and W. R. Vaugh, "Stratospheric Aerosol and Gas Experiment II instrument: A functional description", *Opt. Eng.*, 24 (1985), 307–312.

Mayer, B., and A. Kylling, "Technical note: The libRadtran software package for radiative transfer calculations-description and examples of use", *Atmos. Chem. Phys.*, **5** (2005), 1855–1877.

McClung, F. J., and R. W. Hellarth, "Giant optical pulsations from ruby", *J. Appl. Phys.*, **33** (1962), 828–829.

McConnell, C. L., P. Formenti, E. J. Highwood, and M. A. J. Harrison, "Using aircraft measurements to determine the refractive index of Saharan dust during the DODO Experiments", *Atmos. Chem. Phys.*, **10** (2010), 3081–3098, doi: 10.5194/acp-10-3081-2010.

McCormick, M. P., *Satellite Measurements of Stratospheric Aerosols and Gases: SAM II and SAGE*, 9th International Laser Radar Conference, Munich, Germany, July 2–5, 1979.

McCormick, M. P., P. Hamill, T. J. Pepin, W. P. Chu, T. J. Swissler, and L. R. McMaster, "Satellite studies of the stratospheric aerosol", *Bull. Amer. Meteorol. Soc.*, **60** (1979), 1038–1046.

McCormick, M. P., H. M. Steele, P. Hamill, W. P. Chu, and T. J. Swissler, "Polar stratospheric clouds sightings by SAM II", *J. Atmos. Sci.*, **39** (1982), 1387–1397.

McCormick, M. P., D. M. Winker, E. V. Browell, J. A. Coakley, C. S. Gardner, R. M. Hoff, G. S. Kent, S. H. Melfi, R. T. Menzies, C. M. R. Menzies, D. A. Randall, and J. A. Reagan, "Scientific investigations planned for the Lidar In-space Technology Experiment (LITE)", *Bull. Amer. Meteorol. Soc.*, **74:2** (1993), 205–214.

McMillin, L. M., D. Q. Wark, J. M. Siomkajlo, P. G. Abel, A. Werbowetzki, L. A. Lauritson, J. A. Pritchard, D. S. Crosby, H. M. Woolf, R. C. Luebbe, M. P. Weinreb, H. E. Fleming, F. E. Bittner, and C. M. Hayden, "Satellite infrared soundings from NOAA spacecraft." *NOAA Technical Report NESS 65*, (1973), available at http://www.ncdc.noaa.gov/oa/rsad/noaa-tr-ness65.pdf.

Meador, W. E., and W. R. Weaver, "Two stream approximations to radiative transfer in planetary atmospheres: A united description of existing methods and a new improvement", *J. Atmos. Sci.*, **27** (1980), 630–643.

Mekler, Y., H. Quenzel, G. Ohring, and I. Marcus, "Relative atmospheric aerosol content from ERTS observations", *J. Geophys. Res.*, **82** (1977), 967–972.

Mekler, Y., and Y. J. Kaufman, "The effect of Earth's atmosphere on contrast reduction for a nonuniform surface albedo and two-halves field", *J. Geophys. Res.*, **85** (1980), 4067–4083.

Melfi, S. H., J. D. Lawrence Jr, and M. P. McCormick, "Observations of Raman scattering by water vapor in the atmosphere", *Appl. Phys. Lett.*, **15** (1969), 295.

Menyuk, N. and D. K. Killinger, "Assessment of relative error sources in IR DIAL measurement accuracy", *Appl. Opt.*, **22:17** (1983), 2690–2698.

Menzel, W. P., R. A. Frey, H. Zhang, D. P. Wylie, C. C. Moeller, R. E. Holz, B. Maddux, B. A. Baum, K. I. Strabala, and L. E. Gumley, "MODIS global cloud-top pressure and amount estimation: Algorithm description and results", *J. Appl. Meteor. Climatol.*, **47** (2008), 1175–1198.

Menzies, R. T., D. M. Tratt, and W. H. Hunt, "Lidar in-space technology experiment measurements of sea surface directional reflectance and the link to surface wind speed", *Appl. Opt.*, **37** (1998), 5550–5559.

Mergenthaler, J. L., J. L. Kumer, and A. E. Roche, "CLAES observations of the Mt. Pinatubo aerosol", *Geophys. Res. Lett.*, **22** (1995), 3497.

Meszaros, E., and D. M. Whelpdale, "Manual for BAPMoN Station Operators", World Meteorological Organization, Environmental Pollution Monitoring and Research Programme Report 32, WMO/TD 66, 1985.

Mie, G., "Beiträge zur Optik trüber Medien, speziell kolloidaler Metallösungen", *Ann. Physik*, **25** (1908), 377–445.

Mims, S. R., R. A. Kahn, C. M. Moroney, B. J. Gaitley, D. L. Nelson, and M. J. Garay, "MISR stereo-heights of grassland fire smoke plumes in Australia", *IEEE Trans. Geosci. Rem. Sens.*, **48** (2009), 25–35.

Min, Q., and M. Duan, "A successive order of scattering model for solving vector radiative transfer in the atmosphere", *J. Quant. Spectrosc. Radiat. Transfer,* **87** (2004), 243–259.

Minnis, P., D. F. Young, D. R. Doelling, S. Sun-Mack, Y. Chen, and Q. Z. Trepte, "Surface and clear-sky albedos and emissivities from MODIS and VIRS with application to SEVIRI", *Proceedings of the 2002 EUMETSAT Meteorological Satellite Conference*, EUMETSAT, Vol. EUM P 36 (2002), 600–607.

Mishchenko, M. I., and K. Sassen, "Depolarization of lidar returns by small ice crystals: An application to contrails", *Geophys. Res. Lett.*, **25:3** (1998), 309–312.

Mishchenko, M. I., and L. D. Travis, "Light scattering by polydisperse, rotationally symmetric, non spherical particles : Linear polarization", *J. Quant. Spectrosc. Radiat. Transfer*, **51** (1994a), 759–788.

Mishchenko, M. I., and L. D. Travis, "Light scattering by polydispersions of randomly oriented spheroids with size comparable to wavelengths of observation", *Appl. Optics* **33** (1994b), 7206–7225.

Mishchenko, M. I., and L. D. Travis, "Satellite retrieval of aerosol properties over the ocean using polarization as well as intensity of reflected sunlight", *J. Geophys. Res.*, **102,** D14 (1997), 16,989–17,013.

Mishchenko, M. I., and L. D. Travis, "Gustav Mie and the evolving discipline of electromagnetic scattering by particles", *Bull. Amer. Meteorol. Soc.*, **89** (2008), 1853–1861.

Mishchenko, M. I., L. D. Travis, R. A. Kahn, and R. A. West, "Modeling phase functions for dustlike tropospheric aerosols using a shape mixture of randomly oriented polydisperse spheroids", *J. Geophys. Res.*, **102** (1997), 16,831–16,847.

Mishchenko, M. I., I. V. Geogdzhayev, B. Cairns, W. B. Rossow, and A. A. Lacis, "Aerosol retrievals over the ocean by use of channels 1 and 2 AVHRR data: Sensitivity analysis and preliminary results", *Appl. Opt.*, **38** (1999a), 7325–7341.

Mishchenko, M. I., J. W. Hovenier, and L. D. Travis (Editors), "Light scattering by nonspherical particles '98", *J. Quant. Spectrosc. Radiat. Transfer*, **63** (1999b), 127–738.

Mishchenko, M. I., J. W. Hovenier, and L. D. Travis (Editors), *Light Scattering by Nonspherical Particles: Theory, Measurements, and Applications*, Academic Press, San Diego, 1999c.

Mishchenko, M., J. M. Luck, and T. M. Nieuwenhuizen, "Full angular profile of the coherent polarization opposition effect", *J. Opt. Soc. Am. A.*, **17** (2000), 888–891.

Mishchenko, M. I., L. D. Travis, and A. A. Lacis, *Scattering, Absorption and Emission of Light by Small Particles*, Cambridge University Press, 2002.

Mishchenko, M. I., L. D. Travis, and A. A. Lacis, *Multiple Scattering of Light by Particles: Radiative Transfer and Coherent Backscattering*. Cambridge University Press, Cambridge, 2006.

Mishchenko, M. I., I. V. Geogdzhayev, W. B. Rossow, B. Cairns, B. E. Carlson, A. A. Lacis, L. Liu, and L. D. Travis, "Long-term satellite record reveals likely recent aerosol trend", *Science*, **315** (2007), 1543.

Mishchenko, M. I., G. Videen, and M. P. Mengüç (Editors), "X conference on electromagnetic and light scattering by non-spherical particles", *J. Quant. Spectrosc. Radiat. Transfer*, **109** (2008), 1335–1548.

Mishchenko, M. I., V. K. Rosenbush, N. N. Kiselev, D. F. Lupishko, V. P. Tishkovets, V. G. Kaydash, I. N. Belskaya, Y. S. Efimov, and N. M. Shakhovskoy, *Polarimetric Remote Sensing of Solar System Bodies*. Akademperiodyka, Kyiv, 2010 (available as a PDF file at http://www.giss.nasa.gov/staff/mmishchenko/books.html).

Mogili, P. K., K. H. Yang, M. A. Young, P. D. Kleiber, and V. H. Grassian, "Environmental aerosol chamber studies of extinction spectra of mineral dust aerosol components: Broadband IR-UV extinction spectra", *J. Geophys. Res.*, **112** (2007), D21204, doi:10.1029/2007JD008890.

Molina, M., "Heterogeneous chemistry on polar stratospheric clouds", *Atmos. Environ.*, **25** (1991), 2535–2537.

Molina, L. T., and M. J. Molina, "Absolute absorption cross sections of ozone in the 185 to 350 nm wavelength range", *J. Geophys. Res.*, **91** (1986), 14,501–14,508.

Moorthy, K. K., S. N. Beegum, S. S. Babu, A. Smirnov, S. John, K. R. Kumar, K. Narasimhulu, C. B. S. Dutt, and V. S. Nair, "Optical and physical characteristics of Bay of Bengal aerosols during W ICARB: Spatial and vertical heterogeneities in the marine atmospheric boundary layer and in the vertical column", *J. Geophys. Res.*, **115** (2010), D24213, doi: 10.1029/2010JD014094.

Moosmüller, H., R. K. Chakrabarty, and W. P. Arnott, "Aerosol light absorption and its measurements", *J. Quant. Spectrosc. Radiat. Transfer*, **110** (2009), 844–878.

Moré, J. M., "The Levenberg-Marquardt algorithm: Implementation and theory", *Lect. Notes Math.*, **630** (1978), 105–116.

Morel, A., "Optical modeling of the open ocean in relation to its biogenous water content (case I waters)", *J. Geophys. Res.*, **93** (1983), 10,749–10,768.

Moreno, F., O. Muñoz, J. J. López-Moreno, and A. Molina (Editors), "VIII conference on electromagnetic and light scattering by nonspherical particles", *J. Quant. Spectrosc. Radiat. Transfer*, **100** (2006), 1–495.

Moroney, C., R. Davies, and J. P. Muller, "Operational retrieval of cloud-top heights using MISR data", *IEEE Trans. Geosci. Rem. Sens.*, **40** (2002), 1532–1540, doi: 10.1109/TGRS.2002.801150.

Müller, D., K. Franke, F. Wagner, D. Althausen, A. Ansmann, and J. Heintzenberg, "Vertical profiling of optical and physical particle properties over the tropical Indian Ocean with six-wavelength lidar 1. Seasonal cycle", *J. Geophys. Res.*, **106** (2001a) 28,567–28,575.

Müller, D., U. Wandinger, D. Althausen, and M. Fiebig, "Comprehensive particle characterization from three-wavelength Raman-lidar observations: Case study", *Appl. Opt.* **40** (2001b), 4863–4869.

Müller, D., A. Ansmann, F. Wagner, K. Franke, and D. Althausen, "European pollution outbreaks during ACE 2: Microphysical particle properties and single-scattering albedo inferred from multiwavelength lidar observations", *J. Geophys. Res.*, **107** (2002), 4248, doi 10.1029/2001jd001110.

Muñoz, O., and H. Volten, "Experimental light scattering matrices from the Amsterdam Light Scattering Database", *Light Scattering Rev.*, **1** (2006), 3–29.

Murakami, H., K. Sasaoka, K. Hosoda, H. Fukushima, M. Toratani, R. Frouin, B. G. Mitchell, M. Kahru, P. Y. Deschanmps, D. Clark, S. Flora, M. Kishino, S. Saitoh, I. Asanuma, A. Tanaka, H. Sasaki, K. Yokouchi, Y. Kiyomoto, H. Saito, C. Dupouy, A. Siripong, A., Matsumura., and J. Ishizaka, "Validation of ADEOS-II GLI ocean color products using in-situ observations", *J. Oceanogr.*, **62**(3) (2006), 373–393, doi: 10.1007/s10872-006-0062-6.

Murray, E. R., and J. E. van der Laan, "Remote measurement of ethylene using a co_2 differential-absorption lidar", *Appl. Opt.*, **17**(5) (1978), 814–817.

Myhre, G., F. Stordal, M. Johnsrud, D. J. Diner, I. V. Geogdzhayev, J. M. Haywood, B. N. Holben, T. Holzer-Popp, A. Ignatov, R. A. Kahn, Y. J. Kaufman, N. Loeb, J. V. Martonchik, M. I. Mishchenko, N. R. Nalli, L. A. Remer, M. Schroedter-Homscheidt, D.

Tanré, O. Torres, and M. Wang, "Intercomparison of satellite retrieved aerosol optical depth over ocean during the period September 1997 to December 2000", *Atmos. Chem. Phys.*, **5** (2005), 1697–1719, doi: 10.5194/acp-5-1697-2005.

Nadal, F., and F. M. Bréon, "Parameterization of surface polarized reflectance derived from POLDER spaceborne measurements", *IEEE Trans Geosci. Rem. Sens.*, **37** (1999), 1709–1718.

Nadeau, P. H., Relationships between the mean area, volume and thickness for dispersed particles of kaolinites and micaceous clays and their application to surface-area and ion-exchange properties, *Clay Miner.*, **22**, (1987) 351–356.

Nakajima, T., and A. Higurashi, "A use of two-channel radiances for an aerososl characterization from space", *Geophys. Res. Lett.*, **25** (1998), 38153818.

Nakajima, T., M. Tanaka, and T. Yamauchi, "Retrieval of the optical properties of aerosols from aureole and extinction data", *Appl. Opt.*, **22** (1983), 2951–2959.

Nakajima T., G. Tonna, R. Rao, P. Boi, Y. Kaufman, and B. Holben, "Use of sky brightness measurements from ground for remote sensing of particulate polydispersions", *Appl. Opt.* **35** (1996), 2672–2686.

Nalli, N. R., P. Clemente-Colon, P. J, Minnett, M. Szczodrak, V. Morris, E. Joseph, M. D. Goldberg, C. D. Barnet, W. W. Wolf, A. Jessup, R. Branch, R. O. Knuteson, and W. F. Feltz, "Ship-based measurements for infrared sensor validation during Aerosol and Ocean Science Expedition 2004", *J. Geophys. Res.*, **111** (2006), D09S04, doi: 10.1029/2005JD006385

Nelson, D. L., Y. Chen, R. A. Kahn, D. J. Diner, and D. Mazzoni, "Example applications of the MISR INteractive eXplorer (MINX) software tool to wildfire smoke plume analyses", *Proceedings of SPIE* (2008), 7089, doi: 708909.1-708909.11.

Northrup-Grumman Space & Mission Systems Corporation, National polar-orbiting operational environmental satellite system, (NPOESS) VIIRS aerosol optical thickness (AOT and particle size parameter algorithm theoretical basis document (ATBD), CDRL No. A032, Rev F, D43313, Redondo Beach CA (2004-2010), http://jointmission.gsfc.nasa.gov/science/documents.html.

Obland, M. D., C. A. Hostetler, R. A. Ferrare, J. W. Hair, R. R. Rogers, S. P. Burton, A. L. Cook, and D. B. Harper, "Aerosol profile measurements from the NASA Langley Research Center Airborne High Spectral Resolution Lidar", in: *Proceedings of the 24th International Laser Radar Conference*, Boulder, Colorado (2008), 353–356.

O'Neill, N. T., T. F. Eck, B. N. Holben, A. Smirnov, O. Dubovik, and A. Royer, "Bimodal size distribution influences on the variation of Ångström derivatives in spectral and optical depth space", *J. Geophys. Res.*, **106** (2001), 9787–9806.

O'Neill, N. T., T. F. Eck, A. Smirnov, B. N. Holben, and S. Thulasiraman, "Spectral discrimination of coarse and fine mode optical depth", *J. Geophys. Res.*, **108**, D17 (2003), 4559–4573, doi:10.1029/2002JD002975.

O'Neill, N. T., O. Pancrati, K. Baibakov, E. Eloranta, R. L. Batchelor, J. Freemantle, L. J. B. McArthur, K. Strong, and R. Lindenmaier, "Occurrence of weak, sub-micron, tropospheric aerosol events at high Arctic latitudes". *Geophys. Res. Lett.*, **35**(14) (2008) L14814, doi:10.1029/2008GL033733.

Ota, Y., A. Higurashi, T. Nakajima, and T. Yokota, "Matrix formulations of radiative transfer including the polarization effect in a coupled atmosphere-ocean system", *J. Quant. Spectrosc. Radiat. Transfer,* **111** (2010), 878-894.

Palm, S. P., B. Hart, D. Hlavka, J. Spinhirne, A. Mahesh, and E. J. Welton, "An overview of the GLAS real-time atmospheric processing algorithms and results from the analysis of simulated GLAS data sets", *IEEE International Geoscience and Remote Sensing Symposium* (IGARSS 2002), Vols I–VI/*24th Canadian Symposium on Remote Sensing*, (2002), 1376-1378 .

Pappalardo, G., J. Bösenberg, A. Amodeo, A. Ansmann, A. Apituley, L. A. Arboledas, D. Balis, C. Böckmann, A. Chaikovsky, A. Comeron, V. Freudenthaler, G. Hansen, V. Mitev, D. Nicolae, A. Papayannis, M. R. Perrone, A. Pietruczuk, M. Pujadas, J.-P. Putaud, F. Ravetta, V. Rizi, V. Simeonov, N. Spinelli, D. Stoyanov, T. Trickl, and M. Wiegner, "EARLINET-ASOS: Programs and perspectives for the aerosol study on continental scale", *Proc. SPIE* 6367 (2006), 636701, doi:10.1117/12.690717.

Pappalardo, G., U. Wandinger, L. Mona, A. Hiebsch, I. Mattis, A. Amodeo, A. Ansmann, P. Seifert, H. Linné, A. Apituley, L. A. Arboledas, D. Balis, A. Chaikovsky, G. D'Amico, F. De Tomasi, V. Freudenthaler, E. Giannakaki, A. Giunta, I. Grigorov, M. Iarlori, F. Madonna, R. E. Mamouri, L. Nasti, A. Papayannis, A. Pietruczuk, M. Pujadas, V. Rizi, F. Rocadenbosch, F. Russo, F. Schnell, N. Spinelli, X. Wang, and M. Wiegner, "EARLINET correlative measurements for CALIPSO: First intercomparison results", *J. Geophys. Res.* **115** (2010), D00H19, doi10.1029/2009JD012147.

Patterson, E. M., "Optical properties of the crustal aerosol: Relation to chemical and physical characteristics", *J. Geophys. Res.*, **86** (1981), 3236–3246.

Paur, R. J., and A. M. Bass, The ultraviolet cross sections of ozone. II. Results and temperature dependence, in: C. S. Zeferos and A. Ghazi (Eds), *Atmospheric Ozone*, Proceedings of the Quadriennal Ozone Symposium, 1985, 611–616.

Penney, C. M., R. L. St. Peters, and M. Lapp, "Absolute rotational Raman cross sections for N_2, O_2, and CO_2", *J. Opt. Soc. Am.*, **64** (1976), 712–716.

Penning de Vries, M. J. M., S. Beirle, and T. Wagner, "UV aerosol indices from SCIAMACHY: Introducing the SCattering Index (SCI)", *Atmos. Chem. Phys.*, **9** (2009), 9555–9567.

Pepin, T. J., and M. P. McCormick, "Stratospheric aerosol measurement – Experiment MA-007", Apollo-Soyuz Test Project, Preliminary Science Report, NASA-JSC, TM-X-58173: 9.1–9.8 (1976).

Péquignot, E., A. Chédin, and N. A. Scott, "Infrared continental surface emissivity spectra retrieved from AIRS hyperspectral sensor", *J. Appl. Meteor. Climatol.*, **47:6** (2008), 1619–1633.

Petit, R. H., M. Legrand, I. Jankowiak, J. Molinié, J.-L. Mansot, G. Marion, and C. Asselin de Beauville, "Transport of Saharan dust over the Caribbean Islands. Study of an event", *J. Geophys. Res.*, **110** (2005), doi: 10.1029/ 2004JD004748.

Peyridieu, S., A. Chédin, D. Tanré, V. Capelle, C. Pierangelo, N. Lamquin, and R. Armante, "Saharan dust infrared optical depth and altitude retrieved from AIRS: A focus over North Atlantic – Comparison to MODIS and CALIPSO", *Atmos. Chem. Phys.*, **10** (2010a), 1953–1967.

Peyridieu S., A. Chédin, D. Tanré, V. Capelle, C. Pierangelo, N. Lamquin, and R. Armante, "Dust aerosol optical depth and altitude retrieved from hyperspectral infrared observations (AIRS, IASI) and comparison with other aerosol datasets (MODIS, CALIOP, PARASOL)", 2nd International IASI Conference, Sevrier, France, 25–29 January 2010 (2010b).

Phillips, D. L., "A technique for the numerical solution of certain integral equations of the first kind." *J. Assoc. Comput. Mach.*, **9** (1962), 84–97.

Pierangelo, C., A. Chédin, and P. Chazette, "Measurements of stratospheric volcanic aerosol optical depth from NOAA TIROS Observational Vertical Sounder (TOVS) observations", *J. Geophys. Res.*, **109** (2004a), D03207, doi: 10.1029/2003JD003870.

Pierangelo, C., A. Chédin, S. Heilliette, N. Jacquinet-Husson, and R. Armante, "Dust altitude and infrared optical depth from AIRS", *Atmos. Chem. Phys.*, **4** (2004b), 1813–1822.

Pierangelo, C., M. Mishchenko, Y. Balkanski, and A. Chédin, "Retrieving the effective radius of Saharan dust coarse mode from AIRS", *Geophys. Res. Lett.*, **32** (2005a), L20813, doi: 10.1029/2005GL023425.

Pierangelo C., É. Péquignot, A. Chédin, R. Armante, C. J. Stubenrauch, and S. Serrar, "8-year climatology of dust aerosol in the infrared from HIRS", ITSC XIV 2005 14th International TOVS Study Conference, Beijing, China, 25 May–1 June 2005 (2005b).

Pierce, J. R., R. A. Kahn, M. R. Davis, and J. M. Comstock, "Detecting thin cirrus in Multiangle Imaging Spectroradiometer aerosol retrievals", *J. Geophys. Res.*, **115** (2010), D08201, doi: 10.1029/2009JD013019.

Piironen, P., and E. W. Eloranta, "Demonstration of a high-spectral-resolution lidar based on an iodine absorption filter", *Opt. Lett.*, **19:3** (1994), 234–236.

Platnick, S., M. D. King, A. Ackerman, W. P. Menzel, B. A. Baum, J. C. Riédi, and R. A. Frey, "The MODIS cloud products: Algorithms and examples from Terra", *IEEE Trans. Geosci. Rem. Sens.*, **41** (2003), 459–473.

Plass, G. N., and G. W. Kattawar, "Monte Carlo calculations of radiative transfer in the Earth's atmosphere–ocean system. I. Flux in the atmosphere and the ocean", *J. Phys. Oceanogr.*, **2** (1972), 139–145.

Platt, U., "Differential optical absorption spectroscopy (DOAS)", *Chem. Anal. Series*, **127** (1994), 27–83.

Poole, L. R., and M. C. Pitts, "Polar stratospheric cloud climatology based on Stratospheric Aerosol Measurement II observations from 1978 to 1989", *J. Geophys. Res.*, **99**, D6 (1994), 13,083–13,089.

Popp, C., A. Hauser, N. Foppa, and S. Wunderle, "Remote sensing of aerosol optical depth over central Europe from MSG-SEVIRI data and accuracy assessment with ground-based AERONET measurements", *J. Geophys.Res.*, **112**, D24 (2007), D24S11.

Potter, J. F, "The Delta function approximation in radiative transfer theory", *J. Atmos. Sci.*, **27** (1970), 943–949.

Prados, A. I., S. Kondragunta, P. Ciren, and K. R. Knapp, "GOES Aerosol/Smoke product (GASP) over North America: Comparisons to AERONET and MODIS observations", *J. Geophys. Res.*, **112**, D15 (2007), D15201.

Prodi, F., V. Levizzani, M. Sentimenti, T. Colombo, V. Cundari, T. Zanzu, and V. Juliano, "Measurements of atmospheric turbidity from a network of sun-photometers in Italy during ALPEX", *J. Aerosol. Science.*, **15** (1984), 595–613.

Prospero, J. M., P. Ginoux, O. Torres, and S. E. Nicholson, "Environmental characterization of atmospheric soil dust derived from the Nimbus-7 TOMS absorbing aerosol product", *Reviews of Geophysics* (2002), 1002–1032, doi: 10.129/2000RG000095.

Pueschel, R. F., and J. M. Livingston, "Aerosol spectral optical depths: Jet fuel and forest fire smokes", *J. Geophys. Res.*, **95** (1990), 22,417–22,422.

Pueschel, R. F., J. M. Livingston, P. B. Russell, D. A. Colburn, T. P. Ackerman, B. V. Zak, D. A. Allen, and W. Einfeld, "Smoke optical depths: Magnitude, variability and wavelength dependence", *J. Geophys. Res.*, **93** (1988), 8388–8402.

Radiation Commission: *A Preliminary Cloudless Standard Atmosphere for Radiation Computations*. World Climate Programme, WCP-112. WMO/TD-NO.24, 1986.

Rahman, H., B. Pinty, and M. M. Verstraete, "Coupled surface–atmosphere reflectance (CSAR) model 2. Semiempirical surface model usable with NOAA advanced very high resolution radiometer data", *J. Geophys. Res.*, **98**, D11 (1993), 20,791–20,801.

Rahn, K. A., R. D. Borys, and G. E. Shaw, "The Asian source of Arctic haze bands", *Nature*, **268** (1977), 713–715, doi: 10.1038/268713a0.

Randall, C.E., R. M. Bevilacqua, J. D. Lumpe, and K. W. Hoppel, Validation of POAM III aerosols: Comparison to SAGE II and HALOE", *J. Geophys. Res.*, **106**, D21 (2001), 27,525–27,536, doi: 10.1029/2001JD000528.

Rao, C. R. N., and J.-H. Chen, "Post-launch calibration of the visible and near-infrared channels of the advanced very high resolution radiometer on the NOAA-14 spacecraft", *Int. J. Rem. Sens.*, **17** (1996), 2743–2747.

Rao, C. R. N., E. P. McClain, and L. L. Stowe, "Remote sensing of aerosols over the oceans using AVHRR data theory, practice, and applications", *Int. J. Remote Sens.*, **10**:4–5 (1989), 743–749.

Rault, D., and R. Loughman, "Stratospheric and upper tropospheric aerosol retrieval from limb scatter signals", *Proc. SPIE* 6745 (2007), 674509., doi: 10.1117/12.737325.

Rayleigh, Lord, "On the transmission of light through an atmosphere containing many small particles in suspension, and on the origin of the blue of the sky", *Philos. Mag.*, **47** (1889), 375–384.

Reber, C. A., C. E. Trevathan, R. J. McNeal, and M. R. Luther, "The Upper Atmosphere Research Satellite (UARS) Mission", *J. Geophys. Res.*, **98**, D6 (1993), 10,643–10,647.

Redemann, J., R. P. Turco, K. N. Liou, P. B. Russell, R. W. Bergstrom, B. Schmid, J. M. Livingston, P. V. Hobbs, W. S. Hartley, S. Ismail, R. A. Ferrare, and E. V. Browell, "Retrieving the vertical structure of the effective aerosol complex index of refraction from a combination of aerosol in situ and remote sensing measurements during TARFOX", *J. Geophys. Res.*, **105** (2000a), 9949–9970.

Redemann, J., R. P. Turco, K. N. Liou, P. V. Hobbs, W. S. Hartley, R. W. Bergstrom, E. V. Browell, and P. B. Russell, "Case studies of the vertical structure of the direct shortwave aerosol radiative forcing during TARFOX", *J. Geophys. Res.*, **105** (2000b), 9971–9979.

Redemann, J., Q. Zhang, J. Livingston, P. Russell, Y. Shinozuka, A. Clarke, R. Johnson, and R. Levy, "Testing aerosol properties in MODIS Collection 4 and 5 using airborne sunphotometer observations in INTEX-B/MILAGRO", *Atmos. Chem. Phys.*, **9** (2009), 8159–8172.

Reid, E. A., J. S. Reid, M. M. Meier, M. R. Dunlap, S. S. Cliff, A. Broumas, K. Perry, and H. Maring, "Characterization of African dust transported to Puerto Rico by individual particle and size segregated bulk analysis", *J. Geophys. Res.*, **108** (2003), 8591, doi: 10.1029/2002JD002935.

Reid, J. S., H. H. Jonsson, H. B. Maring, A. A. Smirnov, D. L. Savoie, S. S. Cliff, E. A. Reid, J. M. Livingston, M. M. Meier, O. Dubovik, and S. C. Tsay, "Comparison of size and morphological measurements of coarse mode dust particles from Africa", *J. Geophys. Res.*, **108** (2003), 8593, doi: 10.1029/2002JD002485.

Remer L. A., and Y. J. Kaufman, "Dynamic aerosol models: Urban industrial aerosols", *J. Geophys. Res.*, **103** (1998), 13,859–13,871.

Remer, L. A., Y. J. Kaufman, B. N. Holben, A. M. Thompson, and D. McNamara, "Tropical biomass burning smoke aerosol size distribution model", *J. Geophys. Res.*, **103** (1998), 31,879–31,892.

Remer, L. A., Y. J. Kaufman, D. Tanré, S. Mattoo, D. A. Chu, J. V. Martins, R. R. Li, C. Ichoku, R. C. Levy, R. G. Kleidman, T. F. Eck, E. Vermote, and B. N. Holben,

"The MODIS aerosol algorithm, products and validation", *J. Atmos. Sci.*, **62** (2005), 947–973.

Remer, L. A., R. G. Kleidman, R. C. Levy, Y. J. Kaufman, D. Tanré, S. Mattoo, J. V. Martins, C. Ichoku, I. Koren, H. Yu, and B. N. Holben, "Global aerosol climatology from the MODIS satellite sensors", *J. Geophys. Res.*, **113** (2008), D14S07, doi: 10.1029/2007JD009661.

Renard, J.-B., M. Pirre, C. Robert, G. Moreau, D. Huguenin, and J. M. Russell III, "Nocturnal vertical distribution of stratospheric O_3, NO_2 and NO_3 from balloon measurements", *J. Geophys. Res.* **101** (1996), 28,793–28,804.

Renard, J.-B., M. Chartier, C. Robert, G. Chalumeau, G. Berthet, M. Pirre, J. P. Pommereau, and F. Goutail, "SALOMON: A new, light balloonborne UV-visible spectrometer for nighttime observations of stratospheric trace-gas species", *Appl. Opt.* **39** (2000), 386–392.

Robles-Gonzalez, C., and G. de Leeuw, "Aerosol properties over the SAFARI-2000 area retrieved from ATSR2", *J. Geophys. Res.*, **113** D5 (2008), D05206.

Robles-Gonzalez, C., G. de Leeuw, R. Decae, J. Kusmierczyk-Michulec, and P. Stammes, "Aerosol properties over the Indian Ocean Experiment (INDOEX) campaign area retrieved from ATSR-2", *J. Geophys. Res.*, **111**, D15 (2006), D15205.

Roche, A. E., J. B. Kumer, J. L. Mergenthaler, G. A. Ely, W. G. Uplinger, J. F. Potter, T. C. James, and L. W. Sterritt, "The Cryogenic Limb Array Etalon Spectrometer (CLAES) on UARS: Experiment description and performance", *J. Geophys. Res.*, **98**: D6 (1993), 10,763–10,775.

Rodgers, C. D., "Retrieval of atmospheric temperature and composition from remote measurements of thermal radiation", *Rev. Geophys.*, **14** (1976), 609–624.

Rodgers, C. D., *Inverse Methods for Atmospheric Sounding, Theory and Practice*, World Scientific Series on Atmospheric, Oceanic and Planetary Physics, Vol. 2, p. 238, World Scientific, 2000.

Roger, J. C., and E. F. Vermote, "A method to retrieve the reflectivity signature at 3.75 µm from AVHRR data", *Rem. Sens. Environ.*, **64** (1998), 103–114.

Rogers, R. R., J. W. Hair, C. A. Hostetler, R. A. Ferrare, M. D. Obland, A. L. Cook, D. B. Harper, S. P. Burton, Y. Shinozuka, C. S. McNaughton, A. D. Clarke, J. Redemann, P. B. Russell, J. M. Livingston, and L. I Kleinman, "NASA LaRC airborne high spectral resolution lidar aerosol measurements during MILAGRO: observations and validation". *Atmos. Chem. Phys.*, **9**:14 (2009), 4811–4826.

Romme, W. H., and D. G. Despain, "Historical perspective on the Yellowstone Fires of 1988", *Bioscience*, **39** (1989), 696–699.

Rondeaux, G., and M. Herman, "Polarization of light reflected by crop canopies", *Rem. Sens. Environ.*, **38** (1991), 63–75.

Roosen, R. G., and Ronald J. Angione, "Atmospheric transmission and climate: Results from Smithsonian measurements", *Bull. Amer. Meteorol. Soc.*, **65:9** (1984), 950–957, doi: 10.1175/1520-0477(1984)065<0950:ATACRF>2.0.CO;2.

Roosen, R. G., R. J. Angione, and C. H. Clemcke, "Worldwide variation in atmospheric transmission. 1. Baseline results from Smithsonian observations", *Bull. Amer. Meteorol. Soc.*, **54** (1973), 307–316.

Rothe, K.W., "Monitoring of various atmospheric constituents using a C.W. chemical hydrogen/deuterium laser and a pulsed carbon dioxide laser", *Radio Electron. Eng.*, **50** (1980), 567–574.

Rothman, L. S., I. E. Gordon, A. Barbe, D. C. Benner, P. E. Bernath, M. Birk, V. Boudon, L. R. Brown, A. Campargue, J. P. Champion, K. Chance, L. H. Coudert, V. Dana, V. M. Devi, S. Fally, J. M. Flaud, R. R. Gamache, A. Goldman, D. Jacquemart, I. Kleiner, N. Lacome, W. J. Lafferty, J. Y. Mandin, S. T. Massie, S. N. Mikhailenko, C. E. Miller, N. Moazzen-Ahmadi, O. V. Naumenko, A. V. Nikitin, J. Orphal, V. I. Perevalov, A. Perrin, A. Predoi-Cross, C. P. Rinsland, M. Rotger, M. Simeckova, M. A. H. Smith, T. K. Sung, S. A. Tashkun, J. Tennyson, R. A. Toth, A. C. Vandaele, and J. Vander Auwera, "The HITRAN 2008 molecular spectroscopic database", *J. Quant. Spectrosc. Radiat. Transf.*, **110** (2009), 533–572.

Roujean, J.-L., M. Leroy, and P.-Y. Deschamps, "A bidirectional reflectance model of the Earth's surface for the correction of remote sensing data", *J. Geophys. Res.*, **97**, D18 (1992), 20,455–20,468.

Rozanov, A. A., V. V. Rozanov, M. Buchwitz, A. A. Kokanovsky, and J. P. Burrows, "SCIATRAN 2.0: A new radiative transfer model for geophysical applications in the 175–2400 nm spctral range", *Adv. Space Res.*, **36** (2005), 1015–1016.

Rusch, D. W., C. E. Randall, M. T. Callan, M. Horanyi, R. T. Clancy, S. C. Solomon, S. J. Oltmans, B. J. Johnson, U. Koehler, H. Claude, and D. De Muer, "A new inversion for stratospheric aerosol and gas experiment II", *J. Geophys. Res.*, **103** (1998), D7, doi: 10.1029/97JD03625.

Russell, J. M. III, L. L. Gordley, J. H. Park, S. R. Drayson, W. D. Hesketh, R. J. Cicerone, A. F. Tuck, J. E. Frederick, J. E. Harries, and P. J. Crutzen, "The Halogen Occultation Experiment", *J. Geophys. Res.*, **98**, D6 (1993), 10,777–10,797.

Russell, P., J. Livingston, B. Schmid, J. Eilers, R. Kolyer, J. Redemann, S. Ramirez, J.-H. Yee, W. Swartz, R. Shetter, C. Trepte, A. Risley, Jr., B. Wenny, J. Zawodny, W. Chu, M. Pitts, J. Lumpe, M. Fromm, C. Randall, K. Hoppel, and R. Bevilacqua, "Aerosol optical depth measurements by airborne Sun photometer in SOLVE II: Comparisons to SAGE III, POAM III and airborne spectrometer measurements", *Atmos. Chem. Phys.*, **5** (2005), 1311–1339.

Russell, P. B., J. M. Livingston, J. Redemann, B. Schmid, S. A. Ramirez, J. Eilers, R. Khan, A. Chu, L. Remer, P. K. Quinn, M. J. Rood, and W. Wang, "Multi-grid-cell validation of satellite aerosol property retrievals in INTEX/ITCT/ICARTT 2004", *J. Geophys. Res.,* **112** (2007), D12S09, doi:10.1029/2006JD007606.

SAGE III Algorithm Theoretical Basis Document: Solar and Lunar Algorithm, Earth Observing System Project Science Office web site online availble at: http://eospso.gsfc. nasa.gov, 2002.

Saha, A., N. T. O'Neill, E. Eloranta, R. S. Stone, T. F. Eck, S. Zidane, D. Daou, A. Lupu, G. Lesins, M. Shiobara, and L. J. B. McArthur, "Pan-Arctic sunphotometry during the ARCTAS-A campaign of April 2008", *Geophys. Res. Lett.,* **37** (2010), L05803, doi:10.1029/2009GL041375.

Salisbury, J. W., and D. M. d'Aria, "Emissivity of terrestrial materials in the 8–14 μm atmospheric window", *Rem. Sens. Environ.,* **42** (1992), 83–106.

Sano, I., S. Mukai, M. Yamano, T. Takamura, T. Nakajima, and B. Holben, "Calibration and validation of retrieved aerosol properties based on AERONET and SKYNET", *Adv. Space Res.,* **32** (2003), 2159–2164.

Sano, I, M. Tanabe, T. Kamei, M. Nakata and S. Mukai, "Carbonaceous aerosol over Siberia and Indonesia with GOSAT/CAI", in: *Remote Sensing of the Atmosphere and Clouds III, Proc. of SPIE 7859* (2006), 785906, doi: 10.1117/12.869625.

Sano, I., S. Mukai, M. Mukai, B. Holben and I. Slutsker, "Estimation algorithm for aerosol properties from CAI on GOSAT", in: R. Picard, H. K. Schäfer, A. Comeron, E. I. Kassianov, and C. J. Mertens (Eds) *Remote Sensing of Clouds and the Atmosphere XIV,* Proceedings of the SPIE, 7475 (2009), 74751E-74751E-8, doi:10.1117/12.830155.

Santer, R., and M. Herman, "Particle size distributions from forward scattered light using the Chahine inversion scheme", *Appl. Opt.,* **22** (1983), 2294–2301.

Sasano Y., E. V. Browell, and S. Ismail, "Error caused by using a constant extinction/backscattering ratio in the lidar solution", *Appl. Opt.,* **24** (1985), 3929–3932.

Satheesh, S. K., O. Torres, L. A. Remer, S. S. Babu, V. Vinoj, T. F. Eck, R. G. Kleidman, and B. N. Holben, "Improved assessment of aerosol absorption using OMI–MODIS joint retrieval", *J. Geophys. Res.,* **114** (2009), D05209, doi: 10.1029/2008JD011024.

Sayer, A. M., G. E. Thomas, and R. G. Grainger, "A sea surface reflectance model for (A) ATSR, and application to aerosol retrievals", *Atmos. Meas. Tech.,* **3** (2010), 813–838.

Sayer, A., N. Hsu, C. Bettenhausen, Z. Ahmad, B. Holben, A. Smirnov, G. E. Thomas, and J. Zhang, "SeaWiFS Ocean Aerosol Retrieval (SOAR): Algorithm, validation, and comparison with other datasets", *J. Geophys. Res.* **117**, D3 (2012), doi:10.1029/2011JD016599.

Schmid, B., P. R. Spyak, S. F. Biggar, C. Wehrli, J. Sekler, T. Ingold, C. Matzler, and N. Kampfer, "Evaluation of the applicability of solar and lamp radiometric calibrations of a precision sun photometer operating between 300 and 1025 nm", *Appl. Opt.,* **37** (1998), 3923–3941.

Schmid, B., J. Redemann, P. B. Russell, P. V. Hobbs, D. L. Hlavka, M. J. McGill, B.N. Holben, E. J. Welton, J. R. Campbell, O. Torres, R. A. Kahn, D. J. Diner, M.C. Helmlinger, D.A. Chu, C. Robles-Gonzalez and G. de Leeuw, "Coordinated airborne, spaceborne, and ground-based measurements of massive thick aerosol layers during the dry season in southern Africa", *J. Geophys. Res.*, **108** (2003), 8496, doi: 10.10290/2002JD002297.

Schmid, B., R. Ferrare, C. Flynn, R. Elleman, D. Covert, A. Strawa, E. Welton, D. Turner, H. Jonsson, J. Redemann, J. Eilers, K. Ricci, A. G. Hallar, M. Clayton, J. Michalsky, A. Smirnov, B. Holben, and J. Barnard, "How well do state-of-the-art techniques measuring the vertical profile of tropospheric aerosol extinction compare?" *J. Geophys. Res.* (2006), **111**, doi:10.1029/2005jd005837.

Schneider, J., D. Balis, C. Böckmann, J. Bösenberg, B. Calpini, A. P. Chaikovsky, A. Comeron, P. Flamant, V. Freudenthaler, A. Hågård, I. Mattis, V. Mitev, A. Papayannis, G. Pappalardo, J. Pelon. M. R. Perrone, D. P. Resendes, N. Spinelli, T. Trickl, G. Vaughan, and G. Visconti, "A European aerosol research lidar network to establish an aerosol climatology (EARLINET)", *J. Aerosol Sci.*, **31** (2000), 592–593.

Schoeberl, M. R., A. R. Douglass, E. Hilsenrath, P. K. Bhartia, R. Beer, J. W. Waters, M. R. Gunson, L. Froidevaux, J. C. Gille, J. J. Barnett, P. F. Levelt, and P. DeCola, "Overview of the EOS Aura Mission", *IEEE Trans. Geosci. Rem. Sens.*, **44:5** (2006), 1066–1074.

Scott, N. A., and Chédin, A., "A fast line-by-line method for atmospheric absorption computations: The automatized atmospheric absorption atlas", *J. Appl. Meteor.*, **20** (1981), 802–812.

Seftor, C., N. Hsu, J. Herman, P. Bhartia, O. Torres, W. Rose, D. Schneider, and N. Krotkov, "Detection of volcanic ash clouds from Nimbus 7/total ozone mapping spectrometer", *J. Geophys. Res.*, **102** (1997), 16,749–16,759.

Sellers, P., F. Hall, H. Margolis, B. Kelly, D. Baldocchi, G. Denhartog, J. Cihlar, M. G. Ryan, B. Goodison, P. Crill, K. J. Ranson, D. Lettenmaier, and D. E. Wickland, "The boreal ecosystem – Atmosphere study (Boreas) – an overview and early results from the 1994 field year", *Bull. Amer. Meteorol. Soc.*, **76**(9) (1995), 1549–1577, doi: 10.1175/1520-0477(1995)076<1549:TBESAO>2.0.CO;2.

Shaw, G. E., *An Experimental Study of Atmospheric Turbidity Using Radiometric Techniques*, Dissertation, 170 pp, University of Arizona, 1971.

Shaw G. E., "Error analysis of multi-wavelength sun photometry", *Pure and Applied Geophysics*, **114** (1976), 1–14.

Shaw, G. E., "Inversion of optical scattering and spectral extinction measurements to recover aerosol size spectra", *Appl. Opt.*, **18** (1979), 988–993.

Shaw, G. E., "Transport of Asian desert aerosol to the Hawaiian Islands", *J. Appl. Meteor.*, **19** (1980), 1254–1259.

Shaw, G. E., "Solar spectral irradiance and atmospheric transmission at Mauna Loa Observatory", *Appl. Opt.*, **21** (1982), 2006.

Shaw, G. E., "The Arctic Haze phenomenon", *Bull. Amer. Meteorol. Soc.*, **76** (1995), 2403–2413.

Shaw, G. E., J. A. Reagan, and B. M. Herman, "Investigations of atmospheric extinction using direct solar radiation measurements made with multiple wavelength radiometer", *J Appl. Meteor.*, **12** (1973), 374.

Shettle, E. P., and R. W. Fenn, "Models for the aerosols of the lower atmosphere and the effect of humidity variations on their optical parameters". AFGL Project 7670, AFGL-TR-0214. Environmental Research Papers No.676, 1979.

Shimizu, H., S. A. Lee, and C. Y. She, "High spectral resolution lidar system with atomic blocking filters for measuring atmospheric parameters", *Appl. Opt.*, **22:9** (1983), 1373–1381.

Shipley, S. T., D. H. Tracy, E. W. Eloranta, J. T. Sroga, F. L. Roesler, and J. A. Weinman, "High spectra resolution lidar to measure optical scattering properties of atmospheric aerosols. 1: Theory and instrumentation", *Appl. Opt.*, **22**(23) (1983), 3716–3724.

Sicard, M., P. R. Spyak, G. Brogniez, M. Legrand, N. K. Abuhassan, C. Pietras, and J.-P. Buis, "Thermal-infrared field radiometer for vicarious cross-calibration: Characterization and comparisons with other field instruments", *Opt. Eng.*, **38** (1999), 345–356.

Siewert, C. E., "On the phase matrix basic to the scattering of polarized light", *Astron. Astrophys.*, **109** (1982), 195–200.

Siewert, C., "A discrete-ordinates solution for radiative-transfer models that include polarization effects", *J. Quant. Spectrosc. Radiat. Transfer,* **64** (2000), 227–254.

Sinyuk, A., O. Dubovik, B. Holben, T. F. Eck, F-M Breon, J. Martonchik, R. Kahn, D. J. Diner, E. F. Vermote, J.-C. Roger, T. Lapyonok, and I. Slutsker, "Simultaneous retrieval of aerosol and surface properties from a combination of AERONET and satellite", *Rem. Sens. Env.*, **107** (2007), 90–108.

Smirnov, A., B. N. Holben, Y. J. Kaufman, O. Dubovik, T. F. Eck, I. Slutsker, C. Pietras, and R. N. Halthore, "Optical properties of atmospheric aerosol in maritime environments", *J. Atmos. Sci.*, **59** (2002), 501–523.

Smirnov, A., B. N. Holben, I. Slutsker, D. M. Giles, C. R. McClain, T. F. Eck, S. M.Sakerin, A. Macke, P. Croot, G. Zibordi, P.K. Quinn, J. Sciare, S. Kinne, M. Harvey, T. J. Smyth, S. Piketh, T. Zielinski, A. Proshutinsky, J. I. Goes, N. B. Nelson, P. Larouche, V F. Radionov, P. Goloub, K. Krishna Moorthy, R. Matarrese, E. J. Robertson, and F. Jourdin, "Maritime Aerosol Network as a component of Aerosol Robotic Network", *J. Geophys. Res.*, **114** (2009), D06204, doi: 10.1029/2008JD011257.

Smirnov, A., B. N. Holben, D. M. Giles, I. Slutsker, N. T. O'Neill, T. F. Eck, A. Macke, P.Croot, Y. Courcoux, S. M. Sakerin, T. J. Smyth, T. Zielinski, G. Zibordi, J. I. Goes, M. J. Harvey, P. K. Quinn, N. B. Nelson, V. F. Radionov, C. M. Duarte, R. Losno, J. Sciare, K. J. Voss, S. Kinne, N. R. Nalli, E. Joseph, K. Krishna Moorthy, D. S. Covert, S. K. Gulev, G. Milinevsky, P. Larouche, S. Belanger, E. Horne, M. Chin, L. A. Remer,

R. A. Kahn, J. S. Reid, M. Schulz, C. L. Heald, J. Zhang, K. Lapina, R. G. Kleidman, J. Griesfeller, B. J. Gaitley, Q. Tan and T. L. Diehl, "Maritime aerosol network as a component of AERONET – first results and comparison with global aerosol models and satellite retrievals", *Atmos. Meas. Tech.*, **4** (2011), 583–597.

Sneep, M., J. F. de Haan, P. Stammes, P. Wang, C. Vanbauce, J. Joiner, A. P. Vasilkov, and P. F. Levelt, "Three-way comparison between OMI and PARASOL cloud pressure products", *J. Geophys. Res.*, **113** (2008), D15S23, doi: 10.1029/2007JD008694.

Sokolik, I. N., "The spectral radiative signature of wind-blown mineral dust: Implications for remote sensing in the thermal IR region", *Geophys. Res. Lett.*, **29:24**, (2002), 2154, doi: 10.1029/2002GL015910.

Sokolik, I. N., and G. Golitsyn, "Investigation of optical and radiative properties of atmospheric dust aerosols", *Atmos. Environ.*, **37A** (1993), 2509–2517.

Sokolik, I. N. and O. B. Toon, "Incorporation of mineralogical composition into models of the radiative properties of mineral aerosol from UV to IR wavelengths", *J. Geophys. Res.*, **104:D8** (1999), 9423–9444.

Sokolik I. N., A.V. Andronova, and T. C. Jonhson, "Complex refractive index of atmospheric dust aerosols". *Atmos. Envir.*, **16** (1993), 2495–2502.

Sokolik, I. N., O. B. Toon, and R. W. Bergstrom, "Modeling the radiative characteristics of airborne mineral aerosols at infrared wavelengths", *J. Geophys. Res.*, **103** (1998), 8813–8826.

Sorensen, C. M., "Light scattering by fractal aggregates: A review", *Aerosol Sci. Technol.*, **35** (2001), 648–687.

Spinhirne, J. D., "Micro pulse lidar", *IEEE Trans. Geosci. Rem. Sens.*, **31** (1993), 48–55.

Spinhirne, J. D., and M. D. King, "Latitudinal variation of spectral optical thickness and columnar size distribution of the El Chichón stratospheric aerosol layer", *J. Geophys. Res.*, **90** (1985), 10,607–10,619.

Spinhirne, J. D., and S. P. Palm, "Space based atmospheric measurements by GLAS", in: A. Ansmann (Ed.), *Advances in Atmospheric Remote Sensing with Lidar*, Springer, Berlin, pp. 213–217, 1996.

Spinhirne, J. D., M. Z. Hansen, and L. O. Caudill, "Cloud top remote sensing by airborne lidar", *Appl. Opt.*, **21:9** (1982), 1564–1571.

Spinhirne, J. D., J. A. R. Rall, and V. S. Scott, "Compact eye safe lidar systems", *Rev. Laser Eng.*, **23** (1995), 112–118.

Spinhirne, J. D., S. P. Palm, W. D. Hart, D. L. Hlavka, and E. J. Welton, "Cloud and aerosol measurements from the GLAS space borne lidar: Initial results", *Geophys. Res. Lett.*, **32** (2005), L22S03, doi: 10.1029/2005GL023507.

Stamnes, K., S. C. Tsay, W. Wiscombe, and K. Jayaweera, "Numerically stable algorithm for discrete-ordinate-method radiative transfer in multiple scattering and emitting layered media", *Appl. Opt.*, **27** (1988), 2502–2509.

Stammes, P., J. F. de Haan, and J. W. Hovenier, "The polarized internal radiation field of a planetary atmosphere", *Astron. Astrophys.*, **22** (1989), 239–259.

Stammes, P., "Spectral radiance modelling in the UV-Visible range", in: W. L. Smith and Y. M. Timofeyev (Eds), *IRS 2000: Current Problems in Atmospheric Radiation*, A. Deepak Publications, Hampton, VA (2001), 385–388.

Steele, H. M., and R. P. Turco, "Retrieval of aerosol size distributions from satellite extinction spectra using constrained linear inversion", *J. Geophys. Res.*, **102** (1997), 16,737–16,747,.

Stokes, G. G., "On the composition and resolution of streams of polarized light from different sources", *Trans. Cambridge Philos. Soc.*, **9** (1852), 399–423.

Stowe, L. L., R. Hitzenberger, and A. Deepak, "Report on Experts Meeting on Space Observations of Tropospheric Aerosols and Complementary Measurements", WCRP-48, WMO/TD-No. 389, World Meteorological Organization, 1990.

Stowe, L. L., R. M. Carey, and P. P. Pellegrino, "Monitoring the Mt. Pinatubo aerosol layer with NOAA/11 AVHRR data", *Geophys. Res. Lett.*, **19:2** (1992), 159–162.

Stowe, L. L., A. M. Ignatov, and R. R. Singh, "Development, validation, and potential enhancements to the second-generation operational aerosol product at the National Environmental Satellite, Data, and Information Service of the National Oceanic and Atmospheric Administration", *J. Geophys. Res.*, **102** (1997), 16,923–16,934.

Stowe, L. L., H. Jacobowitz, G. Ohring, K. Knapp, and N. Nalli, "The Advanced Very High Resolution Pathfinder Atmosphere (PATMOS) climate dataset: Initial analysis and evaluations", *J. Climate,* **15** (2002), 1243–1260.

Stratton, J. A., *Electromagnetic Theory*, McGraw Hill, New York, 1941.

Stricker, N. C. M., A. Hahne, D. L. Smith, J. Delderfield, M. B. Oliver, and T. Edwards, "ATSR-2: The evolution in its design from ERS-1 to ERS-2", *ESA Bulletin-European Space Agency*, **83** (1995), 32–37.

Strutt, J. W. (Lord Rayleigh), "On the light from the sky, its polarization and colour", *Philos. Mag.*, **41** (1871), 107–120, 274–279.

Swart, D. P. J., and J. B. Bergwerff, *Abstracts of papers*, Fifteenth International Laser Radar Conference (Tomsk, USSR), July 23–27, 1990.

Swartz, W. H., J. H. Yee, C. E. Randell, R. E. Shetter, E. V. Browell, J. F. Burris, T. J. Mcgee, and M. A. Avery, "Comparison of high-latitude line-of-sight ozone column density with derived ozone fields and the effects of horizontal inhomogeneity", *Atmos. Chem. Phys.*, **6** (2006), 1843–1852.

Taflove, A., and S. C. Hagness, Computational Electrodynamics: The Finite-Difference Time-Domain Method. Artech House, Boston, 2000.

Tanaka, M., T. Takamura, and T. Nakajima, "Refractive index and size distribution of aerosols as estimated from light scattering measurements", *J. Climate Appl. Meteor.*, **22** (1983), 1253–1261.

Tanré, D., M. Herman, P.-Y. Deschamps, and A. de Leffe, "Atmospheric modeling for space measurements of ground reflectances, including bidirectional properties", *Appl. Opt.*, **18** (1979), 3587–3594.

Tanré, D., P. Y. Deschamps, C. Devaux, and M. Herman, "Estimation of Saharan aerosol optical thickness from blurring effects in Thematic Mapper data", *J. of Geophys. Res.*, **92** (1988), 15,955–15,964.

Tanré, D., C. Deroo, P. Duhaut, M. Herman, and J. J. Morcrette, "Description of a computer code to simulate the satellite signal in the solar spectrum: the 5S code", *Int. J. Rem. Sens.*, **11** (1990), 659–668.

Tanré, D., M. Herman, and Y. Kaufman, "Information on the aerosol size distribution contained in the solar reflected spectral radiances", *J. Geophys. Res.*, **101** (1996), 19,043–19,060.

Tanré, D., Y. J. Kaufman, M. Herman, and S. Mattoo, "Remote sensing of aerosol properties over oceans using the MODIS/EOS spectral radiances", *J. Geophys. Res.*, **102:** D14 (1997), 16,971–16,988.

Tanré, D., F. M. Bréon, J. L. Deuzé, O. Dubovik, F. Ducos, P. François, P. Goloub, M. Herman, A. Lifermann, and F. Waquet, "Remote sensing of aerosols by using polarized, directionnal and spectral measurements within the A-Train: The PARASOL mission", *Atmos. Meas. Tech.*, **4** (2011), 1383–1395.

Tarantola, A., *Inverse Problem Theory: Methods for Data Fitting and Model Parameter Estimation*, Elsevier, Amsterdam, 1987, 500 pp.

Taylor, F. W., C. D. Rodgers, J. F. Whitney, S. T. Werrett, J. J. Barnett, G. D. Peskett, P. Venters, J. Ballard, C. W. P. Palmer, R. J. Knight, P. Morris, T. Nightingale, and A. Dudhia, "Remote Sensing of Atmospheric Structure and Composition by Pressure Modulator Radiometry From Space: The ISAMS Experiment on UARS", *J. Geophys. Res.*, **98:**D6 (1993), 10,799–10,814.

Textor, C., M. Schulz, S. Guibert, S. Kinne, Y. Balkanski, S. Bauer, T. Berntsen, T. Berglen, O. Boucher, M. Chin, F. Dentener, T. Diehl, R. Easter, H. Feichter, D. Fillmore, S. Ghan, P. Ginoux, S. Gong, A. Grini, J. Hendricks, L. Horowitz, P. Huang, I. Isaksen, I. Iversen, S. Kloster, D. Koch, A. Kirkevåg, J. E. Kristjansson, M. Krol, A. Lauer, J. F. Lamarque, X. Liu, V. Montanaro, G. Myhre, J. Penner, G. Pitari, S. Reddy, Ø. Seland, P. Stier, T. Takemura, and X. Tie, "Analysis and quantification of the diversities of aerosol life cycles within AeroCom", *Atmos. Chem. Phys.*, **6** (2006), 1777–1813.

Thomas, G. E., and Stamnes, K., *Radiative Transfer in the Earth and the Ocean*, Cambridge University Press, 1999.

Thomas, G. E., C. A. Poulsen, A. M. Sayer, S. H. Marsh, S. M. Dean, E. Carboni, R. Siddans, R. G. Grainger, and B. N. Lawrence, "The GRAPE aerosol retrieval algorithm", *Atmos. Meas. Tech.*, **22** (2009), 679–701.

Thomason, L. W., L. R. Poole, and T. R. Deshler, "A global climatology of stratospheric aerosol surface area density as deduced from SAGE II: 1984–1994", *J. Geophys. Res.*, **103** (1997), 8967–8976.

Thomason, L. W., J. M. Zawodny, S. P. Burton, and N. Iyer, "The SAGE II algorithm: Version 6.0 and on-going developments", 8th Scientific Assembly of the International Association of Meteorology and Atmospheric Sciences, Innsbruck Austria, July 10–18 (2001).

Thomason, L. W., L. R. Poole, and C. E. Randall, "SAGE III aerosol extinction validation in the Arctic winter: Comparisons with SAGE II and POAM III", *Atmos. Chem. Phys.*, **7** (2007), 1423–1433.

Thomason L. W., S. P. Burton, B.-P. Luo, and T. Peter, "SAGE II measurements of stratospheric aerosol properties at non-volcanic levels", *Atmos. Chem. Phys.*, **8** (2008), 983–995.

Thomason, L.W., J. R. Moore, M. C. Pitts, J. M. Zawodny, and E. W. Chiou, "An evaluation of the SAGE III version 4 aerosol extinction coefficient and water vapor data products", *Atmos. Chem. Phys.*, **10** (2010), 2159–2173.

Tikhonov, A. N., "On the solution of incorrectly stated problems and a method of regularization", *Dokl. Acad. Nauk SSSR*, **151** (1963), 501–504.

Tilstra, L. G., and P. Stammes, "Earth reflectance and polarization intercomparison between SCIAMACHY onboard Envisat and POLDER onboard ADEOS-2", *J. Geophys. Res.*, **112** (2007), D11304, doi: 10.1029/2006JD007713.

Tilstra, G., O. N. E. Tuinder, and P. Stammes, "GOME-2 absorbing aerosol index: Statistical analysis, comparison to GOME-1 and impact of instrument degraduation, in: *Proc. 2010 EUMETSAT Meteorological Satellite Conference*, EUMETSAT, 2010, p. 57.

Timofeyev, Y. M., A. V. Polyakov, H. M. Steele, and M. J. Newchurch, "Optimal eigenanalysis for the treatment of aerosols in the retrieval of atmospheric composition from transmission measurements", *Appl. Opt.*, **42** (2003), 2635–2646.

Toriumi R., H. Tai, and N. Takeuchi, "Tunable solid-state blue laser differential absorption lidar system for no_2 monitoring", *Opt. Eng.*, **35:8** (1996), 2371–2375.

Torres, O., and P. K. Bhartia, "Impact of tropospheric aerosol absorption on ozone retrieval from buv measurements", *J. Geophys. Res.*, **104** (1999), 21,569–21,578.

Torres, O., J. R. Herman, P. K. Bhartia, and Z. Ahmad, "Properties of Mount Pinatubo aerosols as derived from Nimbus 7 Total Ozone Mapping Spectrometer", *J. Geophys. Res.*, **100** (1995), 14,043–14,055.

Torres, O., P. K. Bhartia, J. R. Herman, Z. Ahmad, and J. Gleason, "Derivation of aerosol properties from satellite measurements of backscattered ultraviolet radiation: Theoretical basis", *J. Geophys. Res.*, **103** (1998), 17,099–17,110.

Torres, O., P. K. Bhartia, J. R. Herman, A. Sinyuk, and B. Holben, "A long term record of aerosol optical thickness from TOMS observations and comparison to AERONET measurements", *J. Atmos. Sci.*, **59** (2002), 398–413.

Torres, O., P. K. Bhartia, A. Sinyuk, E. J. Welton and B. Holben, "Total Ozone Mapping Spectrometer measurements of aerosol absorption from space: Comparison to SAFARI 2000 ground-based observations", *J. Geophys. Res.*, **110** (2005), D10S18, doi: 10.1029/JD004611.

Torres, O., A. Tanskanen, B. Veihelman, C. Ahn, R. Braak, P. K. Bhartia, P. Veefkind, and P. Levelt, "Aerosols and surface UV products from OMI observations: An overview", *J. Geophys. Res.*, **112** (2007), D24S47, doi: 10.1029/2007JD008809.

Torres, O., H. Jethva, and P. K. Bhartia, "Retrieval of aerosol optical depth above clouds from OMI observations: Sensitivity analysis and case studies", *J. Atmos. Sci.*, **116** (2011), doi: 10.1175/JAS-D-11-0130.1.

Torricella, F., E. Cattani, M. Cervino, R. Guzzi, and C. Levoni, "Retrieval of aerosol properties over the ocean using Global Ozone Monitoring Experiment measurements: Method and applications to test cases", *J. Geophys. Res.*, **103**:D10 (1999), 12,085–12,098.

Tucker, C. J., "Red and photographic infrared linear combinations for monitoring vegetation", *Rem. Sens. Environ.*, **8** (1979), 127.

Turner, D. D., "Ground-based infrared retrievals of optical depth, effective radius, and composition of airborne mineral dust above the Sahel", *J. Geophys. Res.*, **113** (2008), D00E03, doi: 10.1029/2008JD010054.

Turner, D. D., R. A. Ferrare, and L. A. Brasseur, "Average aerosol extinction and water vapor profiles over the southern great plains", *Geophys. Res. Lett.*, **28**:23 (2001), 4441–4444.

Turquety, S., J. Hadji-Lazaro, C. Clerbaux, D. A. Hauglustaine, S. A. Clough, V. Cassé, P. Shlüssel, and G. Mégie, "Trace gas retrieval algorithm for the Infrared Atmospheric Sounding Interferometer", *J. Geophys. Res.*, **109** (2004), D21301, doi: 10.1029/2004JD0048.

Twomey, S., "On the numerical solution of Fredholm integral equations of the first kind by the inversion of the linear system produced by quadrature", *J. Assoc. Comput. Mach.*, **10** (1963), 97–101.

Twomey, S., "Comparison of constrained linear inverse and an iterative nonlinear algorithm applied to the indirect estimation of particle size distributions". *J. Comp. Phys.*, **18** (1975), 188–200.

Twomey, S., "The influence of pollution on the shortwave albedo of clouds", *J. Atmos. Sci.*, **34** (1977a), 1149–1152.

Twomey, S., *Introduction to the Mathematics of Inversion in Remote Sensing and Indirect Measurements*, Elsevier, 1977b.

Uthe, E. E., and W. B. Johnson, *Lidar Observations of the Lower Tropospheric Aerosol Structure during BOMEX*. Final report, Tech. Report SRI Project 7929, Fallout Studies Branch, Division of Biology and Medicine, U.S. Atomic Energy Commission, Washington, D.C., January 1971.

Val Martin, M., J. A. Logan, R. Kahn, F-Y. Leung, D. Nelson, and D. Diner, "Fire smoke injection heights over North America constrained from the Terra Multi-angle Imaging SpectroRadiometer", *Atmos. Chem. Phys.*, **10** (2010), 1491–1510.

Vanbauce, C., B. Cadet, and R. T. Marchand, "Comparison of POLDER apparent and corrected oxygen pressure to ARM/MMCR cloud boundary pressures", *Geophys. Res. Letters*, **30** (2003), 1212, doi: 10.1029/2002GL016449.

Van de Hulst, A. C., *A New Look at Multiple Scattering*, NASA Institute for Space Studies, New York, 1963.

Van de Hulst, H. C., *Multiple Light Scattering*. Academic Press, New York, 1980, 2 volumes.

Van de Hulst, H.C., *Light Scattering by Small Particles*. John Wiley & Sons Inc., New York, 1957. Also: Dover, New York, 1981.

Vanderbilt, V. C., L. Grant, and S. L. Ustin, Polarization of light by vegetation, in: R. B. Myneni and J. Ross (Eds), *Photon-Vegetation Interactions: Applications in Optical Remote Sensing and Plant Ecology*, Springer-Verlag, New York, 1990, 190–228.

Van der Mee, C. V. M., and J. W. Hovenier, "Expansion coefficients in polarized light transfer", *Astron. Astrophys.*, **228** (1990), 559–568.

Veefkind, J. P., G. de Leeuw, and P. A. Durkee, "Retrieval of aerosol optical depth over land using two–angle view satellite radiometry during TARFOX", *Geophys. Res. Lett.*, **25:16** (1998), 3135–3138.

Veefkind, J. P., G. de Leeuw, P. A. Durkee, P. B. Russell, P. V. Hobbs, and J. M. Livingston, "Aerosol optical depth retrieval using ATSR-2 and AVHRR data during TARFOX", *J. Geophys. Res.*, **104:D2** (1999), 2253–2260.

Veefkind, J. P., G. de Leeuw, P. Stammes, and R. B. A. Koelemeyert, "Regional distribution of aerosol over land, derived from ATSR-2 and GOME", *Rem. Sens. Environ.*, **74** (2000), 377–386.

Veihelmann, B., P. F. Levelt, P. Stamnes and J. P. Veefkind, "Simulation study of the aerosol information content in OMI spectral reflectance measurements", *Atmos. Chem. Phys.*, **7** (2007), 3115–3127.

Vergé-Dépré, G., M. Legrand, C. Moulin, A. Alias, and P. Francois, "Improvement of the detection of desert dust over the Sahel using Meteosat IR imagery", *Ann. Geophys.*, **24** (2006), 1–9.

Vermeulen, A., C. Devaux, and M. Herman, "Retrieval of the scattering and microphysical properties of aerosols from ground-based optical measurements including polarization, I. Method". *Appl. Opt.*, **39** (2000), 6207–6220.

Vermote, E., and Y. J. Kaufman, "Absolute calibration of AVHRR visible and near infrared channels using ocean and cloud views", *Int J. Remote Sensing*, **16** (1995), 2317–2340.

Vermote, E., N. El Saleous, and B. N. Holben, Aerosol retrieval and atmospheric correction in: *Advances in the Use of NOAA AVHRR for Land Applications*, D'Souza, G., Belward, A.S and Malingreau, J.P (Eds.), Kluwer AcademicPress, Dordrecht, (1996), 93–124.

Vermote, E. F., D. Tanré, J. L. Deuzé, M. Herman, and J. J. Morcrette, "Second simulation of the satellite signal in the solar spectrum, 6S: An overview", *IEEE Trans. Geosci. Rem. Sens.*, **35** (1997), 675–686.

Veselovskii, I., A. Kolgotin, V. Griaznov, D. Müller, U. Wandinger, and D. N. Whiteman, "Inversion with regularization for the retrieval of tropospheric aerosol parameters from multiwavelength lidar sounding", *Appl. Opt.*, **41** (2002), 3685–3699.

Videen, G., Q. Fu, and P. Chýlek (Editors), "Light scattering by non-spherical particles", *J. Quant. Spectrosc. Radiat. Transfer*, **70** (2001), 373–831.

Videen, G., Y. S. Yatskiv, and M. I. Mishchenko (Editors), "Photopolarimetry in remote sensing", *J. Quant. Spectrosc. Radiat. Transfer*, **88** (2004), 1–406.

Vidot, J., R. Santer, and O. Aznay, "Evaluation of the MERIS aerosol product over land with AERONET", *Atmos. Chem. and Phys.*, **8:24** (2008), 7603–7617.

Vigroux, E., "Contribution à l'étude expérimentale de l'absorption de l'ozone", *Ann. Phys.*, **8** (1953), 709–762.

Viollier, M., D. Tanré, and P. Y. Deschamps, "An algorithm for remote sensing of water color from space", *Boundary Layer Meteorol.*, **18** (1980), 247–267.

Vogelmann, A., P. Flatau, M. Szczodrak, K. Markowicz, and P. Minnett, "Observations of large aerosol infrared forcing at the surface", *Geophys. Res. Lett.*, **30** (2003), 1655, doi: 10.1029/2002GL016 829.

Volten, H., O. Munoz, E. Rol, J. F. de Haan, W. Vassen, and J. W. Hovenier, "Scattering matrices of mineral aerosol particles at 441.6 and 632.8 nm", *J. Geophys. Res.*, **106** (2001), 17,375–17,401.

Volz, F. E., "Turbidity at Uppsala from 1909 to 1922 from Sjostroms solar radiation measurements", *Sver. Meteoro. Hydrolog. Inst.*, Rept 28, Stockholm, (1968), 100–104.

Volz, F. E., "Some results of turbidity networks", *Tellus*, **21** (1969), 625–630.

Volz, F. E., "Infrared refractive index of atmospheric aerosol substances", *Appl. Opt.*, **11** (1972), 755–759.

Volz, F. E., "Infrared optical constants of ammonium sulfate, Sahara dust, volcanic pumice, and fly ash", *Appl. Opt.*, **12** (1973), 564–568.

von der Gathen, P., "Aerosol extinction and backscatter profiles by means of a multi-wave-length Raman lidar: A new method without a priori assumptions", *Appl. Opt.*, **34:3** (1995), 463–466.

von Hoyningen-Huene, W., A. A. Kokhanovsky, M. W. Wuttke, M. Buchwitz, S. Noël, K. Gerilowski, J. P. Burrows, B. Latter, R. Siddans, and B. J. Kerridge, "Validation of SCIAMACHY top-of-atmosphere reflectance for aerosol remote sensing using MERIS L1 data", *Atmos. Chem. Phys.*, **7** (2007), 97–106.

Voshchinnikov, N. V., and G. Videen (Editors), "IX conference on electromagnetic and light scattering by non-spherical particles", *J. Quant. Spectrosc. Radiat. Transfer*, **106** (2007), 1–621.

Wagner, S. C., Y. M. Govaerts, and A. Lattanzio, "Joint retrieval of surface reflectance and aerosol optical depth from MSG/SEVIRI observations with an optimal estimation approach: 2. Implementation and evaluation", *J. Geophys. Res.*, **115** (2010), D02204, doi: 10.1029/2009JD011780.

Wald, A. E., Y. J. Kaufman, D. Tanré, and B.-C. Gao, "Daytime and nighttime detection of mineral dust over desert using infrared spectral contrast", *J. Geophys. Res.*, **103** (1998), 32,307–32,313.

Walthall, C. L., J. M. Norman, J. M. Welles, G. Campbell, and B. L. Blad, "Simple equation to approximate the bidirectional reflectance from vegetative canopies and bare soil surfaces", *Appl. Opt.*, **24** (1985), 383–387.

Wandinger, U., D. Müller, C. Böckmann, D. Althausen, V. Matthias, J. Bösenberg, V. Weiß, M. Fiebig, M. Wendisch, A. Stohl, and A. Ansmann, "Optical and microphysical characterization of biomass-burning and industrial-pollution aerosols from multi-wavelength lidar and aircraft measurements", *J. Geophys. Res.*, **107**, (2002), 8125, doi: 10.1029/2000JD000202.

Wang, J., S. A. Christopher, F. Brechtel, J. Kim, B. Schmid, J. Redemann, P. B. Russell, P. Quinn, and B. N. Holben, "Geostationary satellite retrievals of aerosol optical thickness during ACE-Asia", *J. Geophys. Res.*, **108:D23** (2003), 8657, doi: 10.1029/2003JD003580.

Wang, M., and H. Gordon, "Retrieval of water-leaving radiance and aerosol optical thickness over the oceans with SeaWiFS: A preliminary algorithm", *Appl. Opt.*, **33** (1994), 443–452.

Wang, P. H., M. P. McCormick, L. R. McMaster, W. P. Chu, T. J. Swissler, M. T. Osborn, P. B. Russell, V. R. Oberbeck, J. Livingston, J. M. Rosen, D. J. Hofmann, G. W. Grams, W. H. Fuller, and G. K. Yue, "SAGE II aerosol data validation based on retrieved aerosol model size distribution from SAGE II aerosol measurements", *J. Geophys. Res.*, **94** (1989), 8381–8393.

Waquet, F., B. Cairns, K. Knobelspiesse, J. Chowdhary, L. D. Travis, B. Schmid, and M. I. Mishchenko, "Polarimetric remote sensing of aerosols over land", *J. Geophys. Res.*, **114** (2009a), D01206, doi: 10.1029/2008JD010619.

Waquet, F., J. Riedi, L. C. Labonnote, P. Goloub, B. Cairns, J. L. Deuze, and D. Tanré, "Aerosol remote sensing over clouds using A-train observations", *J. Atm. Sci.*, **66:8** (2009), 2468-2480.

Waquet, F., J. Riedi, L. Labonnotte, F. Thieuleux, F., Ducos, Ph. Goloub, and D. Tanré, "Aerosols remote sensing over clouds using the A-train observations", A-Train Symposium, New Orleans, USA, 25–28 October 2010.

Washington, R., M. Todd, N. J. Middleton, and A. S. Goudie, "Dust-storm source areas determined by the total ozone monitoring spectrometer and surface observations", *Annals of the Association of American Geographers*, **93:2** (2003), 297–313.

Waterman, P. C., "Symmetry, unitarity, and geometry in electromagnetic scattering", *Phys. Rev. D.*, **3** (1971), 825–839.

Weaver, C. J., J. Joiner, and P. Ginoux, "Mineral aerosol contamination of TIROS Operational Vertical Sounder (TOVS) temperature and moisture retrievals", *J. Geophys. Res.*, **108:D8** (2003), 4246, doi: 10.1029/2002JD002571.

Weinzierl, B., A. Petzold, M. Esselborn, M. Wirth, K. Rasp, K., Kandler, L. Schutz, P. Koepke, and M. Fiebig, "Airborne measurements of dust layer properties, particle size distribution and mixing state of Saharan dust during SAMUM" 2006", *Tellus B*, **61** (2009), 96–117.

Weitkamp, C. (Ed.), *Lidar: High Spectral Resolution Lidar*, Springer, New York, 2005.

Wells, K. C., J. V. Martins, L. A. Remer, S. M. Kreidenweis and G. L. Stephens, "Critical reflectance derived from MODIS: Application for the retrieval of aerosol absorption over desert regions", *J. Geophys. Res.*, **117** (2012), D03202, doi:10.1029/2011JD016891.

Welton, E. J., J. R. Campbell, J. D. Spinhirne, and V. S. Scott, "Global monitoring of clouds and aerosols using a network of micro-pulse lidar systems", in: U. N. Singh, T. Itabe, and N. Sugimoto (Eds), *Lidar Remote Sensing for Industry and Environmental Monitoring,* Proc. SPIE, 4153, 2001, 151–158.

Welton, E. J., T. A. Berkoff, S. A. Stewart, L. Belcher, J. R. Campbell, B. N. Holben, and S.-C. Tsay, "The NASA Micro Pulse Lidar Network (MPLNET): Summary of the last 10 years, current status, and future plans", in: *Proceedings of the 25th International Laser Radar Conference*, St. Petersburg, 2010, 875–878.

West, R. A., "Optical properties of aggregate particles whose outer diameter is comparable to the wavelength", *Appl. Opt.*, **30** (1991), 5316–5324.

Whiteman, D. N., "Examination of the traditional Raman lidar technique. I. Evaluating the temperature-dependent lidar equations", *Appl. Opt.*, **42:15** (2003), 2571–2592.

Whiteman, D. N., S. H. Melfi, and R. A. Ferrare, "Raman lidar system for the measurement of water vapor and aerosols in the Earth's atmosphere", *Appl. Opt.*, **31:16** (1992), 3068–3082.

Wiegner, M., J. Gasteiger, S. Groß, F. Schnell, V. Freudenthaler, and R. Forkel, "Characterization of the Eyjafjallajφkull ash-plume: Potential of lidar remote sensing", *Physics and Chemistry of the Earth, Parts A/B/C,* **45-46** (2012), 79–86.

Wilson, D. I., S. J. Piketh, A. Smirnov, B. N. Holben, and B. Kuyper, "Aerosol optical properties over the South Atlantic and Southern Ocean during the 140th cruise of the M/V S.A. Agulhas", *Atm. Res.*, **98** (2010), 285–296.

Winker, D. M., M. A. Vaughan, A. Omar, Y. Hu, K. A. Powell, Z. Liu, W. H. Hunt, and S. A. Young, "Overview of the CALIPSO mission and CALIOP data processing algorithms", *J. Atmos. Ocean. Tech.*, **26** (2009), 2310–2323.

Wiscombe, W. J., and A. Mugnai, "Single scattering from nonspherical Chebyshev particles: A compendium of calculations. NASA Ref. Publ. NASA RP-1157", National Aeronautics and Space Administration: Washington, D.C., 1986.

WMO, *Report on the Measurements of Atmospheric Turbidity in BAPMoN* (TD No. 603), GAW Report No. 94, WHO, 1990.

Wriedt, T. (Editor), "VII electromagnetic and light scattering by non-spherical particles: Theory, measurement and applications", *J. Quant. Spectrosc. Radiat. Transfer*, **89** (2004), 1–460.

Yamamoto, G., and M. Tanaka, "Determination of aerosol size distribution from spectral attenuation measurements", *Appl. Opt.*, **8** (1969), 447–453.

Yang, P., and K.-N. Liou, "Light scattering and absorption by nonspherical ice crystals", *Light Scattering Rev.*, **1** (2006), 31–71.

Yang, P., K. N. Liou, M. I. Mishchenko, and B.-C. Gao, "Efficient finite-difference time-domain scheme for light scattering by dielectric particles: Application to aerosols", *Appl. Opt.*, **39** (2000), 3727–3737.

Yang, S., P. Ricchiazzi, and C. Cautier, "Modified correlated k-distribution method for remote sensing", *J. Quant. Spectrosc. Radiat. Transfer*, **64** (2000), 585–608.

Yoshioka, M., N. Mahowald, J.-L. Dufresne, and C. Luo, "Simulation of absorbing aerosol indices for African dust", *J. Geophys. Res.*, **110** (2005), D18S17, doi: 10.1029/2004JD005276.

Young, A., "Airmass and refraction", *Appl. Opt.*, **33** (1994), 1108–1110.

Yu, H., Y. J. Kaufman, M. Chin, G. Feingold, L. A. Remer, T. L. Anderson, Y. Balkanski, N. Bellouin, O. Boucher, S. Christopher, P. DeCota, R. Kahn, D. Koch, N. Loeb, M. S. Reddy, M. Schulz, T. Takemura, and M. Zhou, "A review of measurement-based assessments of the aerosol direct radiative effect and forcing", *Atmos. Chem. Phys.*, **6** (2006), 613–666.

Yu, H., L. A. Remer, M. Chin, H. Bian, R. G. Kleidman, and T. Diehl, "A satellite-based assessment of transpacific transport of pollution aerosol", *J. Geophys. Res.*, **113** (2008), D14S12, doi: 10.1029/2007JD009349.

Yuan, T., L. A. Remer, and H. Yu, "Microphysical, macrophysical and radiative signatures of volcanic aerosols in trade wind cumulus observed by the A-Train", *Atmos. Chem. Phys.*, **11** (2011), 7119–7132.

Yue, G. K., "Retrieval of aerosol size distributions and integral properties from simulated extinction measurements at SAGE III wavelengths by the linear minimizing error method", *J. Geophys. Res.*, **105** (2000), 14719–14736.

Yurkin, M. A., and A. G. Hoekstra, "The discrete dipole approximation: An overview and recent developments", *J. Quant. Spectrosc. Radiat. Transfer*, **106** (2007), 558–589.

Zender, C., D. Newman, and O. Torres, "Spatial heterogeneity in aeolian erodibility: Uniform, topographic, geomorphologic, and hydrologic hypotheses", *J.Geophys. Res.*, **108** (2003), 4543, doi: 10.1029/2002JD003039.

Zhang, J., and J. S. Reid, "A decadal regional and global trend analysis of the aerosol optical depth using a data-assimilation grade over-water MODIS and Level 2 MISR aerosol products", *Atmos. Chem. Phys.*, **10** (2010), 10,949–10,963.

Zhang, J., J. S. Reid, D. L. Westphal, N. L. Baker, and E. J. Hyer, "A system for operational aerosol optical depth data assimilation over global oceans", *J. Geophys. Res.*, **113** (2008), D10208, doi: 10.1029/2007JD009065.

Zhu, L., J. V. Martins, and L. A. Remer, "Biomass burning aerosol absorption measurements with MODIS using the critical reflectance method", *J. Geophys. Res.*, **116**, (2011), D07202, doi: 10.1029/2010JD015187.

Zubko, E., H. Kimura, H., Y. Shkuratov, K. Muinonen, T. Yamamoto, H. Okamoto, and G. Videen, "Effect of absorption on light scattering by agglomerated debris particles", *J. Quant. Spectrosc. Radiat. Transfer.*, **110** (2009), 1741–1749.

Index